Nanomaterials for Spectroscopic Applications

Nanomaterials for Spectroscopic Applications

edited by
Kaushik Pal

CRC Press
Taylor & Francis Group
Boca Raton London New York

CRC Press is an imprint of the
Taylor & Francis Group, an **informa** business

Nanomaterials for Spectroscopic Applications

First published 2021 by Jenny Stanford Publishing Pte. Ltd.

Published 2023 by CRC Press
Taylor & Francis Group
6000 Broken Sound Parkway NW, Suite 300
Boca Raton, FL 33487-2742

ISBN 13: 978-981-4877-69-5 (hbk)

Visit the Taylor & Francis Web site at
http://www.taylorandfrancis.com

and the CRC Press Web site at
http://www.crcpress.com

British Library Cataloguing-in-Publication Data
A catalogue record for this book is available from the British Library.

Contents

Preface

Electrical, optical, dielectric, electro-optical, and thermal properties are among the most fascinating and useful properties of nanomaterials and have been extensively investigated by using a variety of spectroscopic/microscopic advances. Moreover, a fundamental understanding of optical/electrical properties and related spectroscopic or microscopic investigations is essential for anyone interested in learning about semiconductor nanomaterials, insulators, or novel metals. This is partly because optical properties are closely related to other properties and functionalities (e.g. electronic, magnetic, and thermal) that are of fundamental importance to many technological applications, such as energy conversion, chemical analysis, biomedicine, optoelectronics, and smart display technologies. Recent years have witnessed the sheer growth of nanomaterial fabrication concepts and nanotechnology-based innovations in diverse sensors. The fabrication and characterization of nanosystems is an evergreen domain in nanoscience. The quality measurements of newly designed polymeric stuffs demand systematic and unexplored characterization protocols. As an outcome of this, the research in this regard has attracted considerable attention of the scientific community worldwide. Materials science and advanced nanotechnology have now emerged as specific subjects being taught in universities and practiced in industries for obtaining new products with specific features.

This book paves the way for the improvement of unique and more capable novel analytical tools to gain prominence advances in their special physicochemical and optical properties, including size-tunable emission and bioanalytical as well as exclusive switchable device implementation. Drawing upon case studies from manufacturing and structural monitoring and involving chemical and biological advances, it suggests an approach that frames relevant technical issues to expedite the consideration of novel material fabrication and develop remarkable strategies. New systems are vital to our awareness of our surroundings and provide safety, security, and surveillance, as well as enable the monitoring of our health and environment. With contributions from materials scientists, nanotechnologists, and biologists across the globe, the book is for upper-level postgraduate students, doctoral research scholars, and postdocs with some basic knowledge of materials science and nanotechnology and provides the first systematic coverage of cost-effective nanomaterial-processing techniques for spectroscopic advancements. We hope this book will be an asset for scholars, scientists, engineers, and academicians in nanotechnology and sensor applications. The goal is to have a self-contained smart sensor fabrication that is cost-effective, reliable, self-monitoring, and reconfigurable and can operate indefinitely. The overall approach can achieve distributed novel mechanisms feeding information from multiple locations to improve the overall understanding of system conditions. The book provides the end-user with critical data in a rapid, reliable, robust, and efficient manner with a seamless interface to applications.

Certainly, the coverage is not complete, but the book provides a unique glimpse of what has been achieved and what remains to be explored in functional nanomaterials.

Dr. Kaushik Pal
PhD; DSc (Malaysia); Marie-Curie (Europe);
CAS Fellow (China)

Chapter 1

An Emerging Avenue of Functional Graphene Aggregated Liquid Crystalline Hybrid Nanocomposite Diverse Attempts: Key Challenges and Novel Premises

Asiya S.I.[a] and Kaushik Pal[b]

[a]*Bharath Institute of Higher Education and Research (BIHER), 173 Agharam Road, Selaiyur, Chennai 600073, Tamil Nadu, India*
[b]*University Federal Rio de Janeiro, Cidade Universitária, Rio de Janeiro - RJ, 21941-901, Brazil*
kaushikphysics@gmail.com

Recently, semiconductor nanomaterials dopped liquid crystal design and as well as cultivates highly centralized on numerous approaches towards the development of innovative and assorted applications in the area of Nanoscience and Nanotechnology. Emerging nanomaterials have in recent times been employed to a major extent to revise the physical properties of liquid crystals by dopping ultra-thin graphene nanostructures. The self-organized liquid crystalline optical material is exploited to order templates and orient nanomaterials. Moreover, nanomaterials controllable growth dynamics of nanorods, nanowires, nanospikes, nanoparticles, nanoflakes and graphene oxide (GO) homogeneously blended into lyotropic liquid

Nanomaterials for Spectroscopic Applications
Edited by Kaushik Pal

ISBN 978-981-4877-69-5 (Hardcover), 978-1-003-16033-5 (eBook)
www.jennystanford.com

crystal phases when isotropic host solvents are present. These results contribute to a variety of entirely promising applications, most of them also discussed in this exclusive chapter. Hence, the architectures of functional nanomaterials with nanometer scale dimension beneficial with the treating of material on an industrial scale. Experimental motivation behind nanofabrication is to create nanoscale structures that form part of a system, device, or component in huge amounts and cost-effective, also known as top-down and bottom-up attempts. Exploration of data, however, produces an incoherent interpretation of the favorable effects of nanostructures on the physical properties of LC hosts. The brand new work must be credited to an excellent outcome of significant nanomaterials size and shape dependent extensive morphology, dispersibility and novel functionalization.

1.1 Introduction

The liquid crystal (LC) is one of the precious materials are quite different, yet share some common features, for example, their beauty, distinguished optical properties and a competition of high prices per gram. Recently, keeping in pace with topical science and modern technology, already LCs have entered into the fascinating domains of Nanoscience and Nanotechnology. This chapter explores most recent and significant innovations in 'Liquid Crystal Nanoscience' embracing contemporary nanomaterials different kinds of structural varieties, e.g., nanowires, nanorods, nanospikes, nanoflakes, and nanoparticles.

Indeed, LCs are soft materials that show both order and mobility on a molecular, supra molecular level and as well as macroscopic level. They are well known as active components in liquid crystal displays (LCDs), but many other commercial applications have also been explored. Liquid crystals are self-paced anisotropic fluids that are thermodynamically located phase that exist between crystalline (ordered) and the isotropic liquid phase (disordered), presenting the liquid fluidity as well as long-range lattice order seen in crystal-like solids [1–3]. The peculiar state of phase is known as 'mesophases.' The homogeneous

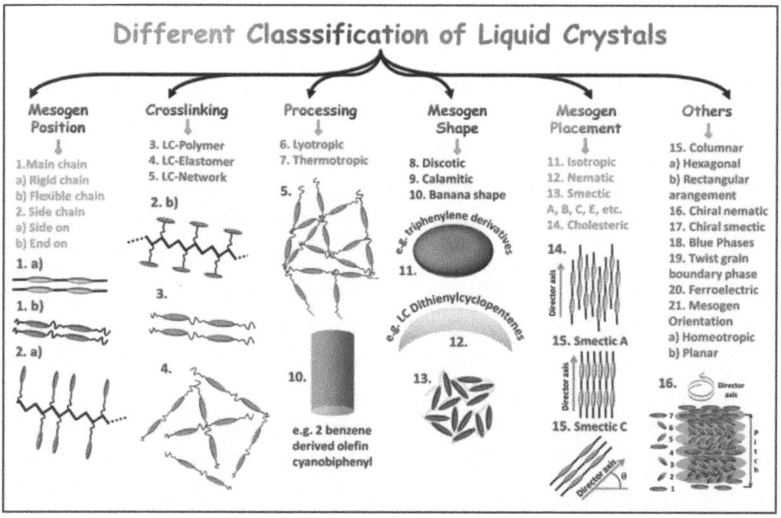

Figure 1.1 Classification of various phases and different kinds of liquid crystal structures, i.e., discotic, calamitic and bent-core shape. A schematic illustration of different thermotropic LC phases observed on heating from crystalline state, e.g., Smectic A to Smectic C to nematic and cholesteric. Reproduced with permission from Rastogi et al. [8], copyright © 2019 Elsevier.

distribution of 0-, 1- and 2-dimensional nanomaterials/graphene into liquid crystalline materials for the improvement of properties, liquid-crystalline hybrid nanocomposite matrix phase behavior of nanomaterials themselves, self-assembly and alignment of nanomaterials in liquid-crystalline media, and the high yield controllable synthesis of nanomaterials by using liquid crystals as 'templates' or 'precursors' have been highlighted and explored in this chapter [2]. Even predicted that, the 'fourth sate of matter' is most likely to play more prominent positions in Nanotechnology and advanced Material Science in the years to come. Normally, the two common range of LCs are 'thermotropic LCs' and 'lyotropic LCs' [4–7]. Based on the shape of their constituent molecules, e.g. thermotropic LCs are further differentiated into calamitic due to 'rod-like,' discotic for disk-like and sanidic for brick-like or lath-like, bent-core for banana-like molecules Fig. 1.1 [8].

Controllable growth dynamics of nanomaterials is the promising technology which will soon conquer all manufacturing technologies by providing the random synthesis of functional nanomaterials capable of meeting all present and future industrial challenges. 1D nanomaterials have gained the focus of researchers owing to their exceptional properties and possible applications in mesoscopic research and in the nanodevice synthesis. Electrospinning is the most adaptable of all reported 1D nanofabrication method and is a fairly inexpensive method that result in the production of long and continuous fibers with nanoscale diameters. Materials below 100 nm are usually referred to as nanomaterials in one, two or three dimensions and they also have special properties that vary from those of bulk counterparts. It has the potential to customize individual nanostructures and to control the functional hybrid composite morphology. The mechanism is extremely robust to handle all sorts of functional nanomaterials, that include organic and inorganic polymers, in different forms, such as solutions, emulsions, mixtures, or melts, to produce complex structures, including porous, hollow, core–shell, etc.

This chapter offers an outline of the latest development in the research area of Nanoscience by designing nanomaterials that can be self-assembled in several directions into functional super-structures. Though assembly of nanostructures was caused by different methods, the usage of LC as a coordinating medium to cause nanoparticles to assemble themselves is an active tool. A variety of LC nanoparticle hybrid structures, which are the most widely utilized material in the fabrication of such hybrid matrixes, were well investigated.

1.2 Liquid Crystalline Materials by Nanodopants

Outstanding electro-optic properties of the liquid crystal has been well applicable for commercialized display. However, the novel advancement in smart technologies demanded improved visual requirements which the traditional LC dispersion failed to meet. Innovative nanomaterials were explored in the quest for liquid crystal displays with less driving voltage, improved color contrast ratio and shorter response time. LCs with enhanced physical properties were also required for the advancement in other research and development fields such as novel optical systems, advanced materials, Nanotechnology and biosciences. The implementation of hybrid nanocomposites obtained by dispersing nanodopants into high performing liquid crystal materials were an excellent invention in the field of nanotechnology. The electro-optic (EO) responses and physical properties of LCs can be improved with the help of nanodoping materials such as: ferromagnetic, ferroelectric, numerous semiconducting materials and carbon nanoparticles.

Inorganic nanoparticles, such as metallic nanoparticles, quantum dots (QDs) and oxides, are also responsible in extemporizing the optical, electrical and electro-optic characteristics of LCs. It has been observed that the electrostatic interaction between liquid crystal molecules and metal nanoparticles such as gold nanoparticles could dramatically enhance the alignment of LC molecules thereby improving the physical and electro-optical characteristics [9–15].

1.3 Carbon Based Materials Graphene Dopant Liquid Crystals

Carbon based nanomaterial has been identified as one of the most extensively used dopant for manipulating the physical properties of LC material. Owing to their unusual physical and chemical properties, carbon nanomaterials comprising fullerene, carbon nanotube, graphene and graphene oxides, have drawn considerable interest of researchers in the field of Nanotechnology.

Carbon allotrope of graphene, which is a monolayer of sp^{2-} hybridized carbon atoms arranged into a 2D honeycomb lattice [16], plays an integral part as smart electronic materials since its discovery in 2004. Graphene has been extensively utilized in a wider range of applications such as electronics, energy storage, biomedical engineering, new composite materials, or waste water treatment

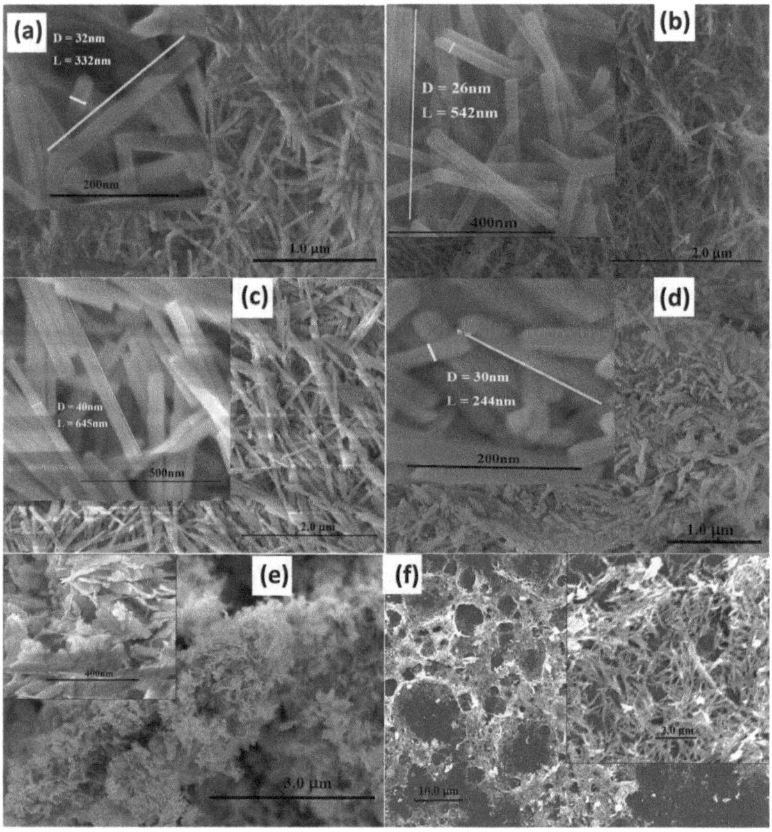

Figure 1.2 Typical FE-SEM micrograph of hydrothermal assisted process fabricated cadmium sulfide (CdS) controllable growth of structural morphologies as investigated (inset images) (a) nanowires at 1 h (D ~ 32 nm, L ~ 332 nm) at micrographic scale 200 nm, (b) nanowires at 2 h (D ~ 26 nm, L ~ 54 nm) at micrographic scale 400 nm scale, (c) nanowires at 3 h (PEG –.35 gm/D ~ 40 nm, L ~ 645 nm) at micrographic 500 nm scale, (d) nanowires at 24 h (D ~ 30 nm, L ~ 244 nm) at micrographic 200 nm scale, (e) nanoflakes at 1 week micrographic 400 nm scale, and (f) uniformly distributed nanowires at 2 h dispersed with ferroelectric liquid crystal (Felix) attributes porous surface structures. Reproduced with permission from Pal et al. [17, 18, 20], copyright © 2011, 2013, 2016 Elsevier.

1.4.2 Fabrication of Titanium Dioxide (TiO_2) Nanostructures Assembled LC Matrix

In this typical synthesis procedure, ordinary metallic titanium powder (Merck) of 0.3 gm was added into 30 ml of 10 M NaOH (sodium hydroxide) and then stirred for a transparent solution. This solution was inserted into a 50 ml capacity stainless steel lined autoclave and the entire system was placed inside an oven 119°C for 19 h. The bluish precipitation that resulted from cooling the system at room temperature was filtered and collected. It was then washed with 0.1 M HCl (35% concentrated hydrochloric acid, Merck) and distilled water for numerous times. The end product was then dried at 60°C for 4 h. This was annealed at 650°C for 1 h in a heating chamber for oxidation after collecting powder like samples. To maintain the chemical methodology dealing only with bulk anatase TiO_2 powder [19], 40 ml mixture of 10 M NaOH and 5 M KOH was stirred vigorously in a magnetic stirrer (Remi) for 15 min before the transparent solution emerges. A 0.2 gm of anatase titanium (IV) oxide powder (Sigma Aldrich) was mixed with previous precursor of NaOH and KOH (sodium and potassium hydroxide pellets) in a 100 ml beaker contained deionized distilled water. The solution which resulted was then stirred for 25 min and after properly stirring the solution, it was moved into an autoclave lined in teflon stainless steel with a capacity of 100 ml and the entire system was then inserted into a heating chamber. Following the similar experimental conditions, the white precipitation obtained was filtered and then washed with 0.1 M HCl solution; 150 ml deionized distilled water and 50 ml ethanol for numerous times till the samples are free from impurities. Sodium-potassium titanate nanowires turned into hydrogen titanate nanowires and finally obtained desired pure TiO_2 nanowires, due several time of acid washing with deionized distilled water. The desired sample was dried out at 60°C for 4 h in an oven. This final sample was annealed at 650°C for 1 hr in furnace and pure TiO_2 nanocabbage obtained [21], when it cooled naturally at room temperature and confirmed by FE-SEM micrographs displayed in Figs. 1.3a–d. Ultra-thin hydrogen-titanate nanowires were obtained by anatase TiO_2 of diameter 1.5 nm and length approximate 750 nm as identified

Figure 1.3 FE-SEM micrograph of anatase TiO_2 (a) hydrogen-titanate nanowires, (b) nanoparticles, (c) nanocabbages, and (d) coagulated nanoparticles. Reproduced with permission from Pal et al. [21, 22], copyright © 2012, 2013 Elsevier.

in Fig. 1.3a. By applied annealing process in regular laboratory oven that reveals gradually formulated nanoparticles of diameter 40 nm corresponding to Fig. 1.3b. Typical nanocabbages formed during control of chemical methodologies in Fig. 1.3c, and further coagulated nanoparticles formed in Fig. 1.3d.

The controllable growth strategies of TiO_2 nanostructures summarizes as following Table 1.1.

While, fabricated TiO_2 nanorods, nanowires [22] and nanocabbage [23] were harvested in high yield fabrication and chemical reduction process from sodium hydrogen titanates with general chemical formula $Na_yH_{2-y}Ti_nO_{2n+1}{\cdot}xH_2O$. By acid washing of obtained nanomaterials results in ion exchange to produce the layered hydrogen titanates $H_2Ti_nO_{2n+1}{\cdot}xH_2O$, which exhibits features similar to $H_2Ti_3O_7$, $H_2Ti_4O_9{\cdot}H_2O$, and other members of the hydrogen-titanate family. After maintaining an oxidation process for several times and previously prepared sample confirms into unique crystalline pure TiO_2 nanostructure formation. We predicted from

Table 1.1 Growth dynamics of TiO_2 nanostructures chemical assisted hydrothermal route [22]

Morphology of nanostructures	Bulk quality of TiO_2 Powder	Hydroxide solvent	Temp./°C	Diameter of nanostructures
Nanowires	(0.3 gm) metal	10M of 30ml	119°C	11 nm (diameter) 232 nm (length)
Nanowires	—	—	Annealed at 650°C	30 nm (diameter) 120 nm (length)
Nanowires anatase	(0.2 gm)	10M of 30ml NaOH and KOH	119°C	1.5 nm (diameter) 653 nm (length)
Nanocabbage	—	—	Annealed at 650°C	42 nm (diameter)

our earliest publication, i.e., there is really a disadvantage of using bulk metal TiO_2 powder rather than using of TiO_2 (IV) anatase bulk powder for nanostructure synthesis confirmation through FE-SEM images investigation. The above discussed titanium nanorods synthesis and dehydration reaction mechanism process as follows

(1) $NaOH + TiO_2 = Na_4TiO_3 + H_2O$
(2) Na_4TiO_3 (annealed at 650°C for 1 h) $\rightarrow Na_XH_{2-X}Ti_3O_7$
(3) $H_2Ti_2O_4(OH)_2 \rightarrow H_2O + H_2Ti_2O_5$ (Hydrogen titanate nanowires)
(4) $H_2Ti_2O_5$ (annealed at 650°C 1 h) $\rightarrow H_2O + 2TiO_2$ (Pure TiO_2 nanowires from bulk TiO_2 anatase powder)

1.4.3 Controllable Synthetic Mechanism of Zinc Oxide (ZnO) Nanostructures Assembled LC Matrix

Typical set of experiments introduced a cationic surfactant cetyltrimethylammonium bromide (CTAB) of 0.8 gm to prepare ZnO nanoparticles. 20 ml CTAB aqueous solution strongly stirred in a beaker for 15 min. Occasionally, 0.65 gm of $Zn(Ac)_2$ dissolved into 2 M of 10 ml NaOH, which were previously stirred in separate beaker. Now, aqueous solution of CTAB dropwise added with zinc acetate solution for several times at continuous stirring. After the dissolution of CTAB, which resulted in a white aqueous solution

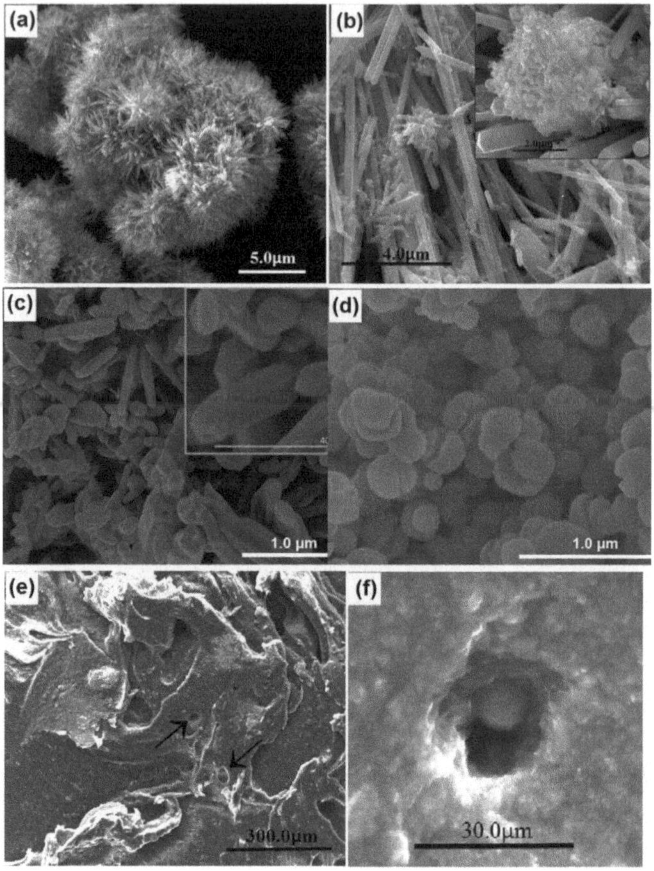

Figure 1.4 FE-SEM images of zinc oxide (ZnO) performed under the controllable chemical acceleration process (a) nanospikes, (b) nanocapsules coagulated with hexagonal nanorods, (c) nanoflakes, (d) nanoparticles, (e) porous surface due to nanoparticle dispersed anti-ferroelectric liquid crystals (W-182), and (f) high magnification images nanoparticles inside porous cavity. Reproduced with permission from Pal et al. [19, 23], copyright © 2012, 2013 Elsevier.

obtained, and it transferred into stainless steel coated teflon-lined autoclave. The autoclave was placed for firing at 120°C for 20 h in a regular laboratory oven. The stainless steel coated jacketed vessel of autoclave was allowed to cool to room temperature naturally. The resultant sample was filtered off and washed with deionized distilled water and ethanol for several times until the produced impurity washout and separated from the impurity ingredients. The final products were dried in a hot vacuum chamber at 65°C for 4 h [19, 23]. Furthermore, similar hydrothermal treatment we obtained nanoflakes like structures [19] reported recently (kept all chemical ingredients remain same except 0.75 gm of CTAB and [Zn(Ac)$_2$]. FE-SEM investigates the various morphological and structural analysis of ZnO nanospikes, nanocapsules coagulated with nanorods, nanoflakes, nanoparticles as depicted in Figs. 1.4a,b,c and d, respectively. While, surface porosity attributes when ZnO nanoparticles came into contact of anti-ferroelectric liquid crystal (AFLC; W-182) as show in Fig. 1.4e, and as well as captured higher magnification where nanoparticle appeared inside the porous cavity as depicted in Fig. 1.4f.

Different growth dynamics of ZnO nanostructures also summarized in following Table 1.2.

The chemical element Zn existed in $[Zn(OH)_4]^{2-}$ as a negatively charged tetrahedrons that were formed in presence of CTAB is a cationic surfactant. When CTAB is dissolved in water or ethanol, it

Table 1.2 High yield fabrication of ZnO series of nanostructural morphologies under chemical conditions. Reproduced with permission from Pal et al. [23], copyright © 2013 Elsevier

Nanostructures and energy band gap	Complexion reagent	Zn(Ac)$_2$	NaOH solvent	Temp./°C	Time duration
Micro-flowers grown by nanospikes, (E_g = **1.66 eV**)	Hydrazine monohydrate (5 ml)	8.0 gm	3.75 M of 40 ml DIW	60°C	2 h
Nanocapsules bundle with nanorods (E_g = **1.53 eV**)	Hydrazine monohydrate (10 ml)	8.0 gm	5 M of 30 ml DIW	60°C	2 h
Nanoparticles (E_g = **2.46 eV**)	CTAB (0.8 gm in 20 ml DIW)	0.65 gm	2 M of 10 ml DIW	120°C	20 h
Nanoflakes (E_g = **2.85 eV**)	CTAB (0.75 gm in 20 ml DIW)	0.75 gm	2.5 M of 10 ml NaOH	120°C	20 h

will ionize into CTA^+ and Br^- ZnO reaction, as a polar crystal, has a polar axis and possesses a positive face and a negative face on the crystal due to the asymmetrical distribution of Zn atoms and O atoms along its polar axis, and the positive face (0 0 0 1) is occupied by Zn atoms while the negative face (0 0 0 1) is distributed by O atoms.

In typical reaction mechanism, while CTA^+ was positively charged with a tetrahedral head and a hydrophobic tail. It was found from the B^- assisted hydrothermal process, CTA^+ and $[Zn(OH)_4]^{2-}$ ion pairs were formed initially by electrostatic interaction. The CTAB could accelerate the ionization of $[Zn(OH)_4]^{2-}$ as it was a strong-acid weak-base salt. The CTA^+ and $[Zn(OH)_4]^{2-}$ ion pairs formed a combination of CTAB and ZnO. It was inferred that, the CTAB aggregated in between the ZnO crystallites during hydrothermal crystallization and on washing the material with ethanol, the particle like structures of ZnO were formed. Variation of the growth mechanism of ZnO nanostructures [19, 23] formation obtained through FE-SEM imaging performance.

1.5 LC-Aggregated Self-Assembly Nanomaterials and CNTs/Graphene

Nanomaterials possess unique physical and chemical properties which vary greatly from both bulk material and individual atoms and molecules. However, single nanomaterial entities are not adequate for direct implementation of any functional device. Their vast range of application depends on the generation of macroscopic assemblies of nanomaterials with controlled localization and orientation. The most common technique to accomplish this is to pursue a 'bottom-up' methodology where the functional nanomaterials are built and organized into a specific structure by self-organization, which involves the desired capacity to control nanomaterial interactions. As a promising approach for producing numerous functional materials

with complex hybrid architectures, the unusual ability of LC material to self assemble into long-range organized super-structures has been extensively exploited. The chemical reactions executed in aqueous lyotropic LCs have successfully resulted in obtaining various mesoporous materials [24]. Furthermore, thermotropic LCs may also act as an effective tool for modeling 'top-down' approaches that exhibit a long-range or even spatial order and high sensitivity to different external stimuli, which are supportive of LCs composed of anisotropic molecules. During homogeneous dispersion into an liquid crystal hybrid matrix, anisotropic torques and forces would be exerted on the micro-additives. This hybrid composite can also steer them on a macroscopic scale, or even align their position with different micro-scale locations. The efficient fabrication of complex periodic super-structures composed of microparticles, were also recorded in the recent decades [25–30]. Owing to the release of elastic energy produced by the volume occupied by the nanoparticles, nanoparticles localized in the distorted LC regions can effectively reduce the systems free energy. The nanoparticles appear to be drawn spontaneously into the distortion core in the presence of a distortion gradient [31–34]. The core, bounded by highly elastically distorted areas are typically used as a model for the bottom-up assemblage of nanoparticles [35]. In addition, in the case of anisotropy, surface anchoring energy is the critical factor in determining the alignment of nanoparticles. The particles tends to coordinate with tangential anchoring around the nematic director while the particle orientation inclines to be perpendicular to the director for radial anchoring. In order to couple the LC-mediated elastic interaction between nanoparticles and the long-range drag power, the distortion gradient of director area, nanoparticles assemblies with ordered mesostructures in LCs are required to obtain. It will support the uses of the optoelectronics, photonics and smart electronic sectors [36–43].

On the other hand, carbon nanotubes (CNTs) are outstanding for their extraordinary anisotropic thermal, electrical, optical and mechanical characteristics as well as their exceptional thermal, electrical, optical and mechanical characteristics as well as their unique thermal and chemical stability and have engrossed a great deal of interest in several applications such as sensors, electronic devices, conductive and transparent films, as well as robust, lightweight and conductive fibers. The organized arrangement and configuration of CNTs in lyotropic LCs that were tested over several years has led to the noteworthy advancement of electronics, new materials and energy storage devices as per the introduction of macroscopically arranged CNT assemblies [44–46]. The precise orientation of an appropriate host mesophase under an exterior magnetic field [47] is due to the fact that the CNTs distributed in films of polymerizable lyotropic LC surfactants can be lined up in the film thickness direction, and CNT-dopped LCs can help remove filaments from uniformly spaced CNTs [48]. The effective dispersion of CNTs inside cylindrical micelles was investigated because of the physical absorption of the nonplanar trails of the surfactants. Single walled carbon nanotubes (SWCNTs) are also structured into ordered nanostructures with both lamellar and gyroid nanostructures following complete photo polymerization cycle [49]. Carbon nanotubes form lyotropic nematic LC phases when they are suspended at a sufficiently high concentrated in an isotropic liquid medium for their typical anisotropic shape [50–52]. Simultaneously, these smooth and homogeneous carbon nanotube.

Dispersion with liquid crystalline (CNT-LC) samples can also be produced with the aid of exterior forces such as magnetic field [53], electric field [54], and especially shear [55]. CNTs are particularly tough to disperse in any medium, and owing to the intense attraction of van der Waals forces between CNTs, the long term stability of the dispersion is often weak. To overcome this difficulty and to attain a complete homogeneous suspension of CNTs numerous chemical methodologies have been identified [6–60].

Apart from this, aqueous GO dispersions are expected to self-assemble into lyotoropic LCs as GO has an incredibly high aspect ratio. Investigations dealing with liquid crystallinity of aqueous GO dispersions reveals that the graphene dispersed in chlorosulfonic acid [61] was recorded as the first significant study for graphene liquid crystal. Graphene in its intrinsic form is barely soluble in water and organic solvents, which seriously restricts its possible usage in LCs. As an alternative, GO for its tunable amphiphilic properties has been found to diffuse well soluble in water [62].

Intensive research on the liquid crystalline properties of GO dispersions are very encouraging in the area of Composite Science and Technology. The liquid crystalline properties of GO dispersions

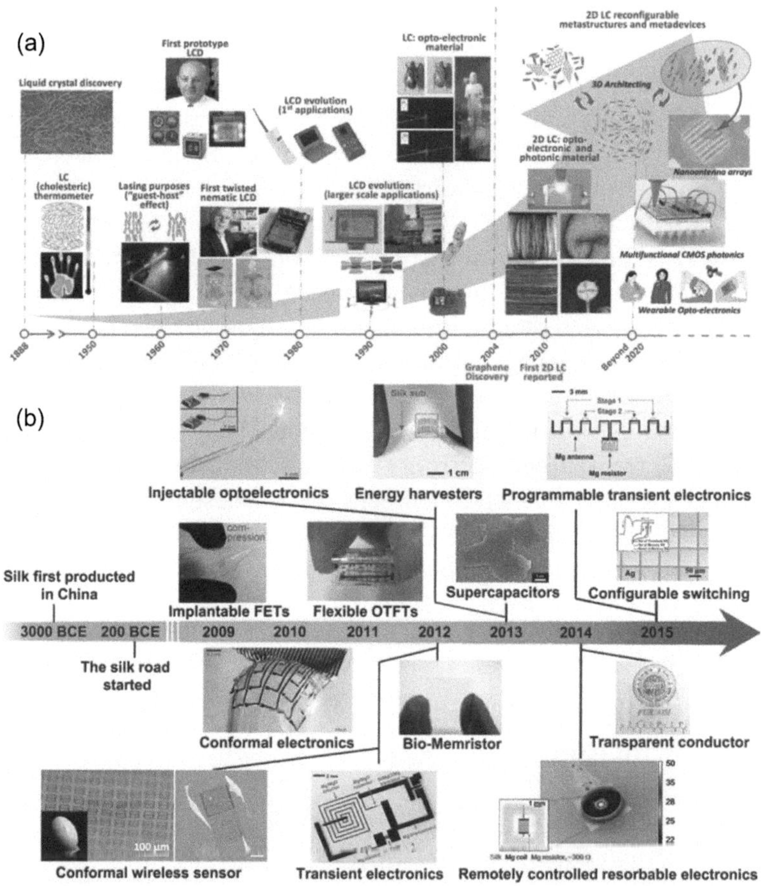

Figure 1.5 (a) Historical timeline of liquid crystals phase applications from the discovery of past decades 1988 by Friedrich Reinitzer to latest design of LCD and photonics applications (reproduced with permission from Mitov [67], copyright © 2014 Wiley-VCH) and (b) a brief timeline of the silk 'electronic road map' of graphene based liquid crystalline device implementations (reproduced with permission from Zhu et al. [68], copyright © 2015 Wiley-VCH).

are highly dependent on the concentration, which investigated earlier [63, 64], and aspect ratio of GO flakes [65]. Higher concentration of GO can result in a 'pseudo' lamellar phase [63] and a tunable structural color can be obtained by tailoring the concentration of the host solution [64], whereas a broad aspect ratio of GO flakes will effectively minimize the critical concentration for the spontaneous development of the nematic phase [66]. In addition, interesting physical properties of liquid crystalline has lead to the development of several applications Fig. 1.5a, from old discovery 1988 to latest day of moderation by 'Friedrich Reinitzer' [67] and extraordinary smart electronic materials 'Graphene-Liquid Crystals' based possible application on the advantageous for improving biocompatible smart electronic devices for next generation usages. The potential application of silk fibroin is highlighted in versatile electronic devices as both passive and active components as depicted in (Fig. 1.5b) [68], such as GO-LC gels with 'shape memory' effect [69], nanocomposites [70–72], transparent electrodes [73], photoluminescence imaging [74], conductive fibers [75–78], sensitive strain sensors [79], flame-retardant nanocoatings [80], and membranes [81], optical devices [80, 82] and biomedical applications [83], have been anticipated and some of them have been realized.

1.6 Photo Refractive ITO-LC Cell Manufacturing Principle

1.6.1 Liquid Crystal Phase Transition and Switchable Device Implementations

A switchable viewing angle liquid crystal smart display by means of an optically isotropic liquid crystal was proposed. A pixel and a sub-pixel are the two main parts in which the pixels are divided into in the process of developing optical devices where the image contents are displayed by the main pixels and the sub-pixels are responsible for controlling the viewing angle. The birefringence is caused by the transverse electric field in the main pixel, hence the images formed are not sensitive to the viewing angle as the liquid crystal re-orientation are in the same plane. In the sub-pixels, longitudinal electric fields cause birefringence, which indicates that the liquid crystal's optic axis is at right angle to the substrates. As a consequence, the liquid crystal's phase retardation induces light leakage in the oblique viewing directions and the amount of leakage can be regulated by the voltage applied. Based on the voltage applied, the light leakage could be managed thereby maintaining zero retardation in the normal direction while the image is being displayed. The viewing angle of the liquid crystal display is switchable with the current operating theory as depicted in Fig. 1.6. In the process of heating and cooling cycles of LC materials into loaded photo-refractive cells confined geometric orientation, (EO) textural phase patterns were produced and the sample was retained for 1 min at its isotropic temperature to gain thermal stability. The cooling process was executed with a scan rate of 10°C/min. The molecular orientation was diagrammatically represented using the recorded values of equilibrium transition temperatures and the resultant enthalpy values of the mesogens of the homologous series. The textural phase variance appeared with temperature under cross polarizers. The thermotropic liquid crystal implies an enantiotropic Isotropic (I), Smectic A (Sm A), Nematic (N) and crystalline (Cr) phase, which is characterized by heating, Cr - Sm A phase transitions were illustrated in Fig. 1.6c–f.

During cooling the LC cell the phase transition from 'Cr' of smooth 'multi-colored marbel' patterns as indicated in Fig. 1.6c was identified at temperature 90°C. Free standing threaded 'N' phase pattern of textures identified at 195°C (Fig. 1.6d), rather Newtown's ring of schariline patterns of 'Sm A' were obtained at 163.5°C in Fig. 1.6e. Hence, shiny droplet of patterns of 'Isotropic' phase investigated corresponding at 110.5°C.

The temperature variance textural phase sequences (Cr $\leftrightarrow$ Iso) of diagrammatic representation liquid crystalline sample under cross polarizer identified as follows

$$(\text{Iso } 125.5 \leftrightarrow \text{Sm A } 163.5 \rightarrow \text{N } 195 \leftrightarrow \text{Cr } 90)$$

Moreover, to realize the controllable viewing angle, high yield contrast ration, optically isotropic liquid crystal blend, under two domains in-plane switching cell approaches have been developed for electro-optic novel switchable device identified by a transversal electric field reveals a viewing angle that is relatively wide and symmetrical as schematically demonstrated in Fig. 1.7. Variation in the POM textural phase pattern often shows switching states 'ON' and 'OFF' states by the use of new device modulation of two electrode system defined as 'smart switchable device.'

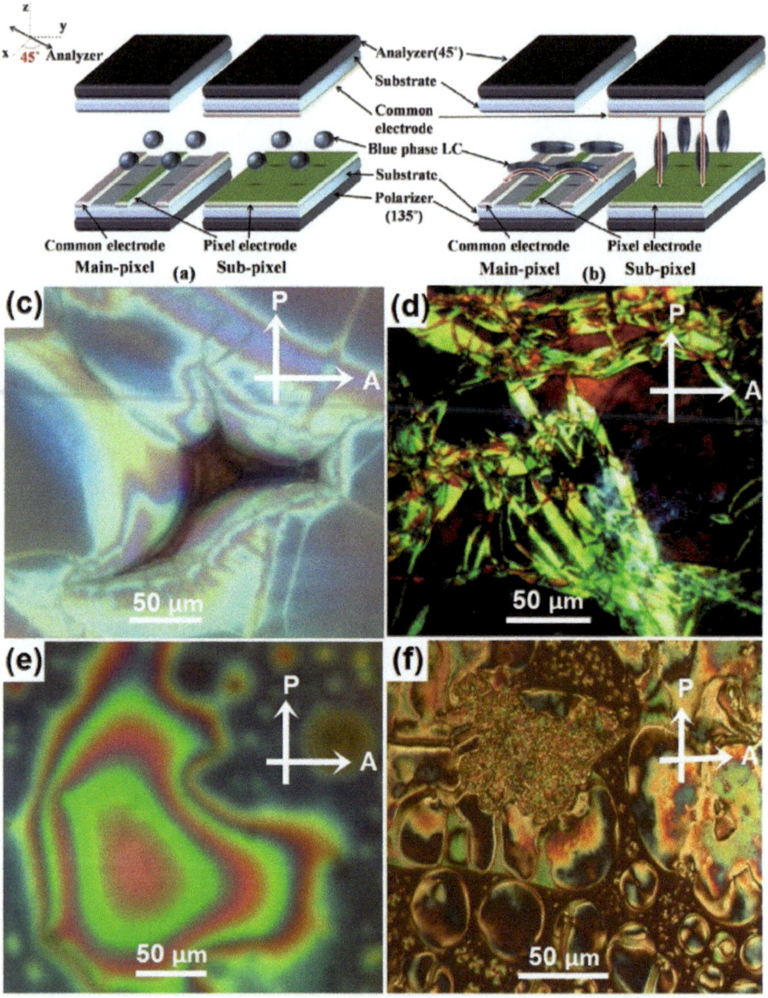

Figure 1.6 Implemented LCD device configuration of the viewing angle switchable device with an optically isotropic LC: (a) voltage-off state, and (b) voltage-on state. The temperature variance typical collection of textural phase studies by optical micrograph under cross polarizer with 20× magnification of liquid crystal compound (c) a multi-colored marbel patterns of crystalline phase appeared at 90°C, (d) free standing threaded 'N' phase pattern of textures identified at 195°C, (e) Newtown's ring of schariline patterns of 'Sm A' phase identified at 163.5°C, and (f) shiny nematic droplet phase patterns were identified at 125.5°C.

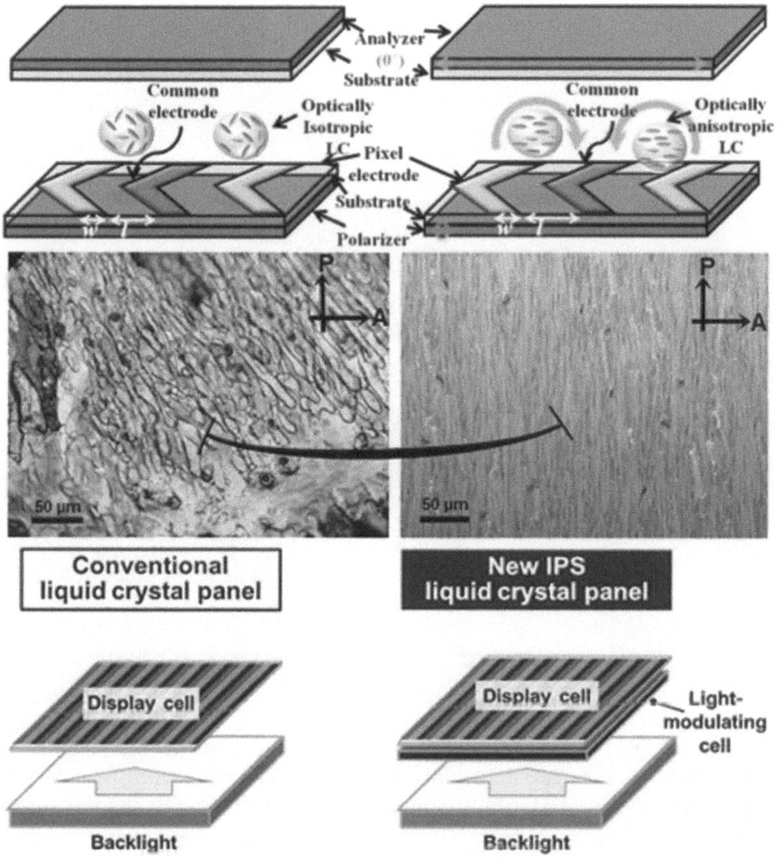

Figure 1.7 Schematic illustration of confined geometry of photo-refractive ITO LC cell unidirectionally rubbed for uniformly aligned liquid crystalline molecules with common electrode system by using spacer 4 μm, and in-plane switching (IPS) of textural phase patterns applied to latest liquid crystal display technology in moderation of OLED.

1.6.2 Graphene Deposited Liquid Crystalline Nanocomposite Phase Transition

Additionally, the surface stabilized GO-LC moves from a planer to the homeotropic alignment of GO containing ITO cells geometry and the molecular orientation is diagrammatically shown in Fig. 1.8a–c. Indeed, GO dispersions are an excellent demonstration of constant

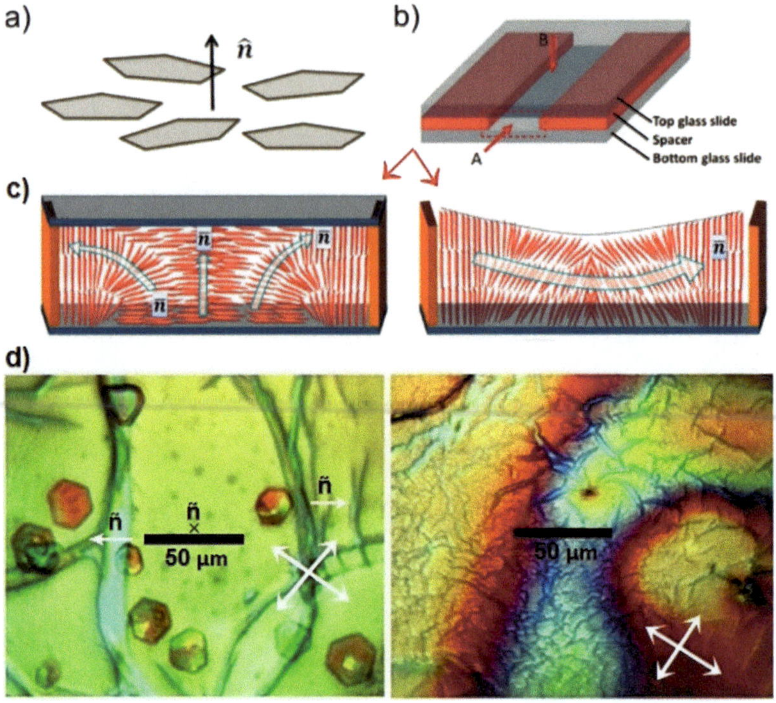

Figure 1.8 (a) Schematic diagram of the orientation of GO-flakes with respect to the director, ñ, (b) decoration of in-plane switching of ITO cells, (c) confining geometry used to photograph GO-LC dispersions under cross polarizing optical microscopy (POM), and (d) ideal and homogeneous GO-LC dispersion of high aspect ratio hexagonal crystal patterns of nematic droplet and Newton's ring patterns attributes beneficial for functionalization.

and homogeneous dispersions into thermotropic LCs of high aspect ratio hexagonal crystal-like nematic droplets and Newton's ring patterns are attributed due to the functionalization of graphene with liquid crystalline materials, as identified under cross polarizers as depicted in Fig. 1.8d. Although, the distinctive stability of these GO-LC dispersions opens up a range of solutions-based processing as well as spontaneous self-assembly into nematic and lamellar phases with improved concentration to help in the fabrication of unique tri-dimensional architectures.

1.7 Electro-optical (EO) Phase Transition of Switchable Sensor Modulation

To date, significant investigations have shown remarkable progress in the electro-optical efficiency of thermotropic LCs blended with nanomaterials/graphene oxide (GO) [82–87] for the manipulation of the configuration and applications of switchable sensors. The main aim of the progress could be to couple electrical dipoles on the surface treatment of the nanostructures with those of mesogens and to capture free ionic contamination by GO. This lowering of the composites isotropic transformation temperatures leads to a disordered mesophases of composites. A non-stabilizing relations among the graphene functionalities is driven by the surface functionalized graphene oxide (GO).

Comparing the melting point of hexadecanethiol functionalized graphene oxide and thiophenol functionalized graphene oxide discloses that van der Waals interaction among the alkyl chains of LC and that of GO results in a low stabilization, that is lacking in LC-GO composites and it is due to the fact that the hexadecanethiol functionalized graphene oxide melts a higher temperature. In comparison, steric phenyl group interruption may also be responsible for mesophases disordering. The synthesized graphene oxide when dispersed into deionized water displays a non-homogeneous chocolate-like milky look into naked eyes which is a indication of nematic liquid crystal as seen in Fig. 1.9. For the development of dispersed LC phases, a small concentration (0.05–0.6 wt%) aqueous dispersion of GO was immobilized for more than 3 weeks.

Figure 1.9a clearly indicates that the upper part symbolizes the isotropic phase, while the lower dense chocolate portion displayed optical birefringence formation, which supports optically active nematic LC of biphasic behavior (where both isotropic and nematic phases coexist).

In the bottommost phase of the sample, the phase consists of both dark and bright brushes.

Figure 1.9b is a standard graphene oxide dispersion nematic schlieren texture of 0.3 wt% concentration representing the local

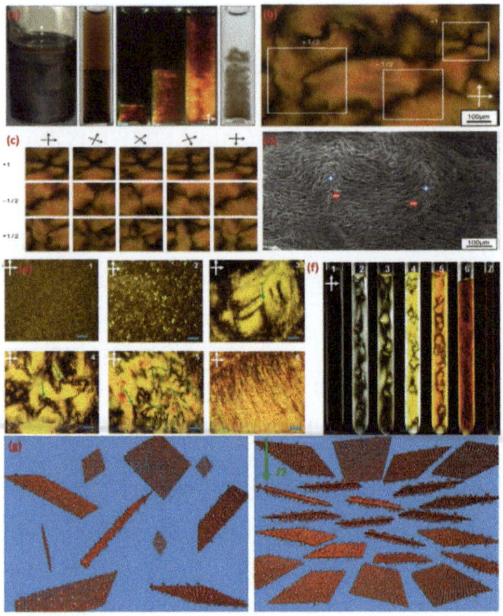

Figure 1.9 (a) (Left to right) 0.5 wt% GO dispersion exhibiting a milky appearance; phase-separated 0.2 wt% dispersion three weeks after preparation; three phase-separated dispersions (0.05, 0.2, 0.5 wt%); coagulated 0.01 wt% dispersion upon adding 50 mm NaCl. (b) Typical nematic schlieren texture of a 0.3 wt% dispersion with 1 = 2 disclinations and a +1 disclination. (c) Disclination morphologies of GOLC upon successive rotations of crossed polarizers. (d) SEM image of a graphene oxide liquid crystal in a freeze-dried sample (0.5 wt%). Blue and red symbols indicate +1/2 and −1/2 disclinations, respectively (reproduced with permission from Kim et al. [62], copyright © 2011 Wiley-VCH). Polarizing optical microscopy (POM) images between crossed polarizers of (e) GO aqueous dispersions in planar cells, (f) GO aqueous dispersions in test tubes (g) representation for isotropic (left) and nematic (right) phases of GO aqueous dispersions (reproduced with permission from Xu and Gao [63], copyright © 2011 American Chemical Society).

orientation of graphene oxide platelets, where the rotation operator of the brushes determined ± 1/2 disclinations (Fig. 1.9c). The local orientation of GOLC phases explained in Fig. 1.9d, is a Cryo-SEM (scanning electron microscopy) image of GOLC with 0.5 wt% concentration. It also discloses the distinctive disclinations morphology with continuous nematic phase. It has also been proven that polydisperse hard-disc modal is responsible for the concentration values and impurities present in the dispersion results in the distortion of the LC phase. Researches were also concentrated on the consequence of the outside stimulus on the macroscopic alignment of GOLC. Application of a magnetic field (H, 0.25 T) to the GO-LC device shows that the segregated LC orientational domains are separated by disclinations, then slowly reoriented and then integrated into a wide domain [62]. In the surveys performed by Xu and Gao about the aqueous phase nematic liquid crystals of graphene oxide dispersions and their self-dependent phase behavior [63], they attempted to establish the isotropic-nematic solid phase diagram versus mass fraction (GO) and salt concentration (NaCl). The dispersion of GO sheet with thickness of 2.1 μm and polydispersity index of 83% indicate the evolution from isotropic (the phase from at 0.025 wt%) to nematic phase (maximum stable phase concentration at 0.5 wt%). POM textures of GOLC as displayed in Fig. 1.9e indicates the appearance of isolated birefringence and the transition from isotropic to nematic phase starting at a lower f_m values (maximum mass fraction) at 2.5×10^{-4}. Upon increasing the value of f_m birefringence extents the entire dispersion along with the *vivid-schlieren* texture of nematic liquid crystals. Figure 1.9f, shows that the birefringence were not present at the dispersion with the value of f_m at 1×10^{-4} (tube 1), but as the value of f_m increases to 2.5×10^{-4}, the birefringence starts appearing and the primary development of nematic phase (tube 2) is indicated by the thread like textures, at higher concentration it branches with more compactness to the optical textures (from tubes 3 to 7) and as the f_m value increases to 5×10^{-3}, colorful birefringence (tube 5) start appearing that signals the development of a uniform nematic phase. A schematic illustration in Fig. 1.9g displays lamellar (I) and Nematic (N) phases.

1.7.1 Homogeneously Dispersed Graphene-Liquid Crystal in Organic Solvents

It is obvious that, after dispersal in water, big graphene sheets and flakes form lyotropic nematic LC phase. In order to extend this survey Wallace et al. recently demonstrated the development

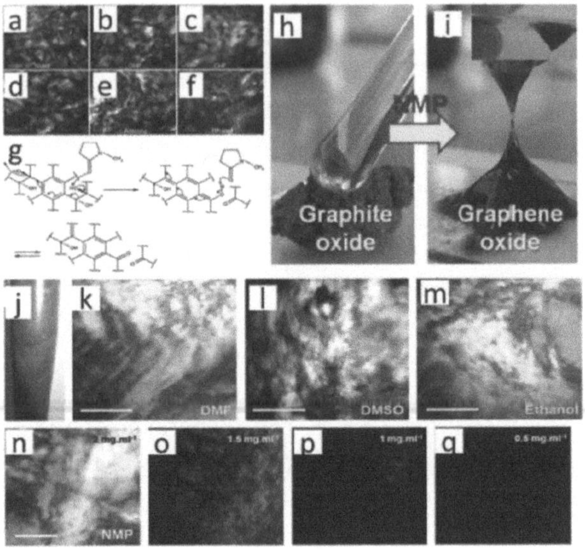

Figure 1.10 (a–f) POM images of LCGO in various organic solvents at a GO concentration of 2.5 mg/mL. (a) Water, (b) DMF, (c) CHP, (d) THF, (e) acetone, and (f) ethanol (reproduced with permission from Jalili et al. [88], copyright © 2013 American Chemical Society). (g) Chemical transformations induced by the reaction of graphite oxide and NMP or other polar aprotic solvents, (h) graphite oxide flakes (i) a gel-like dispersion of GO formed upon addition of NMP, (j) 0.2 wt% GO dispersion in DMF showing anisotropic texture, (k–n) optical micrographs of GO in organic solvents (2 mg ml^{-1}) demonstrating lyotropic LC phases, and (o–q) micrographs of GO dispersions at lower concentrations (0.5–1.5 mg ml^{-1}) showing isotropic and biphasic textures. Reproduced with permission from Gudarzi et al. [89], copyright © 2013 Elsevier.

of colloidal GOLC phase amphiphilic self-organized in a variety of organic solvents like: ethanol, acetone, tetrahydrofuran (THF), *N*, *N*-dimethylformamide (DMF), *N*-cyclohexyl-2-pyrrolidone (CHP) by the dispersal of extremely huge graphene sheets [88]. Since these solvents are used in the processing of different polymers, the dispersal and development of LC phase in graphene sheets offers a novel platform for the fabrication of self-organized functionalized GOLC composites.

Improvement of birefringence and lyotropic LC phase (Fig. 1.10a–f) are due to the organic phase dispersal of large sheets of graphene as indicated by the POM micrographs textural patterns. These organic phase dispersions can remain stable for many months without undergoing phase degradation owing to the adaptive nature of solvent in-between graphene sheets to balance the steric and repulsive force that allows graphene to develop supramolecular interactions such as hydrogen bonding. In addition, this also compensates for the lack of rotational entropy that exists between two adjacent graphene sheets in the self-organized phase and also with confined solvent molecules and electrostatic repulsion. Further, these GOLC can incorporate liquid crystallinity in the dispersion of single layer carbon nanotubes (SWCNT) dispersions.

Further attempts by Sharif et al. recommended organic solvent selection for the development of graphene oxide monolayers sheets through exfoliation [89]. It is essential for the production of nematic lyotropic LC phase in huge sheets of graphene with sufficiently high aspect ratio and this procedure often needed an incredible amount of ultrasonication in organic solvents to exfoliate graphene. Researches also confirmed the development of lyotropic nematic liquid crystalline phase in 0.1 mg/mL graphene dispersal in different organic solvents such as DMF, NMP (N-methyl-2-pyrrolidone) and proposed that chemical transformations are caused by these polar aprotic organic solvents, involving the ionization of GO sheets devoid of any sonication and mechanical agitation as indicated in Fig. 1.10g. However, after the aprotic solvent has been applied to the GO, it form yellow dark brown viscous 'gel'-structure (Fig. 1.10h,i) at concentrations above 0.5 wt%. The graphene flakes without any difficulty disperse in protic solvents like ethanol, methanol etc., by employing solvent exchange mode without the need of sonication resulting in the sudden development of nematic lyotropic GOLC at low concentrations (0.1 wt%) (Fig. 1.10j,n). In such GO dispersions the isotropic–biphasic transition was observed at a concentration range of 0.025–0.3 wt% (Fig. 1.10o–q). Increasing the pH value of the solution in protic solvents did not obstruct the formation of liquid crystal phase. While, manufacturing of 3D assemblaged nanocomposites with strong electrical conductivity up to 1050 S m^{-1}, the usage of these assemblies in combination with the polymer matrix is endorsed.

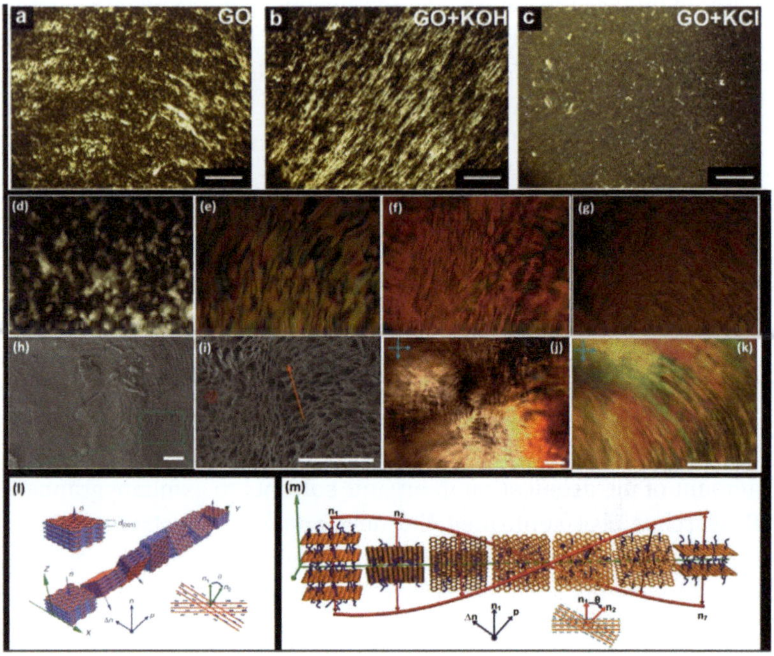

Figure 1.11 POM images of (a) GO, (b) (GO + KOH), and (c) GO + KCl suspensions, respectively; Scale bar = 400 μm (reproduced with permission Li et al. [90], copyright © 2016 Wiley-VCH). (d–g) POM images of GO LCs with successively increased volume fractions (φs) of 0.23, 0.38, 0.60, 0.76%. (h) Top-view SEM images show the fracture morphology of GO CLCs at volume fraction φ 0.98% confined in a circular cavity. (i) The screw dislocation of neighboring GO lamellar blocks with twist vectors and (j, k) POM images between crossed polarizers of GO CLCs in central (j) and lateral (k) domains (reproduced with permission Xu and Gao [91], copyright © 2011 Nature Publishing Group). Schematic illustrations for (l) one pitch of GO CLCs. The vectors (ns) of the lamellar blocks rotate anticlockwise along the helical axis (reproduced with permission Xu and Gao [63], copyright © 2011 Nature Publishing Group). (m) Helical lamellar phase for GO-g-PAN dispersions in DMF. The blue strings indicate grafted PAN chains, and GO sheets are simplified as yellow meshes (reproduced with permission Liu et al. [92] copyright © 2013 American Chemical Society).

1.7.2 Chemical Formulation of GO-LC Hybrid Nanocompounds

Self-organizing large sheets of graphene oxide into tightly frame structure of LC at low concentrations is an important factor to be considered. Subsequently, Shi and co-workers developed a method to produce low concentration liquid crystallinity (3.2 mg/mL) in GO suspension by rising the pH value using potassium hydroxide [90]. Rod-like nematic texture was observed at a GO concentration of 0.5 mg mL^{-1} (0.132 m KOH). As the graphene oxide concentration is raised to 0.25 mg mL^{-1} a separation occurred in birefringence domains that are constantly connected to each other (Fig. 1.11a). Furthermore, at increased concentration of 3.5 mg mL^{-1} well-organized lamellar liquid crystalline properties were developed (Fig. 1.11b). Disordered microstructure as shown in (Fig. 1.11c) were resulted when a 3.5 mg mL^{-1} GO suspension containing 0.132 M KCl characterized to determine the effect of K^+ or OH ions in the texture transformation. In KCl-GO suspension a significant number of flocculated particles precipitated due to the neutralization of the electric double-layer of the colloidal GO sheets by K^+ ions showing that OH^- ions dominated the formation of liquid crystallinity in graphene dispersion.

Gao et al. described a moist-spun, porous, orientational graphene hydrogel films (GHF) for fast-performance supercapacitor electrodes [93] in their studies. Here, graphene oxide liquid crystal was specifically used as free electrode, where GO-LC wet spinning assembly contributes 'open pores' to constantly spin orientational graphene hydrogel films as indicated in (Fig. 1.12a–f).

A peak capacitance of 203 F g^{-1} at 1 A g^{-1} and 67.1% specific capacitance (140 F g^{-1}) at 50 A g^{-1} is maintained that is much higher than graphene gel based devices. The increased capacitance values are implemented by manipulating orientational order and availability of ions that often resulted in higher rate capability and specific capacitance.

Owing to their excellent mechanical and electrical responses and adherence to the manufacturing on versatile substrates, GO-LCs is recycled in several micromechanical sensors. Graphene's electrical resistance varies on strain [94]. Additionally, GO-LC is arranged

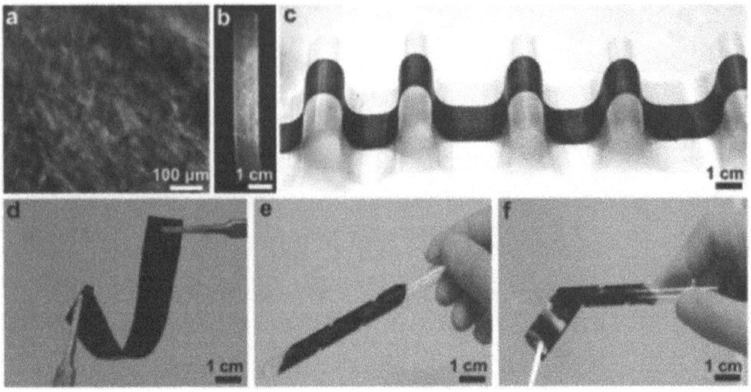

Figure 1.12 (a) POM image of GO liquid crystal aqueous dispersion, image of (b) a GO gel film between crossed polarizers, (c) 1 meter-long GOHF with a polyethylene terephthalate (PET) substrate, and (d–f) flexible, free-standing GHF (reproduced with permission from Kou et al. [93], copyright © 2015 Royal Society of Chemistry).

stably and uniformly into supramolecular assemblages by surface anchoring, complex fluid flow and micro-confinement. GO-LCs has demonstrated hydration-responsive folding and unfolding and offer anisotropic swell on behavior. These systems could be used for materials sensitive to stimuli, control release, application for complex space filling or sealing, nanodevices and microdevices. Nanomaterial dispersed LCs produce composite materials with improved physical characteristics and enhanced electo-optical performance that has high ability to develop looming LC devices and other significant facets of Nanotechnology.

1.8 Conclusions, Outlook and Perspectives

In this chapter, apart from fabrication and self-organization of nanostructures in the LC medium, liquid-crystalline nanomaterials have been effectively built and examined. 'Liquid crystal nanoscience' (LCN) appears to extend the scope of research by using advanced response characteristics which trigger the growth of novel hypotheses and promote new scientific initiatives. This chapter also outlined the impact of multifarious nanomaterials dispersed on LC on the electro-optical and physical characteristics with LC which include molecule orientation, elastic constants, viscosity etc. It is well known that Nanoscience is looking for much easier way of assembling nanomaterials by 'bottom-up' methods. 'Self-assembly' approach is the terrific choice to reach the goal, with an emphasis on streamlining supramolecular and interfacial interactions and LC tend to be powerful supporting medium. Active grouping in LC anisotropic mediums helps in providing improved functional materials that may be applied in switchable nanocomposite devices. Different forms of nanodopants have been identified, such as ferroelectric nanoparticles, noble metallic nanoparticles, semiconductor nanoparticles and carbon nanoparticles, to cause improvements in the physical properties of liquid crystals. These researches demonstrates the favorable effects of nanoparticles on the physical characteristics of liquid crystal hosts. This can be due to the influence of dimensions, shape, dispersibility and functionalization of various nanoparticles. Considerable progress in the fabrication of liquid crystallinity in graphene oxide was observed and numerous discotic nematic LCs of GO and r-GO were also processed at various concentration in aqueous as well as in organic solvents. In depth analysis were also focused on the inconsistency in mesophase properties depending on their size, pH and chirality dependency of GO-LCs and rheological properties. The exclusive chapter aims to provide an overall summary of GOLC development, mesophases properties, physical characteristics, 2D architecture manufacturing for potential software applications, and hybrid systems with other discotic liquid crystals. The study of graphene oxide blended liquid crystals and discotic liquid crystals of 2D materials as well as hybrid composites is novel and growing rapidly from Materials Science and technology. GOLC play a prominent role in advanced functional and optoelectronic materials, transparent electrodes, super capacitors, flexible smart display devices and sensor applications. Despite the centuries-old tradition of LC cultivation, liquid crystal nanoscience (LCN) and liquid crystal graphene nanoscience (LCGN) are relatively novel territory and also expected to evolve in its own ethics as another fascinating space.

Conflict of Interest

For details research work progress on the performance of book chapter contribution all the authors have declared there is no conflict of interest.

References

1. C. Tschierske, *Liquid Crystals: Materials Design and Self-Assembly*, Springer Science & Business Media, 2012.
2. H. K. Bisoyi and S. Kumar, *Chemical Society Reviews*, 2011, **40**, 306–319.
3. T. Kato, N. Mizoshita and K. Kishimoto, *Angewandte Chemie International Edition*, 2006, **45**, 38–68.
4. I. Dierking and S. Al-Zangana, *Nanomaterials*, 2017, **7**, 305.
5. S. Saliba, C. Mingotaud, M. L. Kahn and J.-D. Marty, *Nanoscale*, 2013, **5**, 6641–6661.
6. S. Laschat, A. Baro, N. Steinke, F. Giesselmann, C. Haegele, G. Scalia, R. Judele, E. Kapatsina, S. Sauer and A. Schreivogel, *Angewandte Chemie International Edition*, 2007, **46**, 4832–4887.
7. T. Hegmann, H. Qi and V. M. Marx, *Journal of Inorganic and Organometallic Polymers and Materials*, 2007, **17**, 483–508.
8. P. Rastogi, J. Njuguna and B. Kandasubramanian, *European Polymer Journal*, 2019, 109287.
9. S. K. Prasad, M. V. Kumar, T. Shilpa and C. Yelamaggad, *RSC Advances*, 2014, **4**, 4453–4462.
10. S. Sridevi, S. K. Prasad, G. G. Nair, V. D'Britto and B. Prasad, *Applied Physics Letters*, 2010, **97**, 151913.
11. A. S. Pandey, R. Dhar, S. Kumar and R. Dabrowski, *Liquid Crystals*, 2011, **38**, 115–120.
12. L. A. Holt, R. J. Bushby, S. D. Evans, A. Burgess and G. Seeley, *Journal of Applied Physics*, 2008, **103**, 063712.
13. U. Singh, R. Dhar, R. Dabrowski and M. Pandey, *Liquid Crystals*, 2013, **40**, 774–782.
14. B. Liu, Y. Ma, D. Zhao, L. Xu, F. Liu, W. Zhou and L. Guo, *Nano Research*, 2017, **10**, 618–625.

15. S. Khatua, P. Manna, W.-S. Chang, A. Tcherniak, E. Friedlander, E. R. Zubarev and S. Link, *The Journal of Physical Chemistry C*, 2010, **114**, 7251–7257.
16. K. S. Novoselov, A. K. Geim, S. V. Morozov, D. Jiang, Y. Zhang, S. V. Dubonos, I. V. Grigorieva and A. A. Firsov, *Science*, 2004, **306**, 666–669.
17. K. Pal, U. N. Maiti, T. P. Majumder and S. C. Debnath, *Applied Surface Science*, 2011, **258**, 163–168.
18. K. Pal, X. Yang, M. M. Mohan, R. Schirhagl and G. Wang, *Materials Letters*, 2016, **169**, 37–41.
19. K. Pal, T. P. Majumder, R. Schirhagl, S. Ghosh, S. K. Roy and R. Dabrowski, *Applied Surface Science*, 2013, **280**, 405–417.
20. K. Pal, U. N. Maiti, T.P. Majumder, S.C. Debnath, *Journal of Molecular Structure*, 2013, **1035**, 76–82.
21. K. Pal, J. S. Roy, T. P. Majumder, S. Ghosh, S. K. Roy, R. Schirhagl and R. Dabrowski, *Journal of Luminescence*, 2013, **136**, 278–284.
22. K. Pal, U. N. Maiti, T. P. Majumder, S. C. Debnath, *Journal of Molecular Structure*, 2012, **1016**, 33–38.
23. K. Pal, T. P. Majumder, S. C. Debnath, S. Ghosh and S. K. Roy, *Journal of Molecular Structure*, 2012, **1027**, 36–43.
24. C. Wang, D. Chen and X. Jiao, *Science and Technology of Advanced Materials*, 2009, **10**, 023001.
25. M. Škarabot, M. Ravnik, S. Žumer, U. Tkalec, I. Poberaj, D. Babič, N. Osterman and I. Muševič, *Physical Review E*, 2008, **77**, 031705.
26. I. Muševič, M. Škarabot, U. Tkalec, M. Ravnik and S. Žumer, *Science*, 2006, **313**, 954–958.
27. M. Škarabot, M. Ravnik, S. Žumer, U. Tkalec, I. Poberaj, D. Babič, N. Osterman and I. Muševič, *Physical Review E*, 2007, **76**, 051406.
28. A. Nych, U. Ognysta, M. Škarabot, M. Ravnik, S. Žumer and I. Muševič, *Nature Communications*, 2013, **4**, 1–8.
29. H. Yoshida, K. Asakura, J. Fukuda and M. Ozaki, *Nature Communications*, 2015, **6**, 1–8.
30. J.-C. Loudet, P. Barois and P. Poulin, *Nature*, 2000, **407**, 611–613.
31. D. Voloschenko, O. P. Pishnyak, S. V. Shiyanovskii and O. Lavrentovich, *Physical Review E*, 2002, **65**, 060701.
32. B. Senyuk, J. S. Evans, P. J. Ackerman, T. Lee, P. Manna, L. Vigderman, E. R. Zubarev, J. van de Lagemaat and I. I. Smalyukh, *Nano Letters*, 2012, **12**, 955–963.

33. B. Senyuk and I. I. Smalyukh, *Soft Matter*, 2012, **8**, 8729–8734.
34. R. Jose, G. Skačej, V. Sastry and S. Žumer, *Physical Review E*, 2014, **90**, 032503.
35. J. P. Lagerwall and G. Scalia, *Liquid Crystals with Nano and Microparticles*, World Scientific, 2017.
36. A. L. Rodarte, F. Cisneros, J. E. Hein, S. Ghosh and L. S. Hirst, *Photonics*, 2015, **2**(3), 855–864.
37. L. De Sio, R. Caputo, U. Cataldi and C. Umeton, *Journal of Materials Chemistry*, 2011, **21**, 18967–18970.
38. Q. Liu, Y. Cui, D. Gardner, X. Li, S. He and I. I. Smalyukh, *Nano Letters*, 2010, **10**, 1347–1353.
39. Q. Liu, Y. Yuan and I. I. Smalyukh, *Nano Letters*, 2014, **14**, 4071–4077.
40. J. Milette, S. J. Cowling, V. Toader, C. Lavigne, I. M. Saez, R. B. Lennox, J. W. Goodby and L. Reven, *Soft Matter*, 2012, **8**, 173–179.
41. D. Coursault, J. Grand, B. Zappone, H. Ayeb, G. Lévi, N. Félidj and E. Lacaze, *Advanced Materials*, 2012, **24**, 1461–1465.
42. K. Stratford, O. Henrich, J. Lintuvuori, M. Cates and D. Marenduzzo, *Nature Communications*, 2014, **5**, 1–8.
43. D. Coursault, J.-F. Blach, J. Grand, A. Coati, A. Vlad, B. Zappone, D. Babonneau, G. Lévi, N. Félidj and B. Donnio, *ACS Nano*, 2015, **9**, 11678–11689.
44. G. Scalia, C. Von Bühler, C. Hägele, S. Roth, F. Giesselmann and J. P. Lagerwall, *Soft Matter*, 2008, **4**, 570–576.
45. S. Schymura, E. Enz, S. Roth, G. Scalia and J. P. Lagerwall, *Synthetic Metals*, 2009, **159**, 2177–2179.
46. M. Sousa, S. Cloutier, K. Jian, B. Weissman, R. Hurt and G. Crawford, *Applied Physics Letters*, 2005, **87**, 173115.
47. M. S. Mauter, M. Elimelech and C. O. Osuji, *ACS Nano*, 2010, **4**, 6651–6658.
48. S. Schymura, S. Dölle, J. Yamamoto and J. Lagerwall, *Soft Matter*, 2011, **7**, 2663–2667.
49. C. R. Kasprzak, E. T. Scherzinger, A. Sarkar, M. Miao, D. H. Porcincula, A. M. Madriz, Z. M. Pennewell, S. S. Chau, R. Fernando and M. Stefik, *Macromolecular Chemistry and Physics*, 2018, **219**, 1800197.
50. M. Bravo-Sanchez, T. J. Simmons and M. Vidal, *Carbon*, 2010, **48**, 3531–3542.
51. L. Lu and W. Chen, *ACS Nano*, 2010, **4**, 1042–1048.
52. W. Song and A. H. Windle, *Macromolecules*, 2005, **38**, 6181–6188.

53. B. W. Steinert and D. R. Dean, *Polymer*, 2009, **50**, 898–904.
54. W. Tie, G. H. Yang, S. S. Bhattacharyya, Y. H. Lee and S. H. Lee, *The Journal of Physical Chemistry C*, 2011, **115**, 21652–21658.
55. E. K. Hobbie and D. Fry, *Physical Review Letters*, 2006, **97**, 036101.
56. V. A. Davis, L. M. Ericson, A. N. G. Parra-Vasquez, H. Fan, Y. Wang, V. Prieto, J. A. Longoria, S. Ramesh, R. K. Saini and C. Kittrell, *Macromolecules*, 2004, **37**, 154–160.
57. P. K. Rai, R. A. Pinnick, A. N. G. Parra-Vasquez, V. A. Davis, H. K. Schmidt, R. H. Hauge, R. E. Smalley and M. Pasquali, *Journal of the American Chemical Society*, 2006, **128**, 591–595.
58. A. Pénicaud, P. Poulin, A. Derré, E. Anglaret and P. Petit, *Journal of the American Chemical Society*, 2005, **127**, 8–9.
59. D. Baskaran, J. W. Mays and M. S. Bratcher, *Chemistry of Materials*, 2005, **17**, 3389–3397.
60. R. Haggenmueller, S. S. Rahatekar, J. A. Fagan, J. Chun, M. L. Becker, R. R. Naik, T. Krauss, L. Carlson, J. F. Kadla and P. C. Trulove, *Langmuir*, 2008, **24**, 5070–5078.
61. F. Mirri, A. W. Ma, T. T. Hsu, N. Behabtu, S. L. Eichmann, C. C. Young, D. E. Tsentalovich and M. Pasquali, *ACS Nano*, 2012, **6**, 9737–9744.
62. J. E. Kim, T. H. Han, S. H. Lee, J. Y. Kim, C. W. Ahn, J. M. Yun and S. O. Kim, *Angewandte Chemie International Edition*, 2011, **50**, 3043–3047.
63. Z. Xu and C. Gao, *ACS Nano*, 2011, **5**, 2908–2915.
64. P. Li, M. Wong, X. Zhang, H. Yao, R. Ishige, A. Takahara, M. Miyamoto, R. Nishimura and H.-J. Sue, *ACS Photonics*, 2014, **1**, 79–86.
65. D. Dan, N. Behabtu, A. Martinez, J. S. Evans, D. V. Kosynkin, J. M. Tour, M. Pasquali and I. I. Smalyukh, *Soft Matter*, 2011, **7**, 11154–11159.
66. S. H. Aboutalebi, M. M. Gudarzi, Q. B. Zheng and J. K. Kim, *Advanced Functional Materials*, 2011, **21**, 2978–2988.
67. M. Mitov, *ChemPhysChem*, 2014, **15**, 1245–1250.
68. B. Zhu, H. Wang, W. R. Leow, Y. Cai, X. J. Loh, M. Y. Han and X. Chen, *Advanced Materials*, 2016, **28**, 4250–4265.
69. F. Guo, F. Kim, T. H. Han, V. B. Shenoy, J. Huang and R. H. Hurt, *ACS Nano*, 2011, **5**, 8019–8025.
70. B. Qi, Z. Yuan, S. Lu, K. Liu, S. Li, L. Yang and J. Yu, *Fibers and Polymers*, 2014, **15**, 326–333.
71. S. Lu, S. Li, J. Yu, Z. Yuan and B. Qi, *RSC Advances*, 2013, **3**, 8915–8923.

72. S. Naficy, R. Jalili, S. H. Aboutalebi, R. A. Gorkin III, K. Konstantinov, P. C. Innis, G. M. Spinks, P. Poulin and G. G. Wallace, *Materials Horizons*, 2014, **1**, 326–331.
73. C. Zamora-Ledezma, N. Puech, C. Zakri, E. Grelet, S. E. Moulton, G. G. Wallace, S. Gambhir, C. Blanc, E. Anglaret and P. Poulin, *The Journal of Physical Chemistry Letters*, 2012, **3**, 2425–2430.
74. B. Senyuk, N. Behabtu, B. G. Pacheco, T. Lee, G. Ceriotti, J. M. Tour, M. Pasquali and I. I. Smalyukh, *ACS Nano*, 2012, **6**, 8060–8066.
75. H. Park, K. Lee, Y. Kim, S. Ambade, S. Noh, W. Eom, J. Hwang, W. Lee, J. Huang and T. H. Han, *Science Advances*, 2018, **4**, eaau2104.
76. R. Jalili, S. H. Aboutalebi, D. Esrafilzadeh, R. L. Shepherd, J. Chen, S. Aminorroaya-Yamini, K. Konstantinov, A. I. Minett, J. M. Razal and G. G. Wallace, *Advanced Functional Materials*, 2013, **23**, 5345–5354.
77. Z. Xu and C. Gao, *Accounts of Chemical Research*, 2014, **47**, 1267–1276.
78. I. H. Kim, T. Yun, J. E. Kim, H. Yu, S. P. Sasikala, K. E. Lee, S. H. Koo, H. Hwang, H. J. Jung and J. Y. Park, *Advanced Materials*, 2018, **30**, 1803267.
79. X. Wu, K. Hou, J. Huang, J. Wang and S. Yang, *Journal of Materials Chemistry C*, 2018, **6**, 8717–8725.
80. H. Kim, D. W. Kim, V. Vasagar, H. Ha, S. Nazarenko and C. J. Ellison, *Advanced Functional Materials*, 2018, **28**, 1803172.
81. D. K. Mahalingam, S. Wang and S. P. Nunes, *ACS Applied Nano Materials*, 2018, **1**, 4661–4670.
82. M. Kim, J. H. Park, S. Shahini, J. Yamamoto, S. Campidelli, Y. S. Kim and G. Scalia, *Proceedings of SPIE*, 2018, **10735**, 107350N.
83. X. Huang, J. He, K. Sun, Y. Chen, Z. Zha and C. Zhou, *Carbon*, 2018, **129**, 258–269.
84. S. Asiya, K. Pal, G. S. El-Sayyad, M. A. Elkodous, C. Demetriades, S. Kralj and S. Thomas, *Organic Electronics*, 2019, 105592.
85. K. Pal, S. Sajjadifar, M. A. Elkodous, Y. A. Alli, F. Gomes, J. Jeevanandam, S. Thomas and A. Sigov, *Electronic Materials Letters*, 2019, **15**, 84–101.
86. M. Gökçen, M. Yıldırım and O. Köysal, *The European Physical Journal-Applied Physics*, 2012, **60**, 30104.
87. T. M. Alam and C. J. Pearce, *Chemical Physics Letters*, 2014, **592**, 7–13.
88. R. Jalili, S. Aboutalebi and D. Esra, *ACS Nano*, 2013, **7**, 3981–3990.
89. M. M. Gudarzi, M. H. M. Moghadam and F. Sharif, *Carbon*, 2013, **64**, 403–415.
90. R. Li, M. Zhang, Y. Li, J. Chen, B. Yao, M. Yu and G. Shi, *Physical Chemistry Chemical Physics*, 2016, **18**, 11104–11110.

91. Z. Xu and C. Gao, *Nature Communications*, 2011, **2**, 1–9.
92. Z. Liu, Z. Xu, X. Hu and C. Gao, *Macromolecules*, 2013, **46**, 6931–6941.
93. L. Kou, Z. Liu, T. Huang, B. Zheng, Z. Tian, Z. Deng and C. Gao, *Nanoscale*, 2015, **7**, 4080–4087.
94. K. S. Kim, Y. Zhao, H. Jang, S. Y. Lee, J. M. Kim, K. S. Kim, J.-H. Ahn, P. Kim, J.-Y. Choi and B. H. Hong, *Nature*, 2009, **457**, 706–710.

Chapter 2

Spectroscopic and Microscopic Response of Graphene/CNTs Based Nanomaterials

Saikumar Manchala,[a] Jaison Jeevanandam,[b] and Michael K. Danquah[c]

[a]*Department of Chemistry, National Institute of Technology, Warangal 506004, Telengana, India*
[b]*Department of Chemical Engineering, Faculty of Engineering and Sciences, Curtin University, CDT 250, Miri, Sarawak 98009, Malaysia*
[c]*Chemical Engineering Department, University of Tennessee, Chattanooga, TN 37403, USA*
michael-danquah@utc.edu, jaison.jeevanandam@gmail.com

Carbon is a most common element in the world, which is the 15th abundant element in the crust of the Earth and is the fourth most common element in the universe, with respect to mass, next to helium, hydrogen and oxygen. In general, carbon is present in three distinct allotrope forms or crystallographic arrangements such as diamond, graphite and lonsdaleite or hexagonal diamond. Among these allotropes, graphite is used to fabricate other allotropes of carbon such as C60, C70, C540, amorphous carbon, carbon nano-tubes and graphene. Standalone graphene and carbon nanotubes usually possess enhanced thermal, mechanical, optical, electrical, electronic and biological properties. However, it is noteworthy that there are certain limitations due to the electronic confinement

Nanomaterials for Spectroscopic Applications
Edited by Kaushik Pal

ISBN 978-981-4877-69-5 (Hardcover), 978-1-003-16033-5 (eBook)
www.jennystanford.com

and crystal structure of both these allotropes. Thus, graphene and carbon nanotubes are combined as composites to enhance the already existing properties and reduce the limitations. It is essential to evaluate the properties of the resultant graphene/CNT nanocomposite to alter them according to the desired applications. Therefore, spectroscopies and electron microscopic techniques are utilized for the evaluation of nanocomposite properties. The aim of this chapter is to provide a brief overview of spectroscopic and microscopic responses of graphene/CNT nanocomposites. In addition, the alterations in their properties and analytical responses, depending on the synthesis approaches, were also discussed.

2.1 Introduction

Carbon is a most common element in the world, which is the 15th abundant element in the crust of the Earth [1] and is the fourth most common element in the universe, with respect to mass, next to helium, hydrogen and oxygen [2]. It is a unique element with a melting point of 3652°C [3], density of 2.2 g/cm^{-3} at 20°C [4], boiling point of 4827°C [5], van der Waals radius of 0.091 nm [6], ionic radius of 0.26 nm [7] and consists of three different isotopes [8]. In general, carbon is present in three distinct allotrope forms or crystallographic arrangements [9]. Diamond [10], graphite [11] and lonsdaleite or hexagonal diamond [12] are the three primary allotropes of carbon. Among these allotropes, graphite is used to fabricate other allotropes of carbon such as C60 or Buckminster fullerene or Buckyball [13], C70 [14], C540 [15], amorphous carbon [16], carbon nanotubes (CNTs) [17] and graphene [18]. In these distinct types, graphene is electronically confined in two directions [19], CNTs in one direction [20] and C60, C70 and C540 are carbon atoms arranged in a cage form with electronic confinement in all the directions [21].

Standalone graphene and carbon nanotubes usually possess enhanced thermal [22, 23], mechanical [24, 25], optical [26, 27], electrical [28, 29], electronic [30, 31] and biological properties [32, 33]. However, it is noteworthy that there are certain limitations due to the electronic confinement and crystal structure of both these allotropes [34]. Thus, graphene and carbon nanotubes were combined as composites to enhance the already existing properties and reduce the limitations [35]. Further, they are fabricated into nano sized composites as it will provide a strong interaction between the materials, with high surface to volume ratio and surface charges [36]. Hence, graphene/CNT nanocomposites are gaining much significance in various applications, ranging from industrial to the biomedical field [37]. It is essential to evaluate the properties of the resultant graphene/CNT nanocomposite to alter them according to the desired applications [38]. Therefore, spectroscopies and electron microscopic techniques are utilized for the evaluation of nanocomposite properties [39, 40]. The aim of this chapter is to provide a brief overview of spectroscopic and microscopic responses of graphene/CNT nanocomposites. In addition, the alterations in their properties and analytical responses, depending on the synthesis approaches, were also discussed.

2.2 Graphene and CNT Nanomaterials

In 2004, Novoselov and Geim fortunately isolated and analyzed a monolayer of graphene (MLG), which has led the scientists to elucidate the insights and the possibilities for altering the structure and properties of graphene and its other forms for desired applications. Graphene is a two-dimensional (2D) carbon form, that are constituted in a honeycomb lattice as shown in Fig. 2.1. The unit cell of MLG is made up of two carbon atoms, that are bonded together via sp^2 hybridization with 1.42 Å of bond length and 2.46 Å of lattice constant. In general, the graphene layers are arranged in a staging order of AB (Bernal) to form bilayer graphene (BLG), multilayer graphene (MULG), and graphite. In these graphene layers, carbon atoms exist in one layer will fit in the empty hexagon rings of the other layer in the bottom, where the inter-stacking distance between the layers will be 0.3354 Å. Further, van der Waals attractions are exhibited in between the layers of BLG, MULG, and graphite, which keeps the layers together, due to π-cloud. Both MLG and BLG can be considered as zero-bandgap semiconductors

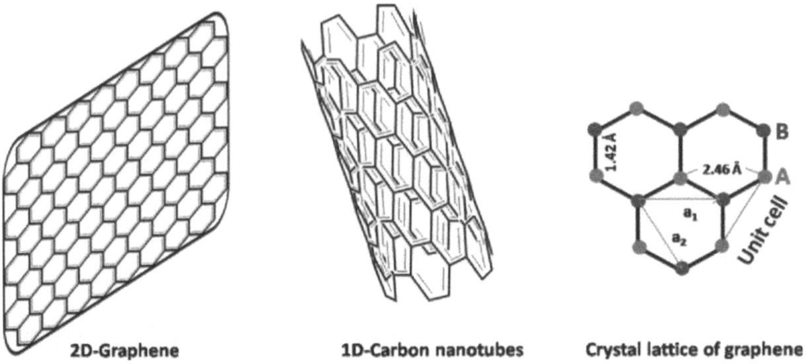

Figure 2.1 Structural representation of the 2D-graphene, 1D-carbon nanotubes, and crystal lattice of graphene.

or semi-metals, which limits their practical applications only to electronics field [41, 42].

On the other hand, CNTs are one-dimensional (1D) tubular carbon forms with a length-to-diameter ratio of 1000:1 and are formed by rolling an MLG, BLG, MULG, and graphite. Hence, CNTs are structurally similar to graphene as illustrated in Fig. 2.1. Generally, CNTs are classified into two types based on the number of walls in CNTs they are single-wall CNTs (SWCNTs) and multiwall CNTs (MWCNTs). SWCNTs (diameter of 0.7–2 nm) are narrower than MWCNTs (diameter of 15 nm) and flexible to be curved. Additionally, CNTs are further classified into three types such as an armchair, zigzag, and chiral CNTs, based on the rolling of MLG or graphite during CNTs formation. Furthermore, the CNTs have the behavior of both metallic and semi-conductive properties with bandgap ranging from 1.8 eV [43, 44]. The structural arrangement of 2D-graphene and 1D-CNTs are reported to be responsible for their exclusive properties to be beneficial in various applications [45].

2.2.1 Synthesis of Graphene

Since, the scientific breakthrough of graphene discovery, several physical and chemical approaches are reported for its synthesis to evaluate and demonstrate their industrial applications. In this section, the most common graphene synthesis approaches are

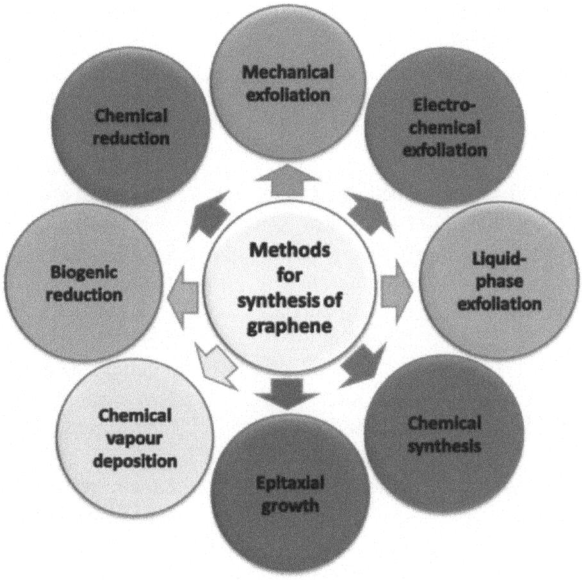

Figure 2.2 Common synthesis approaches to fabricate graphene structures.

discussed as illustrated in Fig. 2.2. Further, the advantages and the disadvantages of those synthesis approaches were mentioned in Table 2.1 [46–48].

2.2.1.1 Micromechanical exfoliation

Micromechanical exfoliation or scotch-tape approach was the first and an oldest physical approach for the synthesis of MLG on the surface of selected substrates. It is an eminent top-down approach, where the longitudinal or external force of $\sim$300 nN/μm^2 was applied on the surface of layered materials to yield 2D-materials. This method generally yields high quality and few micrometers of graphene sheets [49].

2.2.1.2 Electrochemical exfoliation

Electrochemical exfoliation is another distinct method for the synthesis of graphene sheets from graphite rod in an electrochemical

Table 2.1 Advantages and disadvantages of the synthesis approaches used to fabricate graphene

Approach	Advantages	Disadvantages
Mechanical exfoliation	Good quality	Low yield and not a scalable process
Electrochemical	Facile, less time, and economical	DC voltage and harmful electrolytes
Chemical synthesis	Facile and bulk production	Small segments and harmful chemicals
Epitaxial growth	Good quality and large film	Low vacuum and high-temperature conditions and not transferable easily
Chemical vapor deposition	High quality, large film, and reproducibility	Low vacuum and high-temperature conditions
Chemical reduction	Mass production	Expensive, harmful chemical agents, and high defects
Liquid-phase exfoliation	Simple, scalable, good quality, economical, and safe	Low yield, reaggregation, and defects
Biogenic reduction	Mass production, economical, safe, and eco-friendly	High defects

cell, where both anode and cathode were immersed in an electrolyte. In this method, 'graphite' rod will serve as an anode, 'Pt' wire will serve as a cathode, and the solutions of H_3PO_4 or H_2SO_4 will serve as an electrolyte. When applying electric voltage, graphene exfoliation can occur in an electrolyte and the accumulated graphene sheets can be washed and dried for further use [50].

2.2.1.3 Chemical synthesis

In the 20th century, synthetic organic scientists tried to synthesize polycyclic aromatic hydrocarbons (PAHs) via distinct chemical approaches under control reaction conditions, in which a definite length of graphene nanosheets can be prepared. In synthetic organic chemistry, PAHs can be recognized as 2D segments of graphene sheets with sp^2 hybridized carbon atoms [51].

2.2.1.4 Epitaxial growth

The word 'epitaxy' is obtained from the Greek language, where the meaning of 'epi' is'upon' or 'over' and 'taxis' is 'arrangement' or 'order.' This approach develops an epitaxial film of single-crystalline graphene on single-crystalline silicon carbide (SiC). It is considered as an extraordinary approach to synthesize graphene sheets on SiC substrate via higher temperatures of $\geq$ 1100°C. The lateral size of graphene sheets varies with the 'SiC' thickness, as the silicon atoms are desorbed from the surface of the graphene sheet during the synthesis [52].

2.2.1.5 Chemical vapor deposition

Chemical vapor deposition (CVD) is one of the well-known methods to synthesize high quality solid-state materials such as MLG on various substances, including copper, nickel, iridium, ruthenium, rhodium, platinum, palladium and cobalt. In this approach, the volatile compounds will be decomposed or allowed to react together and can be deposited as a thin film of desired product on substrates under controlled pressure and temperature (750–1200°C) conditions in a furnace. Generally, a mixture of hydrogen, methane, argon, and ethylene gas will be used for the synthesis of graphene in this approach [53].

2.2.1.6 Chemical reduction

The chemical reduction method is often employed to synthesize graphene sheets from MLG to MULG from graphene oxide (GO) via Hummer's method. Generally, chemical reduction of GO requires harmful, reducing agents such as sodium borohydride ($NaBH_4$), hydrazine, hydrazine hydrate, and dimethylhydrazine, that are highly toxic to humans and the environment [54].

2.2.1.7 Liquid-phase exfoliation

Recently, liquid-phase exfoliation is considered as one of the best affordable approaches to synthesize graphene sheets (both MLG and MULG) by exfoliating the bulk graphite in a liquid. It uses the sonication process, where mechanical shearing forces are applied to overcome van der Waals forces, that exist between the sheets of graphene in bulk graphite. However, it involves harmful chemical solvents and can lead to low yield and reaggregation of graphene sheets. Currently, researches are focused on developing green solvents and using surfactants such as Cetyl trimethyl ammonium bromide (CTAB), sodium dodecyl sulfate (SDS), and tween-20 for an efficient exfoliation of bulk graphite in different solvents, to overcome the challenges of toxicity and low yield [55].

2.2.1.8 Biogenic reduction

Scientists are currently focusing on alternative methods such as reduction of GO by employing plant extracts, extracts from microbes, sugar molecules, and vitamin C, to overcome the disadvantages of using hazardous chemicals in chemical reduction and liquid-phase exfoliation [56].

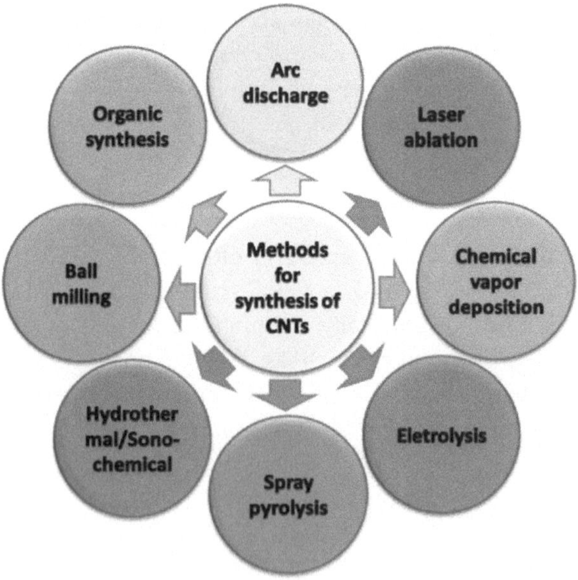

Figure 2.3 Synthesis approaches for the fabrication of CNTs.

2.2.2 Synthesis of Carbon Nanotubes

Similar to graphene, varieties of physical and chemical methods have been employed for the synthesis of CNTs, due to their potential applications. In this section, the common approaches for CNTs synthesis as displayed in Fig. 2.3 were explained. Moreover, the advantages and disadvantages of the distinct approaches, that are employed for the synthesis of CNTs were also described in Table 2.2 [37, 57–63].

2.2.2.1 Arc discharge

Arc discharge is one of the oldest methods, that are generally used for the synthesis of CNTs at higher temperatures (>1700°C) and higher pressures (0.5–100 atm). In this approach, a direct current (DC) arc voltage (10–50 V) was passed through the two graphite electrodes present in an inert gas (He). If pristine graphite

Table 2.2 Advantages and disadvantages of the synthesis approaches to fabricate CNTs

Approach	Advantages	Disadvantages
Arc discharge	Good quality	Defects, high temperatures, DC voltage, and not a scalable process
Laser ablation	High quality	Low vacuum and high temperature conditions
Chemical vapor deposition	High quality	Low vacuum and high-temperature conditions
Electrolysis	Facile, less time, and economical	DC voltage and electrolytes concentration
Spray pyrolysis	High quality	High temperature and not a scalable process
Hydrothermal/ Sono-chemical	Simple and mass production	Harmful chemical agents, high temperature, and high pressures
Ball milling	Mass production, economical, and eco-friendly	High defects, contamination, and high temperatures
Organic synthesis	Simple, good yield, no defects, and selective synthesis of three types of CNTs	Limited length and harmful chemicals

electrodes are employed, MWCNTs are formed on the cathode, and fullerenes are formed as soot in the furnace. Similarly, a graphite electrode consists of the metal catalysts (Co or Fe or Ni) is employed to produce SWCNTs in the soot. Further, doping of metal and non-metal atoms into the SWCNTs and MWCNTs is also possible with the arc discharge approach. However, the synthesized CNTs by this method will have fewer structural defects, compared to other approaches [64].

2.2.2.2 Laser ablation

Laser ablation or photo-ablation is also a conventional method, which was first explained in 1995 by Smalley's research group. This is a significant step in pulse laser deposition (PLD) and an excellent method for the synthesis of SWCNTs. Further, this approach is same as the arc discharge approach, where a short wavelength laser (Nd:YAG) is used to supply energy. Moreover, the material of interest such as CNTs can be extracted from the solid surface by this approach under controlled conditions.

2.2.2.3 Chemical vapor deposition

The CVD method is also an excellent approach for the synthesis of both SWCNTs and MWCNTs, as described in the section 2.2.1.4. However, several modifications have been included to the conventional CVD approach for synthesizing carbon nanomaterials with exclusive properties, which has led to the emergence of several techniques, including plasma-enhanced, thermal, hot filament, microwave, water-assisted, oxygen assisted, liquid pyrolysis, solid pyrolysis, injection assisted, radiofrequency, laser-assisted, aerogel assisted, alcohol assisted, and catalyst assisted CVD (CCVD). It is worthy to note that the CCVD is a widely employed approach for the synthesis of high quality CNTs, compared to other hybrid approaches [65].

2.2.2.4 Electrolysis

It is an economical and an easier approach for the synthesis of CNTs, compared to CVD, laser ablation, and arc discharge. It is an alternative approach for the synthesis of CNTs in bulk quantities, which generally includes electrolysis of carbon or graphite cathode dipped in an alkaline metal chloride electrolyte solution. During electrolysis, carbon components are gradually liberated from the graphite electrode surface into the molten salt that results in the development of CNTs in the liquid phase. After the reaction, the melt is cooled down to 25°C and are washed multiple times with distilled water for the complete cleaning of solidified salt. Finally, the resultant salts are filtered to recover the synthesized CNTs [66].

2.2.2.5 Spray pyrolysis

It is a prominent approach for the synthesis of CNTs, similar to CVD, however, vaporization and pyrolysis of the carbon precursor solutions at a specific time is used in spray pyrolysis. A dissolved/dispersed solution of the precursor is sprayed as an aerosol onto a heated surface in a furnace. Simultaneously, the solvent gets evaporated or decomposed into gaseous substances and the resultant CNTs will remain on the substrate [67].

2.2.2.6 Hydrothermal/sonochemical

CNTs can also be prepared by using hydrothermal approach followed by sonochemical process. In the approach, precursors were initially mixed thoroughly by sonication and the solution can be transferred into the autoclave, where nanomaterials could be synthesized in boiled water under high temperatures and pressures [68, 69].

2.2.2.7 Ball milling

Ball milling is a green and cost-effective mechanical method for the synthesis of CNTs, which readily converts the bulk solids into nanoscale materials. It is a type of grinder that uses different size of balls for the grinding and blending of materials. Further, it can also be used widely for the mass production of CNTs, due to its simplicity and flexibility [70].

Table 2.3 General properties of the 2D-graphene and 1D-CNTs

Property	2D-graphene	1D-CNTs
Thermal conductivity	5000 W/m.K	3000 W/m.K
Electrical conductivity	10^8 S/m	10^6–10^7 S/m
Charge carrier mobility	~230,000 $cm^2/V.s$	~80,000 $cm^2/V.s$
Specific surface area	2630 m^2/g	~200–900 m^2/g
Tensile strength	1100 GPa	~10–500 GPa
Young's modulus of elasticity	1.0 TPa	0.3–1.0 TPa
Density	~2.2 g/cm^3	~1.3 g/cm^3
Transparency	~97.7%	~90.0%
Thermal stability	~500–600°C	>600°C

2.2.2.8 Organic synthesis

In the 21st century, Jasti and Bertozzi synthesized armchair type of SWCNTs by a bottom-up organic approach from macrocyclic molecules such as cycloparaphenylenes via cyclo-addition for the first time. This is an interesting approach, that can be used for the selective synthesis of an armchair, zigzag, and chiral type of CNTs.

2.2.3 Properties of Graphene and Carbon Nanotubes

In general, carbon nanomaterials such as 2D-graphene and 1D-CNTs possess enormous properties, including high thermal conductivity, high electrical conductivity, low density, large surface area, high charge carrier mobility, high tensile strength, chemically inert, thermally stable, high transparency, and high Young's modulus of elasticity as represented in Table 2.3 [43, 71–73]. These extraordinary properties of these carbon allotropes have led to the development of several graphene-based, CNT-based, and graphene/CNT nanocomposites for various applications.

2.3 Graphene/CNT Nanocomposites

Recently, the synthesis of graphene/CNT nanocomposites has received wide attention from the scientific community, due to its variety of applications in optoelectronics, supercapacitors, solar

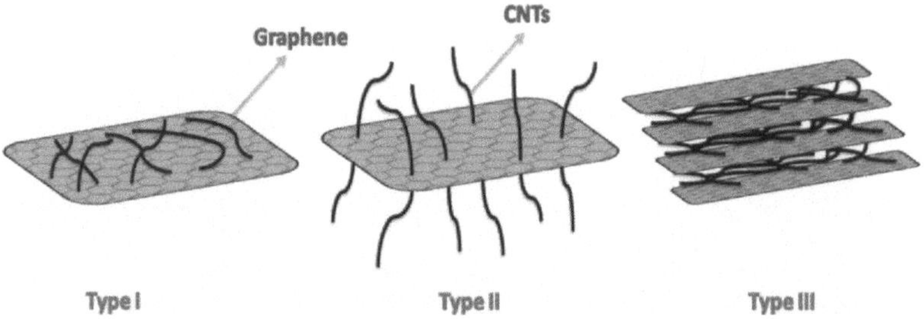

Figure 2.4 Distinct nanoarchitectural types of graphene/CNT nanocomposites.

cells, sensors, fuel cells, lithium-ion batteries, reinforcing agents or fillers, and aerogels [74–78]. Therefore, various methods have been reported to synthesize graphene/CNT nanocomposites.

CNTs are 1D and graphene is in 2D nature. The reinforcement of 1D CNTs with 2D graphene can able to form a variety of topological graphene/CNT nanocomposite structures. These can be divided into three types based on the arrangement of 1D CNTs with the graphene material [74–79], as represented in Fig. 2.4. In type I, CNTs are distributed on the surface of the graphene sheets in horizontal direction. This doesn't require specific conditions to control the orientation of CNTs at the nanoscale level, and hence, most of the graphene/CNT nanocomposites are synthesized in this form. Moreover, the effect on the properties of these nanocomposite types are dependent on the contact resistance and interactions between the CNTs and graphene sheets. In type II, CNTs are oriented perpendicularly with the graphene sheets and there is only one interaction point between the CNTs and graphene sheets, which causes superior out-of-plane electrical and mechanical properties to maintain the outstanding in-plane properties but decreases the interfacial thermal and electrical resistances. Furthermore, the properties of these nanocomposites depend on the nature of C-C bonding between the CNTs and graphene sheets. In type III, CNT is enclosed in the graphene sheets. This type of nanocomposites is difficult to prepare but furnish a good core–shell like dense architectures. The wrapping forms rely on the proportions and interactions between the CNTs and graphene sheets, that are used

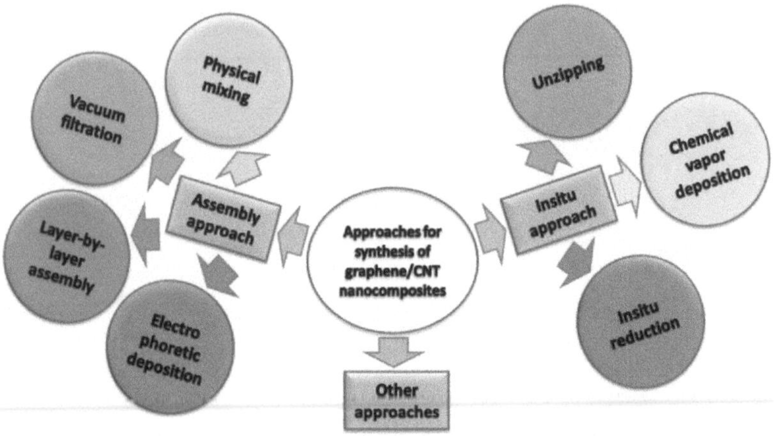

Figure 2.5 Common synthesis approaches of graphene/CNT nanocomposites.

for the fabrication of CNT/graphene nanocomposite. These three types of architectures can lead to the formation of graphene/CNT nanocomposites with enhanced physical and chemical properties, that can be fabricated via various approaches. In recent years, several exclusive approaches are reported for the synthesis of graphene/CNT nanocomposites, which is discussed in the following sections.

2.3.1 Synthesis of Graphene/CNT Nanocomposites

Numerous attempts have been made for the development of different approaches for the synthesis of graphene/CNT nanocomposites to realize their industrial applications. These can be systematically divided into three groups [74, 79, 80], as illustrated in Fig. 2.5. They are assembly methods, in situ method, and other methods. Moreover, to work out a simple, low cost, bulk, and efficient approach to graphene/CNT nanocomposites remains a big challenge.

2.3.1.1 Assembly approach

It is a well-known approach for the synthesis of graphene/CNT nanocomposites, where CNTs and graphene sheets are co-assembled together via π–π interactions or van der Waals interactions by their conjugated aromatic sp^2 hexagonal structure to form hierarchical nanocomposites. In addition, functional groups present on the GO, functionalized GO, and functionalized CNTs can lead to the formation of covalent linkage by chemical assembly or electrostatic interactions [81].

2.3.1.1.1 Physical mixing approach

It is a non-covalent approach, where GO/rGO/graphene can interact with the CNTs through superior π–π interactions due to their extended sp^2 nanostructures. Furthermore, the GO forms stable suspension in water, which is not in the case of CNTs. CNTs agglomerated in the water, when GO suspension is mixed with CNTs, where the CNTs can be dispersed easily with the dispersibility of GO and the CNTs help to avoid the restacking of GO sheets [82].

2.3.1.1.2 Layer-by-layer (LBL) assembly approach

It is a widely employed, productive and versatile approach for the facile synthesis of graphene/CNT nanocomposites. LBL approach is commonly used for the preparation of coatings on the surfaces and thin films of devices. This is a cyclic process in which the oppositely charged monolayers are adsorbed on the surface, either alternatively or sequentially, to form a multi-layer film. By varying the number of layers, the thickness of the films can be controlled in the nanometre range. Further, monolayers of the films bond together via electrostatic forces and other interactions such as hydrogen-bonding, van der Waals forces, covalent and host-guest relation. Moreover, the adhesion property of the films on the surface will be increased, due to the LBL structure and charge-to-charge interactions. Thus, carbon nanomaterials (CNMs) should be initially functionalized with chemical groups that can interact with the other layers [83].

2.3.1.1.3 Vacuum filtration approach

It is a fast filtration method, that are employed for the solid (graphene/CNT nanocomposites) separation from the liquid suspensions, particularly onto a filtration membrane. It has been

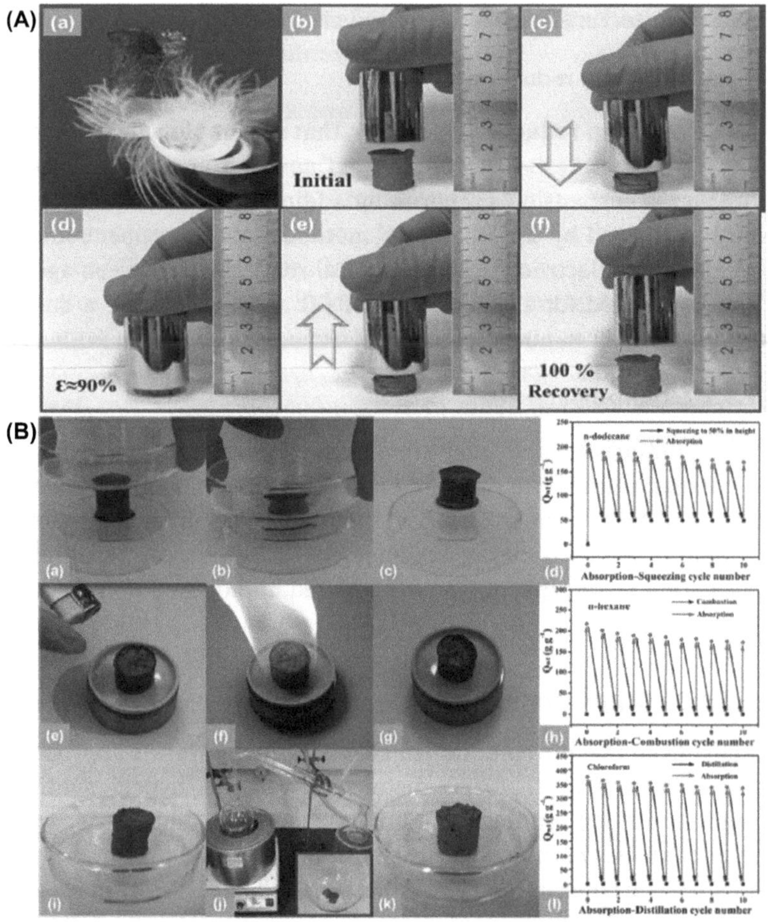

Figure 2.6 (A) As-synthesized GCPCA standing on a feather (2.13 mg/cm^3) with (a) an ultra-low density, and (b–f) images describing the super compressibility. (B) Recycling of n-dodecane absorbed GCPCA through (a–c) squeezing and (d) corresponding absorption-squeezing cycles, (e–g) recycling of n-hexane absorbed GCPCA through combustion and (h) corresponding absorption-combustion cycles, (i–k) recyclability of chloroform absorbed GCPCA through distillation and (l) corresponding absorption-distillation cycles. Reproduced with permission from [78].

in nanocomposites are capable to boost the structural, chemical, electrochemical, catalytic, surface area, mechanical, thermal, electrical, and optical properties of the nanocomposites, where all these properties co-existed together, compared to the pure graphene and CNTs. However, the synthesis approach, functionalization, interfacial adhesion, and weight percentages of the 2D-graphene and 1D-CNTs, that are used in the development of these nanocomposites can influence the properties.

2.3.2.1 Mechanical and thermal properties

The excellent mechanical properties (stiffness, tensile strength, elasticity, and fracture toughness), thermal stability, and low density of 2D-graphene and 1D-CNTs together as a nanocomposite makes them an excellent material to be used as reinforcing agents/fillers and aerogels. The presence of defects can affect the mechanical properties of the graphene/CNT nanocomposites [97–99]. Further, the thermal stability of the 2D-graphene, 1D-CNTs, and graphene/CNT nanocomposites are also very high and mainly depends on the atmosphere and functionalization. Lv et al. (2016) reported the fabrication of superelastic 3-dimensional (3D) graphene/CNT aerogel via freeze-drying approach, followed by a hydrothermal approach and the developed 3D-graphene/CNT aerogel possess extensive properties, such as thermal conductivity of 88.5 W/m.K, low density of 85 mg/cm^3, low thermal interface resistance of 13.6 m.m^{2}K/W, and high stiffness [90].

Similarly, Zhan et al. (2018) developed a strong cabbage-like graphene/MWCNT-PDA (polydopamine) nanocomposite aerogel (GCPCA) with distinct graphene and CNTs proportions. GCPCA-80 possess microporous structure and are employed for the absorption of various organic pollutants from water and it shows remarkable properties including hydrophobicity, ultra-lightweight, low density to stand on feather (4.37–2.13 mg/cm^3), good surface area (84 m^2/g), structural stability, high elasticity, strain (<24%), in situ burning, thermal stability (750°C), repeatable compressibility (10 times with efficiency 90%), good absorption-capacity (533 g/g), and recyclability (501 times for chloroform), as displayed in Fig. 2.6A–B [78].

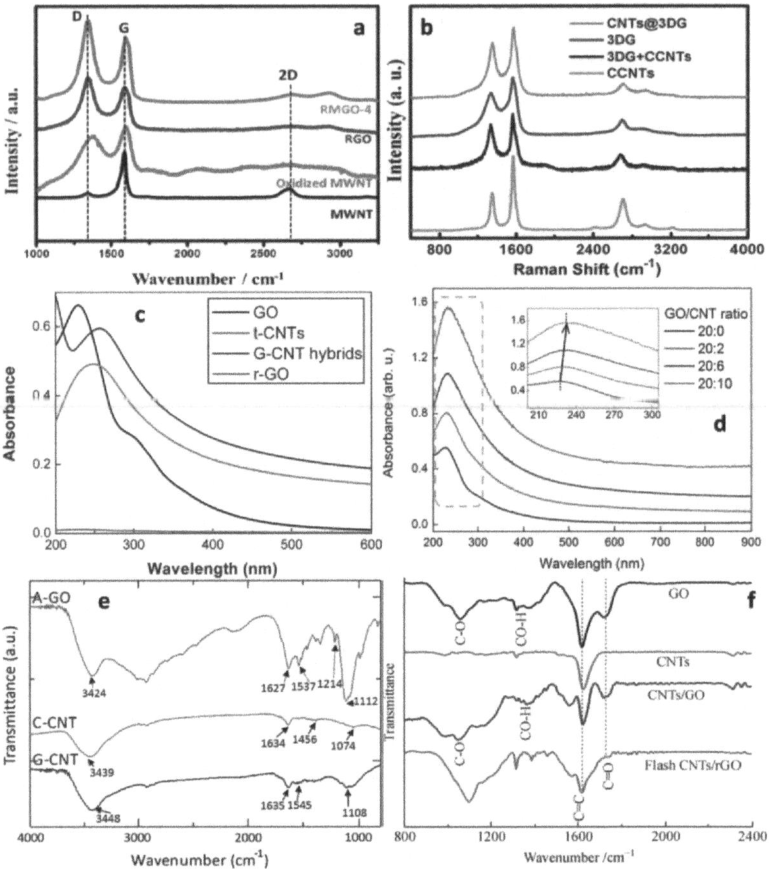

Figure 2.8 Raman spectra of (a) RMGO-4, reproduced with permission from [102], and (b) CNTs@3DG nanocomposites and their components, reproduced with permission from [103], UV-visible spectra of (c) graphene-CNT, reproduced with permission from [104] and (d) GO/CNT nanocomposites and their components, reproduced with permission from [105], and FT-IR spectra of (e) graphene-CNT and (f) rGO/CNT nanocomposites and their components, reproduced with permission from [106, 107].

and graphene/CNT nanocomposites from GO and GO/CNT nanomaterials. Generally, UV-visible spectroscopy of GO reveals that the absorption peak at 228 nm and a small peak at 260 nm, corresponding to the π–π^* and n–π^* transitions of the C=C and C=O bonds. After the reduction of GO/CNT to graphene/CNT nanocomposite, the peak at 228 nm are shifted to higher angles, which indicates the removal of the C=O bonds and increase in the C=C bonds, as displayed in Fig. 2.8c [104]. The intensities of the peaks depend on the composition of GO and CNTs, that are employed in the synthesis of GO/CNT nanocomposite, as exhibited in Fig. 2.8d [105].

2.4.3 FT-IR Spectroscopy

The infrared area of the electromagnetic spectrum includes radiation with wavelengths ranging from 10 to 12500 cm^{-1}, however, the common analytical applications are limited to a portion of the middle IR region extending from 4000 to 400 cm^{-1}. The absorption spectra in the infrared region originate from the transitions between vibrational (along with rotational) levels of a molecule present in its ground electronic state upon irradiation with infrared radiation. It gives the chemical and structural information about the carbon nanomaterials, such as graphene/CNT nanocomposite. Figure 2.8e and f has displayed the FT-IR spectra of the graphene/CNT nanocomposite, that are synthesized by two groups [106, 107]. Commonly, graphene/CNT nanocomposite exhibits several characteristic bands at 1627–1635 cm^{-1}, that are assigned to the C=C bond, whereas the band at 1074–1112 cm^{-1} is attributed to the C–O bond, the broadband at 3400–3500 cm^{-1} corresponding to stretching vibrations of –OH group, the band at 1735 cm^{-1} is assigned to the C=O bond and the band at 1243 cm^{-1} is assigned to the epoxy C–O–C bond. Additionally, the bands at 2924 and 2853 cm^{-1} corresponds to the asymmetric and symmetric stretching frequencies of CH_2 groups. Further, the bands at 1386 and 1038 cm^{-1} are ascribed to the deformation vibrations of OH groups, respectively [108].

2.5 The Microscopic Response of Graphene/CNT Nanocomposites

In addition to spectroscopic studies, microscopic studies, including scanning electron microscope (SEM), field emission (FE) SEM, transmission electron microscope (TEM), and atomic force microscope (AFM) are widely used to describe the distribution, morphology, surface fractures, and microstructures of the graphene/CNT nanocomposites.

2.5.1 Scanning Electron Microscopy

SEM is a powerful technique, that is classified under scanning probe microscopy (SPM), which uses a focused beam of high energy electrons to provide information about the orientation of materials in the sample, its external morphology, surface roughness and elemental composition of the graphene/CNT nanocomposites [109]. Shao et al. (2012) fabricated a type III nanostructure of CNT/GO flexible membrane as displayed in Fig. 2.9a. The cross-sectional images (Fig. 2.9b–c) of this membrane showed that GO layers are aligned horizontally with spaces and the spaces are filled with CNTs in a vertical arrangement [110].

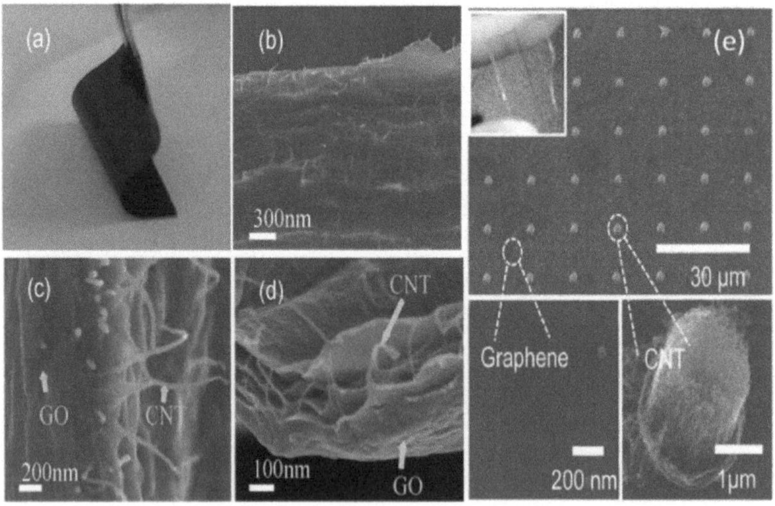

Figure 2.9 A typical (a) photograph, (b–d) cross-sectional SEM images of CNT/GO membranes, reproduced from Ref. [110] with permission from The Royal Society of Chemistry, and (e) SEM images of CNT/graphene nanocomposite on a quartz substrate. Reproduced with permission from [111].

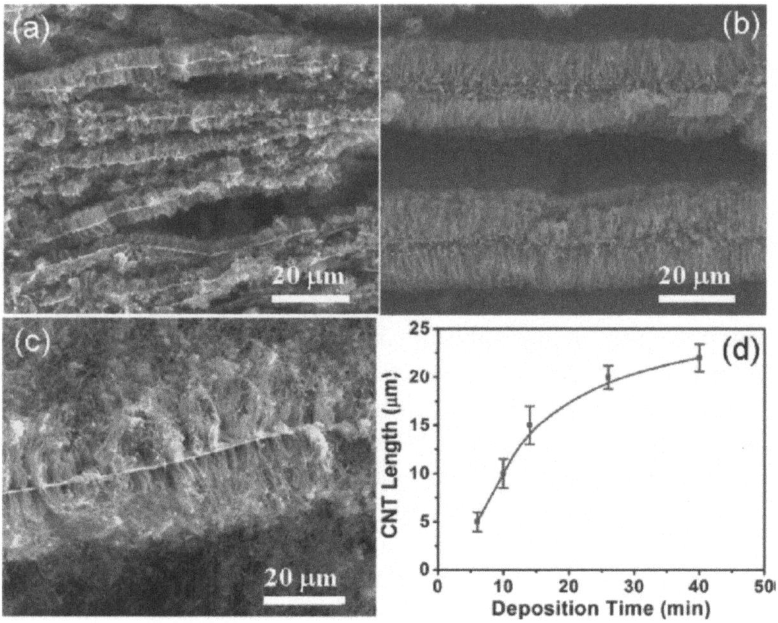

Figure 2.10 (a–c) SEM images of the thermally expanded 3D-pillared CNT/graphene nanocomposite for different pyrolysis times (5, 10, and 30 min), and (d) graph of the VACNT pillar length versus the deposition time. Reproduced with permission from [112].

Further, Kim et al. (2011) reported a type II nanoarchitecture of graphene/CNT nanocomposite transparent film, that are fabricated by CVD approach for future 3D-holography application. The SEM image (Fig. 2.9e) of the transparent film possesses small CNT dot morphology and are systematically grown on graphene with 15 μm of distance. Furthermore, the in situ image revealed that each CNT dot displays a forest of CNTs with 1 μm of height. Finally, the developed transparent electrode exhibited a transmittance of 94% [111]. Moreover, Du et al. (2011) developed a type III nanonetwork of 3D-pillared CNT-graphene. Figure 2.10a–c displayed that the height of the CNT pillars can be extended by increasing the deposition time of CNTs. Likewise, Fig. 2.10c–d showed a constant increase in the length of the CNT pillar up to 20 μm and can increase the volume up to 200 times with the deposition time. Moreover, SEM images in Fig. 2.10a–c demonstrate that the developing CNT pillars are strong to 'push' the layered graphene layers apart from each other. The developed 3D-pillared CNT-graphene electrode proves to be beneficial for supercapacitor applications [112].

2.5.2 Transmission Electron Microscopy

TEM is also a significant SPM technique in which electron beams having the energy of hundreds of KeV, that are transmitted over an ultra-thin sample, where it interacts with the sample to provide morphological, topographical, compositional and crystallographic information of the graphene/CNT nanocomposites [113]. Sridhar et al. (2015) synthesized a nitrogen-doped CNT/graphene nanocomposite and applied them for supercapacitor application. Figure 2.11a–b and 2.11c–d illustrates the SEM and TEM images of the synthesized nanocomposite, respectively, where CNTs with open ends are anchored on the surface of the mesoporous graphene and it is similar to type I nanoarchitecture [95].

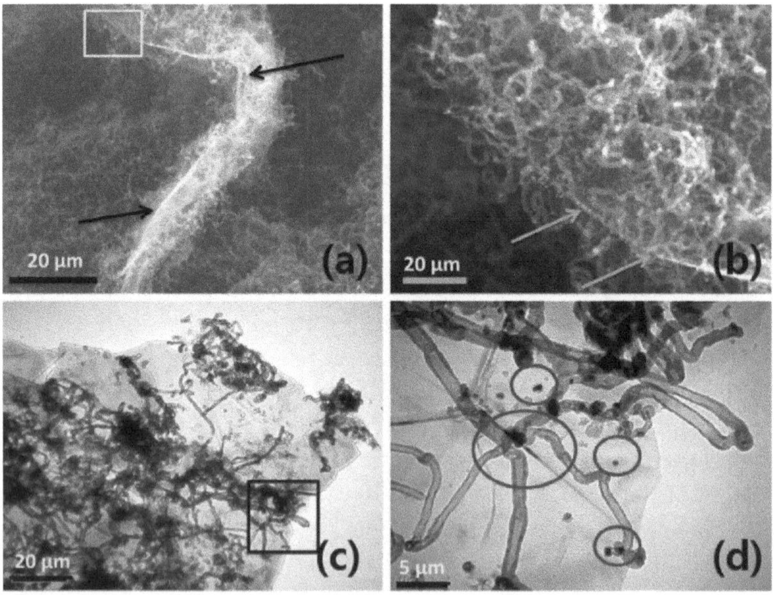

Figure 2.11 (a–b) SEM and (c–d) TEM images of the nitrogen-doped CNT/graphene nanocomposite. Reproduced with permission from [95].

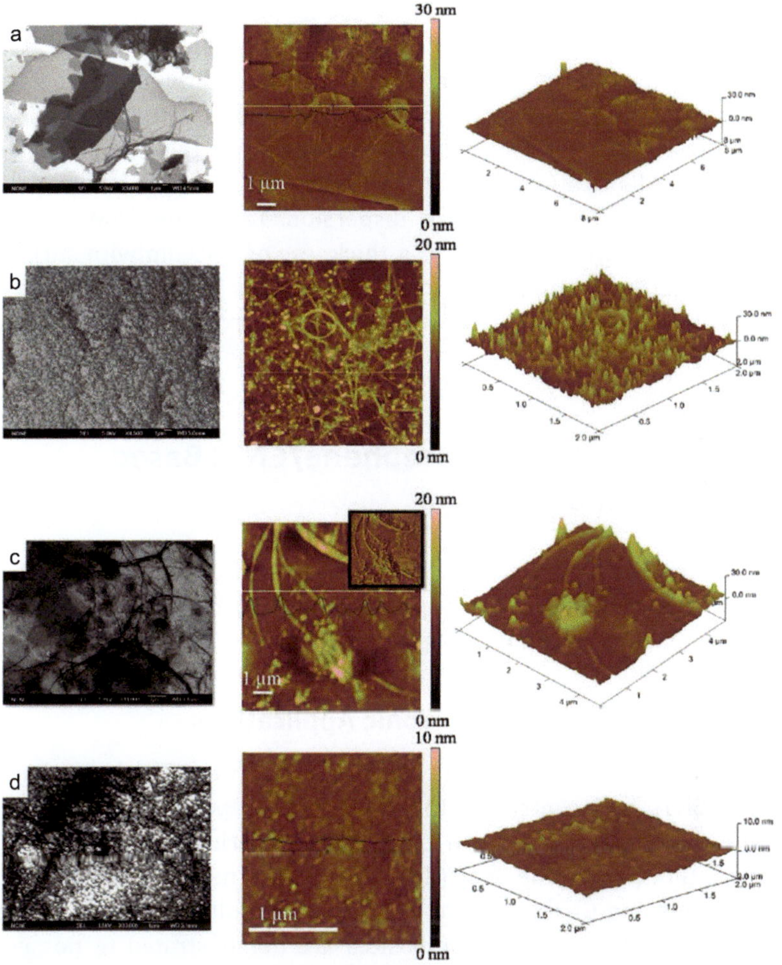

Figure 2.12 AFM images of the (a) graphene, (b) SWCNTs, and graphene/CNT hybrid film (c) before and (d) after sonication. Reproduced with permission from [101].

2.5.3 Atomic Force Microscopy

AFM is also a typical surface imaging technique, where the images can be procured by scanning the samples with a sharp atomic tip. Generally, it is used to determine the topographical morphology, surface roughness, and thickness of the film. Tung et al. (2009)

fabricated a simple and flexible graphene/CNT nanocomposite electronic device. Figure 2.12a–c exhibits the image of pure graphene, CNTs, and graphene/CNT hybrid film. The hybrid film shows a thickness of 5 nm with high roughness due to the presence of CNT bundles. Further, these protruded CNTs are not recommended for the device fabrication, and thus, the hybrid composites are sonicated for 90 min to yield uniform dispersion. Later, the AFM film of sonicated hybrid film displays a thickness of 1.49 nm with surface smoothness as represented in Fig. 2.12d. This study demonstrated that the sonication process can remove the protrusions created by the CNTs [101].

2.6 Applications of Graphene/CNTs Based Nanomaterials

In recent times, graphene/CNTs nanocomposites are widely under extensive research to be used in electrical, electronic, thermal, environmental and biomedical applications.

2.6.1 Electrical and Electronic Applications

The enhanced electrical and electronic properties of both graphene and CNTs will contribute to their superior property, which are under extensive studies to be useful for various applications in the future [79]. Wang et al. (2019) fabricated nanocomposite using graphene and CNTs, that are decorated with ultra-thin manganese oxide. These novel nanocomposites are demonstrated to possess exclusive conductivity, due to their large surface area, and can function as an excellent electron transfer channel with superior lithium ion diffusion. Thus, these hybrid nanocomposites are proven to be beneficial as a potential sulfur hosts for high performance lithium-sulfur batteries [114]. Similarly, Da et al. (2017) utilized carbon nanotubes with graphene oxide as hybrids to form flexible, transparent conducting films, which will be beneficial for organic electronic applications [115]. Likewise, Karteri et al. (2016) fabricated reduced graphene oxide-iron oxide and carbon nanotube-iron oxide nanocomposites with photodiode properties to evaluate their photovoltaic performance. The results revealed that the graphene oxide-iron oxide photodiodes are highly significant as photodiodes, than a CNT hybrid [116]. Further, Wang et al. (2020) recently fabricated a novel reduced graphene oxide-carbon nanotube-poly (3,4-ethylene dioxythiophene): poly (styrene sulfonate) (PEDOT:PSS) nanocomposites that are functionalized with polydopamine to be useful as a flexible transparent electrode for organic light emitting diodes (OLEDs) [117]. Furthermore, these nanocomposites are used to fabricate microbolometer to detect infrared rays [118], flexible electromagnetic interference shielding film [119], supercapacitors [120] and are highly useful in microelectronics [121].

2.6.2 Thermal Applications

The high surface to volume ration and the exclusive crystallographic arrangement of graphene-CNT nanocomposites has provided them with enhanced thermal properties, that can be used in desired thermal applications. Liu et al. (2017) fabricated a novel graphene-carbon nanotube hybrid with polyethylene to form nanocomposites for exhibiting enhanced thermal energy transport between their interface [122]. Further, Shi and Zhu (2020) recently fabricated graphene-CNT nanocomposites with interface regulations to yield particles with exclusive thermal conductivity and anticorrosion properties [123]. Furthermore, Zhang et al. (2019) synthesized interconnected graphite-welded CNT in the form of sponges for the fabrication of stress sensitive, elastic, thermally conductive nanocomposites [124]. Moreover, Zhou et al. (2018) fabricated multiwalled carbon nanotube embedded graphene nanosheets, that are functionalized with silver, to be useful as a high-performance thermal management nanocomposites with enhanced thermal conductivity [125]. In addition, Zhao et al. (2015) synthesized nanocomposite aerogels with graphene and multiwalled carbon nanotube and are proved to possess enhanced electrical conductivity and Seebeck coefficient. These properties of graphene-CNT aerogel has been proposed to be beneficial as a potential thermoelectric material with enhanced thermal conductivity [126]. Recently, Dul et al. (2020) effectively fabricated hybrid graphene nanoplatelets-carbon nanotube nanocomposite via compression molding and fused filament fabrication approach to alter their properties. The results revealed that an alteration in the synthesis approach can change or enhance the thermal conductivity of the nanocomposite [127].

2.6.3 Environmental Applications

It can be noted that these carbon nanocomposites are highly beneficial in wastewater treatment, environmental remediation and photocatalytic degradation of pollutants. Sharma et al. (2015) fabricated a magnetic, three-dimensional, graphene-carbon nanotube iron nanocomposite with the larger adsorption capacity to adsorb and remove metals and organics for effective water purification application, compared to conventional adsorbents. These nanoadsorbents also exhibited enhanced antibacterial activity against *Escherichia coli* and *Staphylococcus aureus*, which can be beneficial in eliminating pathogenic microbes in water [128]. Likewise, Hussein et al. (2018) synthesized a hybrid nanocomposite using graphene and carbon nanotube along with polyaniline for water disinfection application. The resultant nanocomposite showed enhanced ability to inhibit pathogenic bacterial strains such as *E. coli* and *S. aureus*, and can be a better bacterial adsorption material, that can be reused and recycled for several times [129]. Similarly, Ahsan et al. (2020) produced novel nanosized nickel metal organic framework, that are decorated with CNTs and graphene oxide for water remediation application. These nanoframework are proven to possess excellent adsorption capacity via π–π interaction towards toxic methylene blue dye [130]. Further, Hussein et al. (2020) fabricated a hybrid nanocomposite using polymer, such as polyvinyl chloride and polyvinyl alcohol, that are reinforced with CNT-graphene. These novel nanocomposites are proven to be beneficial in the removal of acid red from water samples, obtained from red sea, distilled and tap water [131]. Furthermore, these graphene/CNT nanocomposites are highly beneficial in environmental monitoring applications such as selective silver ionic sensor and heavy metal detection [132]. Recently, Xu et al. (2019) fabricated a novel electrode by using iron oxide-multiwalled CNT, that are scribed with graphene to form nanocomposite and are functionalized with chitosan. These electrodes are utilized in a sensitive electrochemical sensor the simultaneous voltametric detection of lead and cadmium [133].

2.6.4 Biomedical Applications

In general, graphene/CNT nanocomposites are employed as biosensors for the effective diagnosis of diseases or detection of biomolecules. Sharma et al. (2012) demonstrated that the graphene/CNT nanocomposites can be used in magneto-electrochemical immunoassay to amplify signals for the effective detection of analytes. These hybrid nanocomposite based immunoassay are proven to be beneficial in the effective screening of toxins and environmental pollutants in the samples [134]. Similarly, Mani et al. (2014) established the direct myoglobin electrochemistry at reduced graphene oxide-platinum nanoparticles-multiwalled carbon nanotube nanocomposites to demonstrate their effective biosensing ability to sensitively detect nitrite and hydrogen peroxide [135]. Further, Kumar et al. (2015) synthesized graphene-CNT nanohybrid and are immobilized with aptamer as aptasensor for the label free detection of cardiac myoglobin biomarker [136]. Likewise, graphene-CNT nanocomposites functionalized with polymers are widely used as a scaffold with enhanced mechanical and bioactivity for bone regeneration and tissue engineering applications [137]. Furthermore, Ryoo et al. (2010) stated that the graphene-CNT nanocomposites can be used to increase the proliferation of NIH-3T3 fibroblast cells with local adhesion and can be used in the gene transfection applications [138]. Moreover, Fan et al. (2013) demonstrated that the graphene-CNT-iron oxide hybrid nanocomposite can be used for the targeted and controlled delivery of anticancer fluorouracil agent in human liver HepG2 cancer cells for effective cancer treatment and are less toxic towards Chang liver cells [139].

2.6.5 Other Applications

These carbon-based nanocomposites, in combination with other metals, metal oxides and polymers, are also used in various other applications such as fast lithium storage [140], flexible electrodes for hydrogen evolution [141], electrocatalytic applications in hydrogen evolution reactions [142], pseudocapacitive materials [143], catalysts [144] and load transfer agents [145]. They are also used as hybrid films in supercapacitors [146], anode material to improve rate capacities [147], methylene blue adsorption agent [148], asymmetric pseudocapacitors [149], hydrogen peroxide sensors [150] and ultrahigh electrocatalytic agent for methanol oxidation [151]. Thus, extensive research in the future on these nanocomposites will yield more significant particles with highly exclusive properties.

2.7 Conclusion

This chapter provides ample evidences that the synthesis approaches can alter the spectroscopic and microscopic responses of the graphene/CNT nanocomposites and these two analytical approaches are highly beneficial in evaluating their significant properties. However, it is necessary to identify hybrid or novel methods for the synthesis of these nanocomposites, as most of the synthesis approaches utilize toxic chemicals, which may not be recommendable for biological applications. Thus, it will be necessary to fabricate these nanocomposites using biological approaches or functionalize with biological molecules and evaluate their spectroscopic and microscopic responses in the future.

References

1. M. Meybeck, Carbon, nitrogen, and phosphorus transport by world rivers, *American Journal of Science*, 1982, **282**(4), 401–450.
2. T. Henning and F. Salama, Carbon in the universe, *Science*, 1998, **282**(5397), 2204–2210.
3. A. I. Savvatimskiy, Measurements of the melting point of graphite and the properties of liquid carbon (a review for 1963–2003), *Carbon*, 2005, **43**(6), 1115–1142.
4. W. M. Post, W. R. Emanuel, P. J. Zinke and A.G. Stangenberger, Soil carbon pools and world life zones, *Nature*, 1982, **298**(5870), 156–159.

5. S.-H. Jhi, Y.-K. Kwon, K. Bradley and J.-C. P. Gabriel, Hydrogen storage by physisorption: beyond carbon, *Solid State Communications*, 2004, **129**(12), 769–773.
6. S. S. Batsanov, Van der Waals radii of elements, *Inorganic Materials*, 2001, **37**(9), 871–885.
7. M. Rahm, R. Hoffmann and N. W. Ashcroft, Atomic and ionic radii of elements 1–96, *Chemistry A European Journal*, 2016, **22**(41), 14625–14632.
8. D. A. Schoeller, M. Minagawa, R. Slater and I. R. Kaplan, Stable isotopes of carbon, nitrogen and hydrogen in the contemporary North American human food web, *Ecology of Food and Nutrition*, 1986, **18**(3), 159–170.
9. S. K. Tiwari, V. Kumar, A. Huczko, R. Oraon, A. D. Adhikari and G. C. Nayak, Magical allotropes of carbon: prospects and applications, *Critical Reviews in Solid State and Materials Sciences*, 2016, **41**(4), 257–317.
10. S. Kaciulis, A. Mezzi, P. Calvani and D. M. Trucchi, Electron spectroscopy of the main allotropes of carbon, *Surface and Interface Analysis*, 2014, **46**(10–11), 966–969.
11. B. Wunderlich and Y. Jin, The thermal properties of four allotropes of carbon, *Thermochimica Acta*, 1993, **226**, 169–176.
12. P. Németh, L. A. J. Garvie, T. Aoki, N. Dubrovinskaia, L. Dubrovinsky and P. R. Buseck, Lonsdaleite is faulted and twinned cubic diamond and does not exist as a discrete material, *Nature Communications*, 2014, **5**(1), 1–5.
13. A. Hirsch, The era of carbon allotropes, *Nature Materials*, 2010, **9**(11), 868–871.
14. S. B. Singh and A. Singh, The third allotrope of carbon: fullerene an update, *International Journal of ChemTech Research*, 2013, **5**(1), 167–171.
15. G. E. Scuseria, The equilibrium structures of giant fullerenes: faceted or spherical shape? An ab initio Hartree-Fock study of icosahedral C240 and C540, *Chemical Physics Letters*, 1995, **243**(3–4), 193–198.
16. N. Shohoji, L. G. Rosa, J. C. Fernandes, D. Martínez and J. Rodríguez, Catalytic acceleration of graphitisation of amorphous carbon during synthesis of tungsten carbide from tungsten and excess amorphous carbon in a solar furnace, *Materials Chemistry and Physics*, 1999, **58**(2), 172–176.
17. H. Niu, X.-Q. Chen, S. Wang, D. Li, W. L. Mao and Y. Li, Families of superhard crystalline carbon allotropes constructed via cold

compression of graphite and nanotubes, *Physical Review Letters*, 2012, **108**(13), 135501.

18. M. I. Katsnelson, Graphene: carbon in two dimensions, *Materials Today*, 2007, **10**(1–2), 20–27.
19. C. Berger, Z. Song, X. Li, X. Wu, N. Brown, C. Naud, D. Mayou, T. Li, J. Hass and A. N. Marchenkov, Electronic confinement and coherence in patterned epitaxial graphene, *Science*, 2006, **312**(5777), 1191–1196.
20. X. Pan and X. Bao, The effects of confinement inside carbon nanotubes on catalysis, *Accounts of Chemical Research*, 2011, **44**(8), 553–562.
21. J. P. Connerade and R. Semaoune, Relativistic study of the electronic structure and 5d orbital of La confined inside a C60 fullerene cage, *Journal of Physics B: Atomic, Molecular and Optical Physics*, 2009, **33**(5), 869.
22. W. Lee, J. U. Lee, B. M. Jung, J.-H. Byun, J.-W. Yi, S.-B. Lee and B.-S. Kim, Simultaneous enhancement of mechanical, electrical and thermal properties of graphene oxide paper by embedding dopamine, *Carbon*, 2013, **65**, 296–304.
23. J. Hone, M. C. Llaguno, M. J. Biercuk, A. T. Johnson, B. Batlogg, Z. Benes and J. E. Fischer, Thermal properties of carbon nanotubes and nanotube-based materials, *Applied Physics A*, 2002, **74**(3), 339–343.
24. M. Cano, U. Khan, T. Sainsbury, A. O'Neill, Z. Wang, I. T. McGovern, W. K. Maser, A. M. Benito and J. N. Coleman, Improving the mechanical properties of graphene oxide based materials by covalent attachment of polymer chains, *Carbon*, 2013, **52**, 363–371.
25. B. I. Yakobson and P. Avouris, Mechanical properties of carbon nanotubes, in *Carbon Nanotubes: Synthesis, Structure, Properties, and Applications*, M. S. Dresselhaus, G. Dresselhaus and P. Avouris, eds. Springer Berlin Heidelberg, Berlin, Heidelberg, 2001, pp. 287–327.
26. Z. Liu, Y. Wang, X. Zhang, Y. Xu, Y. Chen and J. Tian, Nonlinear optical properties of graphene oxide in nanosecond and picosecond regimes, *Applied Physics Letters*, 2009, **94**(2), 021902.
27. M. F. Lin and K. W. K. Shung, Plasmons and optical properties of carbon nanotubes, *Physical Review B*, 1994, **50**(23), 17744–17747.
28. Z. Lei, L. Lu and X. S. Zhao, The electrocapacitive properties of graphene oxide reduced by urea, *Energy & Environmental Science*, 2012, **5**(4), 6391–6399.
29. P. J. Burke, Luttinger liquid theory as a model of the gigahertz electrical properties of carbon nanotubes, *IEEE Transactions on Nanotechnology*, 2002, **1**(3), 129–144.

30. S. Saxena, T. A. Tyson, S. Shukla, E. Negusse, H. Chen and J. Bai, Investigation of structural and electronic properties of graphene oxide, *Applied Physics Letters*, 2011, **99**(1), 013104.
31. P. G. Collins, K. Bradley, M. Ishigami and A. Zettl, Extreme oxygen sensitivity of electronic properties of carbon nanotubes, *Science*, 2000, **287**(5459), 1801.
32. F. Ordikhani, M. Ramezani Farani, M. Dehghani, E. Tamjid and A. Simchi, Physicochemical and biological properties of electrodeposited graphene oxide/chitosan films with drug-eluting capacity, *Carbon*, 2015, **84**, 91–102.
33. W. Wei, A. Sethuraman, C. Jin, N. A. Monteiro-Riviere and R. J. Narayan, Biological properties of carbon nanotubes, *Journal of Nanoscience and Nanotechnology*, 2007, **7**(4–5), 1284–1297.
34. W. Gu and G. Yushin, Review of nanostructured carbon materials for electrochemical capacitor applications: advantages and limitations of activated carbon, carbide-derived carbon, zeolite-templated carbon, carbon aerogels, carbon nanotubes, onion-like carbon, and graphene, *WIREs Energy and Environment*, 2014, **3**(5), 424–473.
35. H. Yang, Z. Yu, P. Wu, H. Zou and P. Liu, Electromagnetic interference shielding effectiveness of microcellular polyimide/in situ thermally reduced graphene oxide/carbon nanotubes nanocomposites, *Applied Surface Science*, 2018, **434**, 318–325.
36. Y. Ma, Q. Wang, Y. Miao, Y. Lin and R. Li, Plasma synthesis of Pt nanoparticles on 3D reduced graphene oxide-carbon nanotubes nanocomposites towards methanol oxidation reaction, *Applied Surface Science*, 2018, **450**, 413–421.
37. G. Mittal, V. Dhand, K. Y. Rhee, S.-J. Park and W. R. Lee, A review on carbon nanotubes and graphene as fillers in reinforced polymer nanocomposites, *Journal of Industrial and Engineering Chemistry*, 2015, **21**, 11–25.
38. J. H. T. Luong, J. D. Glennon, A. Gedanken and S. K. Vashist, Achievement and assessment of direct electron transfer of glucose oxidase in electrochemical biosensing using carbon nanotubes, graphene, and their nanocomposites, *Microchimica Acta*, 2017, **184**(2), 369–388.
39. G. Bounos, K. S. Andrikopoulos, T. K. Karachalios and G. A. Voyiatzis, Evaluation of multi-walled carbon nanotube concentrations in polymer nanocomposites by Raman spectroscopy, *Carbon*, 2014, **76**, 301–309.

40. M. Martin-Gallego, M. M. Bernal, M. Hernandez, R. Verdejo and M. A. Lopez-Manchado, Comparison of filler percolation and mechanical properties in graphene and carbon nanotubes filled epoxy nanocomposites, *European Polymer Journal*, 2013, **49**(6), 1347–1353.
41. R. Zan, Q. M. Ramasse, R. Jalil and U. Bangert, Atomic structure of graphene and h-BN layers and their interactions with metals, in *Advances in Graphene Science*. IntechOpen, 2013.
42. G. Yang, L. Li, W. B. Lee and M. C. Ng, Structure of graphene and its disorders: a review, *Science and Technology of Advanced Materials*, 2018, **19**(1), 613–648.
43. N. Saifuddin, A. Z. Raziah and A. R. Junizah, Carbon nanotubes: a review on structure and their interaction with proteins, *Journal of Chemistry*, 2012, **2013**, 676815.
44. C. N. R. Rao, R. Voggu and A. Govindaraj, Selective generation of single-walled carbon nanotubes with metallic, semiconducting and other unique electronic properties, *Nanoscale*, 2009, **1**(1), 96–105.
45. M. J. O'Connell, *Carbon Nanotubes: Properties and Applications*. CRC Press, 2018.
46. M. S. A. Bhuyan, M. N. Uddin and M. M. Islam, Synthesis of graphene, *International Nano Letters*, 2016, **6**, 65–83.
47. Z. Wang. *Progress on Preparation of Graphene and Its Application*. IOP Publishing, 2017.
48. S. Manchala, V. Tandava, D. Jampaiah, S. K. Bhargava and V. Shanker, Novel and highly efficient strategy for the green synthesis of soluble graphene by aqueous polyphenol extracts of Eucalyptus bark and its applications in high-performance supercapacitors, *ACS Sustainable Chemistry & Engineering*, 2019, **7**(13), 11612–11620.
49. R. C. Sinclair, J. L. Suter and P. V. Coveney, Micromechanical exfoliation of graphene on the atomistic scale, *Physical Chemistry Chemical Physics*, 2019, **21**(10), 5716–5722.
50. W. Wu, C. Zhang and S. Hou, Electrochemical exfoliation of graphene and graphene-analogous 2D nanosheets, *Journal of Materials Science*, 2017, **52**(18), 10649–10660.
51. M. S. A. Bhuyan, M. N. Uddin, M. M. Islam, F. A. Bipasha and S. S. Hossain, Synthesis of graphene, *International Nano Letters*, 2016, **6**(2), 65–83.
52. X. Xu, Z. Zhang, J. Dong, D. Yi, L. Lin, J. Niu, M. Wu, R. Yin, M. Li and J. Zhou, Ultrafast epitaxial growth of metre-sized single-crystal graphene, *Bulletin of the American Physical Society*, 2018, **63**.

53. J. Nam, D.-C. Kim, H. Yun, D. H. Shin, S. Nam, W. K. Lee, J. Y. Hwang, S. W. Lee, H. Weman and K. S. Kim, Chemical vapor deposition of graphene on platinum: Growth and substrate interaction, *Carbon*, 2017, **111**, 733–740.
54. K. K. H. De Silva, H. H. Huang, R. K. Joshi and M. Yoshimura, Chemical reduction of graphene oxide using green reductants, *Carbon*, 2017, **119**, 190–199.
55. Y. Xu, H. Cao, Y. Xue, B. Li and W. Cai, Liquid-Phase exfoliation of graphene: an overview on exfoliation media, techniques, and challenges, *Nanomaterials*, 2018, **8**(11), 942.
56. N. Gurbani, C.-P. Han, K. Marumoto, R.-S. Liu, R. J. Choudhary and N. Chouhan, Biogenic reduction of graphene oxide: an efficient superparamagnetic material for photocatalytic hydrogen production, *ACS Applied Energy Materials*, 2018, **1**(11), 5907–5918.
57. A. Darabont, P. Nemes-Incze, K. Kertész, L. Tapasztó, A. A. Koós, Z. Osváth, Z. Sárközi, Z. Vértesy, Z. E. Horváth and L. P. Biró, Synthesis of carbon nanotubes by spray pyrolysis and their investigation by electron microscopy, *Journal of Optoelectronics and Advanced Materials*, 2005, **7**(2), 631–636.
58. R. Das, *Nanohybrid Catalyst Based on Carbon Nanotube,* Carbon nanostructures. Springer International Publishing AG, 2017.
59. J. Prasek, J. Drbohlavova, J. Chomoucka, J. Hubalek, O. Jasek, V. Adam and R. Kizek, Methods for carbon nanotubes synthesis, *Journal of Materials Chemistry*, 2011, **21**(40), 15872–15884.
60. S. Manafi, H. Nadali and H. R. Irani, Low temperature synthesis of multi-walled carbon nanotubes via a sonochemical/hydrothermal method, *Materials Letters*, 2008, **62**(26), 4175–4176.
61. D. Kumar, K. Singh, V. Verma and H. S. Bhatti, Low-temperature hydrothermal synthesis and functionalization of multiwalled carbon nanotubes, *Indian Journal of Physics*, 2016, **90**(2), 139–148.
62. S. Abbasloo, M. Ojaghi-Ilkhchi and M. Mozammel, Synthesis of carbon nanotubes by molten salt electrolysis: a review, 1996.
63. H. Lyu, B. Gao, F. He, C. Ding, J. Tang and J. C. Crittenden, Ball-milled carbon nanomaterials for energy and environmental applications, *ACS Sustainable Chemistry & Engineering*, 2017, **5**(11), 9568–9585.
64. C. Somu, A. Karthi, S. Singh, R. Karthikeyan and S. D. N. Ganesh, Synthesis of various forms of carbon nanotubes by arc discharge methods–comprehensive review, *International Research Journal of Engineering and Technology*, 2017, **4**(1).

65. X.-D. Wang, K. Vinodgopal and G.-P. Dai, Synthesis of carbon nanotubes by catalytic chemical vapor deposition, in *Perspective of Carbon Nanotubes*. IntechOpen, 2019.
66. M. Johnson, J. Ren, M. Lefler, G. Licht, J. Vicini, X. Liu and S. Licht, Carbon nanotube wools made directly from CO_2 by molten electrolysis: value driven pathways to carbon dioxide greenhouse gas mitigation, *Materials Today Energy*, 2017, **5**, 230–236.
67. A. Annu, B. Bhattacharya, P. K. Singh, P. K. Shukla and H.-W. Rhee, Carbon nanotube using spray pyrolysis: recent scenario, *Journal of Alloys and Compounds*, 2017, **691**, 970–982.
68. Y. Chen, R. Mu, M. Yang, L. Fang, Y. Wu, K. Wu, Y. Liu and J. Gong, Catalytic hydrothermal liquefaction for bio-oil production over CNTs supported metal catalysts, *Chemical Engineering Science*, 2017, **161**, 299–307.
69. G. J. Price, M. Nawaz, T. Yasin and S. Bibi, Sonochemical modification of carbon nanotubes for enhanced nanocomposite performance, *Ultrasonics Sonochemistry*, 2018, **40**, 123–130.
70. O. Soares, R. P. Rocha, A. G. Gonçalves, J. L. Figueiredo, J. J. M. Órfão and M. F. R. Pereira, Highly active N-doped carbon nanotubes prepared by an easy ball milling method for advanced oxidation processes, *Applied Catalysis B: Environmental*, 2016, **192**, 296–303.
71. Z. Zhen and H. Zhu, Structure and properties of graphene, in *Graphene*. Elsevier, 2018, pp. 1–12.
72. H. Y. Nan, Z. H. Ni, J. Wang, Z. Zafar, Z. X. Shi and Y. Y. Wang, The thermal stability of graphene in air investigated by Raman spectroscopy, *Journal of Raman Spectroscopy*, 2013, **44**(7), 1018–1021.
73. J. C. Goak, S. H. Lee and N. Lee, Effect of purification on the electrical properties of transparent conductive films fabricated from single-walled carbon nanotubes, *Diamond and Related Materials*, 2020, 107815.
74. X. Wu, F. Mu and H. Zhao, Recent progress in the synthesis of graphene/CNT composites and the energy-related applications, *Journal of Materials Science & Technology*, 2020, **55**, 16–34.
75. L. Shahriary, H. Ghourchian and A. A. Athawale, Graphene-multiwalled carbon nanotube hybrids synthesized by gamma radiations: application as a glucose sensor, *Journal of Nanotechnology*, 2014, **2014**, 903872.
76. A. Kumar, K. Sharma and A. R. Dixit, Carbon nanotube- and graphene-reinforced multiphase polymeric composites: review on their properties and applications, *Journal of Materials Science*, 2020, 1–43.

77. S. H. Chae and Y. H. Lee, Carbon nanotubes and graphene towards soft electronics, *Nano Convergence*, 2014, **1**(1), 15.
78. W. Zhan, S. Yu, L. Gao, F. Wang, X. Fu, G. Sui and X. Yang, Bioinspired assembly of carbon nanotube into graphene aerogel with "cabbagelike" hierarchical porous structure for highly efficient organic pollutants cleanup, *ACS Applied Materials & Interfaces*, 2018, **10**(1), 1093–1103.
79. W. Fan, L. Zhang and T. Liu, Graphene-CNT hybrids for energy applications, in *Graphene-Carbon Nanotube Hybrids for Energy and Environmental Applications*. Springer, 2017, pp. 53–90.
80. Z. Yang, M. Liu, C. Zhang, W. W. Tjiu, T. Liu and H. Peng, Carbon nanotubes bridged with graphene nanoribbons and their use in high-efficiency dye-sensitized solar cells, *Angewandte Chemie International Edition*, 2013, **52**(14), 3996–3999.
81. Z. Yuan, X. Xiao, J. Li, Z. Zhao, D. Yu and Q. Li, Self-assembled graphene-based architectures and their applications, *Advanced Science*, 2018, **5**(2), 1700626.
82. X. Li, L. Liang, M. Yang, G. Chen and C.-Y. Guo, Poly(3,4-ethylenedioxythiophene)/graphene/carbon nanotube ternary composites with improved thermoelectric performance, *Organic Electronics*, 2016, **38**, 200–204.
83. W. Chen, P. Liu, Y. Liu, Q. Wang and W. Duan, A temperature-induced conductive coating via layer-by-layer assembly of functionalized graphene oxide and carbon nanotubes for a flexible, adjustable response time flame sensor, *Chemical Engineering Journal*, 2018, **353**, 115–125.
84. T. Kavinkumar and S. Manivannan, Improved dielectric behaviour of graphene oxide-multiwalled carbon nanotube nanocomposite, *Vacuum*, 2018, **148**, 149–157.
85. Y. J. Kwon, Y. Kim, H. Jeon, S. Cho, W. Lee and J. U. Lee, Graphene/carbon nanotube hybrid as a multi-functional interfacial reinforcement for carbon fiber-reinforced composites, *Composites Part B: Engineering*, 2017, **122**, 23–30.
86. H. Cai, K. Han, H. Jiang, J. Wang and H. Liu, Self-standing silicon-carbon nanotube/graphene by a scalable in situ approach from low-cost Al-Si alloy powder for lithium ion batteries, *Journal of Physics and Chemistry of Solids*, 2017, **109**, 9–17.
87. C. Xiong, T. Li, T. Zhao, A. Dang, H. Li, X. Ji, W. Jin, S. Jiao, Y. Shang and Y. Zhang, Reduced graphene oxide-carbon nanotube grown on

107. K. Wang, J. Pang, L. Li, S. Zhou, Y. Li and T. Zhang, Synthesis of hydrophobic carbon nanotubes/reduced graphene oxide composite films by flash light irradiation, *Frontiers of Chemical Science and Engineering*, 2018, **12**(3), 376–382.

108. S. Manchala, V. Tandava, D. Jampaiah, S. K. Bhargava and V. Shanker, A novel strategy for sustainable synthesis of soluble-graphene by a herb delphinium denudatum root extract for use as light-weight supercapacitors, *ChemistrySelect*, 2020, **5**(9), 2701–2709.

109. D. N. Leonard, G. W. Chandler and S. Seraphin, *Scanning Electron Microscopy*, ed. E. N. Kaufmann. Wiley, New Jersey, USA, 2012.

110. J.-J. Shao, W. Lv, Q. Guo, C. Zhang, Q. Xu, Q.-H. Yang and F. Kang, Hybridization of graphene oxide and carbon nanotubes at the liquid/air interface, *Chemical Communications*, 2012, **48**(31), 3706 3708.

111. U. J. Kim, I. L. H. Lee, J. J. Bae, S. Lee, G. H. Han, S. J. Chae, F. Güneş, J. H. Choi, C. W. Baik and S. I. Kim, Graphene/carbon nanotube hybrid-based transparent 2D optical array, *Advanced Materials*, 2011, **23**(33), 3809–3814.

112. F. Du, D. Yu, L. Dai, S. Ganguli, V. Varshney and A. K. Roy, Preparation of tunable 3D pillared carbon nanotube–graphene networks for high-performance capacitance, *Chemistry of Materials*, 2011, **23**(21), 4810–4816.

113. D. H. Anjum. *Characterization of Nanomaterials with Transmission Electron Microscopy*. IOP Publishing, 2016.

114. N. Wang, S. Peng, X. Chen, J. Wang, C. Wang, X. Qi, S. Dai and S. Yan, Construction of ultrathin MnO_2 decorated graphene/carbon nanotube nanocomposites as efficient sulfur hosts for high-performance lithium–sulfur batteries, *RSC Advances*, 2019, **9**(11), 6346–6355.

115. S.-X. Da, J. Wang, H.-Z. Geng, S.-L. Jia, C.-X. Xu, L.-G. Li, P.-P. Shi and G. Li, High adhesion transparent conducting films using graphene oxide hybrid carbon nanotubes, *Applied Surface Science*, 2017, **392**, 1117–1125.

116. İ. Karteri, S. Uruş, H. Özerli and Ş. Karataş, Photovoltaic performance photodiodes based on reduced graphene oxide-Fe_3O_4 and carbon nanotube-Fe_3O_4 nanocomposites, *Materials Today: Proceedings*, 2016, **3**(5), 1297–1302.

117. T. Wang, L.-C. Jing, Q. Zhu, A. Sagadevan Ethiraj, Y. Tian, H. Zhao, X.-T. Yuan, J.-G. Wen, L.-K. Li and H.-Z. Geng, Fabrication of architectural structured polydopamine-functionalized reduced graphene

oxide/carbon nanotube/PEDOT:PSS nanocomposites as flexible transparent electrodes for OLEDs, *Applied Surface Science*, 2020, **500**, 143997.

118. A. K. Sood and E. J. Egerton, Silicon nitride-carbon nanotube-graphene nanocomposite microbolometer IR detector. Google Patents, 2019.
119. S.-H. Lee, D. Kang and I.-K. Oh, Multilayered graphene-carbon nanotube-iron oxide three-dimensional heterostructure for flexible electromagnetic interference shielding film, *Carbon*, 2017, **111**, 248–257.
120. P. K. Adusei, S. N. Kanakaraj, S. Gbordzoe, K. Johnson, D. DeArmond, Y.-Y. Hsieh, Y. Fang, S. Mishra, N. Phan and N. T. Alvarez, A scalable nano-engineering method to synthesize 3D-graphene-carbon nanotube hybrid fibers for supercapacitor applications, *Electrochimica Acta*, 2019, **312**, 411–423.
121. B. Dey, M. W. Ahmad, A. Almezeni, G. Sarkhel, D. S. Bag and A. Choudhury, Enhancing electrical, mechanical, and thermal properties of polybenzimidazole by 3D carbon nanotube@graphene oxide hybrid, *Composites Communications*, 2020, **17**, 87–96.
122. F. Liu, X. Liu, N. Hu, H. Ning, S. Atobe, C. Yan, F. Mo, S. Fu, J. Zhang, Y. Wang and X. Mu, Investigation of thermal energy transport interface of hybrid graphene-carbon nanotube/polyethylene nanocomposites, *Scientific Reports*, 2017, **7**(1), 14700.
123. Q. Shi and A. Zhu, Interface regulation of graphene/carbon nanotube on the thermal conductivity and anticorrosion performance of their nanocomposite, *Progress in Organic Coatings*, 2020, **140**, 105480.
124. F. Zhang, Y. Feng, M. Qin, T. Ji, F. Lv, Z. Li, L. Gao, P. Long, F. Zhao and W. Feng, Stress-sensitive thermally conductive elastic nanocomposite based on interconnected graphite-welded carbon nanotube sponges, *Carbon*, 2019, **145**, 378–388.
125. Y. Zhou, X. Zhuang, F. Wu and F. Liu, High-performance thermal management nanocomposites: silver functionalized graphene nanosheets and multiwalled carbon nanotube, *Crystals*, 2018, **8**(11).
126. L. Zhao, X. Sun, Z. Lei, J. Zhao, J. Wu, Q. Li and A. Zhang, Thermoelectric behavior of aerogels based on graphene and multi-walled carbon nanotube nanocomposites, *Composites Part B: Engineering*, 2015, **83**, 317–322.
127. S. Dul, G. L. Ecco, A. Pegoretti and L. Fambri, Graphene/carbon nanotube hybrid nanocomposites: effect of compression molding and fused filament fabrication on properties, *Polymers*, 2020, **12**(1).

128. V. K. Sharma, T. J. McDonald, H. Kim and V. K. Garg, Magnetic graphene–carbon nanotube iron nanocomposites as adsorbents and antibacterial agents for water purification, *Advances in Colloid and Interface Science*, 2015, **225**, 229–240.
129. M. A. Hussein, R. M. El-Shishtawy, K. A. Alamry, A. M. Asiri and S. A. Mohamed, Efficient water disinfection using hybrid polyaniline/graphene/carbon nanotube nanocomposites, *Environmental Technology*, 2019, **40**(21), 2813–2824.
130. M. A. Ahsan, V. Jabbari, M. A. Imam, E. Castro, H. Kim, M. L. Curry, D. J. Valles-Rosales and J. C. Noveron, Nanoscale nickel metal organic framework decorated over graphene oxide and carbon nanotubes for water remediation, *Science of the Total Environment*, 2020, **698**, 134214.
131. M. A. Hussein, Z. M. Al-Amshany and N. Y. Tashkandi, Hybrid polyvinyl alcohol/polyvinyl chloride nanocomposites reinforced with graphene-carbon nanotube for acid red environmental treatments, *Polymer-Plastics Technology and Materials*, 2020, 1–16.
132. M. A. Hussein, M. M. Alam, A. M. Asiri, Z. M. Al-amshany, K. S. Hajeeassa and M. M. Rahman, Ultrasonic-assisted fabrication of polyvinyl chloride/mixed graphene-carbon nanotube nanocomposites as a selective Ag^+ ionic sensor, *Journal of Composite Materials*, 2019, **53**(16), 2271–2284.
133. Z. Xu, X. Fan, Q. Ma, B. Tang, Z. Lu, J. Zhang, G. Mo, J. Ye and J. Ye, A sensitive electrochemical sensor for simultaneous voltammetric sensing of cadmium and lead based on Fe_3O_4/multiwalled carbon nanotube/laser scribed graphene composites functionalized with chitosan modified electrode, *Materials Chemistry and Physics*, 2019, **238**, 121877.
134. P. Sharma, V. Bhalla, V. Dravid, G. Shekhawat, E. S. Prasad and C. R. Suri, Enhancing electrochemical detection on graphene oxide-CNT nanostructured electrodes using magneto-nanobioprobes, *Scientific Reports*, 2012, **2**, 877.
135. V. Mani, B. Dinesh, S.-M. Chen and R. Saraswathi, Direct electrochemistry of myoglobin at reduced graphene oxide-multiwalled carbon nanotubes-platinum nanoparticles nanocomposite and biosensing towards hydrogen peroxide and nitrite, *Biosensors and Bioelectronics*, 2014, **53**, 420–427.
136. V. Kumar, M. Shorie, A. K. Ganguli and P. Sabherwal, Graphene-CNT nanohybrid aptasensor for label free detection of cardiac biomarker myoglobin, *Biosensors and Bioelectronics*, 2015, **72**, 56–60.

137. M. C. Serrano, M. C. Gutiérrez and F. del Monte, Role of polymers in the design of 3D carbon nanotube-based scaffolds for biomedical applications, *Progress in Polymer Science*, 2014, **39**(7), 1448–1471.
138. S.-R. Ryoo, Y.-K. Kim, M.-H. Kim and D.-H. Min, Behaviors of NIH-3T3 fibroblasts on graphene/carbon nanotubes: proliferation, focal adhesion, and gene transfection studies, *ACS Nano*, 2010, **4**(11), 6587–6598.
139. X. Fan, G. Jiao, L. Gao, P. Jin and X. Li, The preparation and drug delivery of a graphene–carbon nanotube–Fe_3O_4 nanoparticle hybrid, *Journal of Materials Chemistry B*, 2013, **1**(20), 2658–2664.
140. C. Zhu, X. Mu, P.A. van Aken, J. Maier and Y. Yu, Fast Li storage in MoS_2-graphene-carbon nanotube nanocomposites: advantageous functional integration of 0D, 1D, and 2D nanostructures, *Advanced Energy Materials*, 2015, **5**(4), 1401170.
141. S. Peng, L. Li, X. Han, W. Sun, M. Srinivasan, S. G. Mhaisalkar, F. Cheng, Q. Yan, J. Chen and S. Ramakrishna, Cobalt sulfide nanosheet/graphene/carbon nanotube nanocomposites as flexible electrodes for hydrogen evolution, *Angewandte Chemie*, 2014, **126**(46), 12802–12807.
142. M. Khan, A. B. Yousaf, M. Chen, C. Wei, X. Wu, N. Huang, Z. Qi and L. Li, Molybdenum sulfide/graphene-carbon nanotube nanocomposite material for electrocatalytic applications in hydrogen evolution reactions, *Nano Research*, 2016, **9**(3), 837–848.
143. C. J. Hung, P. Lin and T. Y. Tseng, Electrophoretic fabrication and pseudocapacitive properties of graphene/manganese oxide/carbon nanotube nanocomposites, *Journal of Power Sources*, 2013, **243**, 594–602.
144. T. Sun, Z. Zhang, J. Xiao, C. Chen, F. Xiao, S. Wang and Y. Liu, Facile and green synthesis of palladium nanoparticles-graphene-carbon nanotube material with high catalytic activity, *Scientific Reports*, 2013, **3**(1), 1–6.
145. Y. Zhang, X. Zhuang, J. Muthu, T. Mabrouki, M. Fontaine, Y. Gong and T. Rabczuk, Load transfer of graphene/carbon nanotube/polyethylene hybrid nanocomposite by molecular dynamics simulation, *Composites Part B: Engineering*, 2014, **63**, 27–33.
146. D. Yu and L. Dai, Self-assembled graphene/carbon nanotube hybrid films for supercapacitors, *The Journal of Physical Chemistry Letters*, 2010, **1**(2), 467–470.

147. B. Zhang, Q. B. Zheng, Z. D. Huang, S. W. Oh and J. K. Kim, SnO_2–graphene–carbon nanotube mixture for anode material with improved rate capacities, *Carbon*, 2011, **49**(13), 4524–4534.
148. P. Wang, M. Cao, C. Wang, Y. Ao, J. Hou and J. Qian, Kinetics and thermodynamics of adsorption of methylene blue by a magnetic graphene-carbon nanotube composite, *Applied Surface Science*, 2014, **290**, 116–124.
149. C. J. Hung, P. Lin and T. Y. Tseng, High energy density asymmetric pseudocapacitors fabricated by graphene/carbon nanotube/MnO_2 plus carbon nanotubes nanocomposites electrode, *Journal of Power Sources*, 2014, **259**, 145–153.
150. F. Lorestani, Z. Shahnavaz, P. Mn, Y. Alias and N. S. A. Manan, One-step hydrothermal green synthesis of silver nanoparticle-carbon nanotube reduced-graphene oxide composite and its application as hydrogen peroxide sensor, *Sensors and Actuators B: Chemical*, 2015, **208**, 389–398.
151. Y.-S. Wang, S.-Y. Yang, S.-M. Li, H.-W. Tien, S.-T. Hsiao, W.-H. Liao, C.-H. Liu, K.-H. Chang, C.-C.M. Ma and C.-C. Hu, Three-dimensionally porous graphene–carbon nanotube composite-supported PtRu catalysts with an ultrahigh electrocatalytic activity for methanol oxidation, *Electrochimica Acta*, 2013, **87**, 261–269.

Chapter 3

Thermotropic Hydrogen-Bonded Liquid Crystals (HBLCs) Spectral Analysis of Phase Transition: Limitations, Perspectives and Challenges

N. Pongali Sathya Prabu, M. Harikishor, and M. L. N. Madhu Mohan
Liquid Crystal Research Laboratory, Bannari Amman Institute of Technology, Sathyamangalam 638 401, Tamil Nadu, India
mln.madhu@gmail.com

One of the spectroscopic applications of thermotropic hydrogen-bonded liquid crystals (HBLC's) is fabrication of a thermistor using the thermotropic ferroelectric liquid crystalline ingredient. The design, performance and characteristics of this fabricated liquid crystalline device are elucidated. Thermo-resistive profiles are constructed and the characteristics are discussed in various liquid crystalline phases namely Cholesteric and Smectic A*, C*, X*, re-entrant C*, G*. As a representative case, the characteristics of two ferroelectric liquid crystals which exhibit positive and negative temperature coefficient of resistance are detailed. A liquid crystal (S) -1-[4′-(octyloxy carbonyl) phenyl]-4-[2-chloro-(4-benzamido acetyl) phenyl propionato]-benzoate abbreviated as

Nanomaterials for Spectroscopic Applications
Edited by Kaushik Pal

ISBN 978-981-4877-69-5 (Hardcover), 978-1-003-16033-5 (eBook)
www.jennystanford.com

RTFLC yielded negative temperature coefficient thermistor while dextro tartaric acid with p-n decyloxy benzoic acid ferroelectric liquid crystal resulted in positive temperature coefficient thermistor. The characterization of liquid crystal thermistors in various phases is discussed. First time in the history of liquid crystals, a thermistor is derived from a mesogen and is characterized.

3.1 Introduction

The thermistor is a two-terminal device extensively used in electronic circuits. It is a passive transducer. The variation of resistance is manifested with corresponding changes in temperature. Liquid crystal by definition is an organic substance that possesses characteristics of solid and that of liquid [1]. The liquid crystal exhibits mesophases which possess orientational order, positional order or both. The most common mesophases are nematic and Sm. These materials are used widely in display devices. Liquid crystal displays are constructed from one of these mesophases as reported in literature [2–18]. Applications of liquid crystal are not only confined to liquid crystal displays, these materials are used in many other aspects [19–32]. Some typical applications of these materials are in thermographic measurement of skin temperature [23], aerospace thermal mapping applications [24], non-destructive testing [28], optical applications [28–31]. Our recent investigation on these materials enabled us to use liquid crystals in-memory storage [33], light modulation [34], light filtering at infrared region [35, 36], visible region [37–39] and light shuttering action [40].

There are no reports available in the literature about a liquid crystal being used as a thermistor. In this article, we propose the fabrication of a liquid crystal thermistor. Both positive and negative temperature coefficient (PTC, NTC) thermistors can be deduced using various types of liquid crystals. The experimental data points are fitted to an exponential fit derived from a theoreticalpower law. Material constant (β) for individual phases has been calculated. Theoretical molecular modeling of various phases pertaining to the liquid crystal mesogen is provided to validate the NTC and PTC behavior.

3.2 Experimental Details

3.2.1 Material Preparation

To derive a thermistor from an LC, the LC has to be prepared. In the present case, the RTFLC material mentioned above is used. RTFLC is filled in an LC 10 micron cell by standard technique. This compound is reported [4] to exhibit the following mesophases.

$$\text{Iso} \leftrightarrow \text{Colesteric} \leftrightarrow \text{Sm A}^* \leftrightarrow \text{Sm C}^* \leftrightarrow \text{Cryst.}$$

Isotropic to Ch phase transition is reported [4] at 87.2°C, Ch to Sm A* is observed at 77.4°C, and Sm A* to Sm C* is noticed at 49.1°C. The liquid crystal crystallizes at −20°C and the thermal span of Sm A* and Sm C* are 64.1°C and 44.1°C degrees, respectively.

FLC namely tartaric acid-dextro along with decyloxy acid abbreviated as (DTA+10BAO) is reported [41] to exhibit the following phase sequence

$$\text{Iso} \leftrightarrow \text{Ch} \leftrightarrow \text{Sm C}^* \leftrightarrow \text{Sm X}^* \leftrightarrow \text{ReSm C}^* \leftrightarrow \text{Sm G}^* \leftrightarrow \text{Cryst.}$$

Isotropic to Ch phase transition is identified at 144°C, Ch to Sm C* is observed at 135.9°C, Sm C* to Sm X* is noticed at 124.5°C, Sm X* to re-entrant Sm C* is observed at 118°C, the phase transition from re-entrant Sm C* to Sm G* is noticed at 117.1°C Sm G* to crystal is detected at 89.3°C.

3.3 Results and Discussion

3.3.1 Conventional Thermistor

A thermistor is basically composed of carbon material with proper leads to connect it in series to a device. Basically, the variation of resistance with temperature under constant field, pressure and frequency is termed as one of the prominent property of thermistor. Thermistors can be classified into two categories basing on the performance with temperature. A thermistor is governed by the following equation

$$R_t = R_o e^{\beta\left(\frac{1}{T}-\frac{1}{T_0}\right)} \tag{3.1}$$

where R_t – magnitude of the component at a given temp t
R_o – reference resistance at a given temp t_o
T – temp measured
T_o – the standard reference temp
β – constant of material

β is a material constant that varies in each liquid crystalline phase. The magnitudes of β in various phases are calculated and shown in Table 3.1. A positive coefficient thermistor (PCT) is one whose magnitude exponentially increases with an increase in temperature. On the other hand, a negative coefficient thermistor (NCT) is one where the magnitude of the curve decreases exponentially with respect to an increase in temperature.

3.3.2 Theoretical Modeling of Liquid Crystal Analysis

3.3.2.1 Liquid crystal thermistor

A liquid crystal is a substance that exhibits fluid properties like viscosity and crystalline properties like lattice structure. In the liquid crystalline phase, a mesogen exhibit phases either of the positional and orientational ordering or both orderings. Prominent layered phases are Sm C, Sm G, Sm I, etc., while nematic and Ch are less ordered phases. If a liquid crystal is heated it goes to isotropic and when it is cooled the ordering is induced and in the crystal state, the ordering is maximum.

A liquid crystal is essentially an insulator at a crystalline state. In general, ferroelectric liquid crystals (FLC), in the liquid crystalline state behaves as an elegant thermistor, the reasons are listed below:

(1) the dipoles present in FLC paves way for thermal conduction
(2) conjugation (alternate double bonds in the molecular structure) in FLC in turn facilitates the abundance of Π bonds
(3) Π bonds assist in thermal conduction
(4) in a chiral phase, Sm layers are arranged normal to the LC substrates

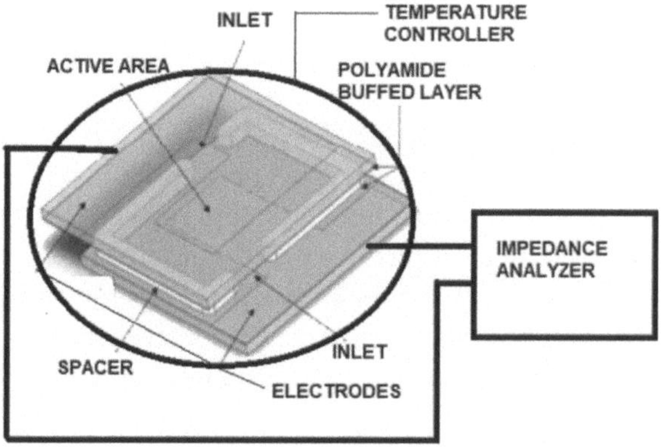

Figure 3.1 Experimental setup of liquid crystal thermistor.

(5) the helicoidal structure which is formed manifests the thermal conduction through the Smectic layers

In the forthcoming sections, liquid crystal exhibiting negative temperature coefficient and positive coefficient will be discussed in detail.

3.3.2.2 Liquid crystal thermistor fabrication

The liquid crystal thermistor fabrication set up is depicted in Fig. 3.1. The liquid crystal is made up of two transparent substrates which are segregated with a 10 micron liner. The indium tin oxide (ITO) coating on the glass plates yields a conductive area of 1 square centimeter. Further, the ITO coated glass plates are treated with a layer of polyamide to form a buffed layer, this procedure helps in the proper alignment of the liquid crystal mesogen. Two electrically conducting silver leads are taken from the LC cell for the measurement of resistance. The LC cell is filled through any one of the inlets with the above liquid crystal compound under vacuum by standard technique and is placed in a temperature controller. This arrangement is mounted on the stage of a polarizing microscope between analyzer and polarizer. The incident incandescent light is polychromatic in nature. The LC cell is connected to the port of impedance analyzer Wayne Kerr 6500B.

Liquid crystalline compounds tartaric acid- dextro and decyoxy acid (benzoic) abbreviated as (DTA+10BAO) are taken to its isotropic state by elevating the temperature through a programmed mode and is observed through the eyepiece of the microscope for confirmation. It is taken to its clearing point and is held for 2 min for thermal equilibrium attainment. The sample is then slowly cooled at the rate of 0.1°C till a Ch phase with schlieren texture is observed. The measurement of the resistance with temperature in the total thermal phase is recorded. Similarly, the experiment is repeated in the thermal ranges of all other phases namely Sm C*, Sm X* and re-entrant Sm C*.

The sample RTFLC is subjected to heat so that it reaches the clearing temperature and is brought to 49.1°C at a programmed sequence. The inducement of the Sm C* phase is noticed through a polarizing microscope. The broken focal conic texture is a token of evidence for the formation of the Sm C* phase. The measurement of the resistance with temperature in the total thermal phase is recorded.

3.3.3 Liquid Crystal Exhibiting Positive Temperature Coefficient

As a representative case, the positive temperature coefficient characteristics of an FLC namely tartaric acid- dextro with decyloxy acid (benzoic) abbreviated as (DTA+10BAO) is discussed. The transition temperatures corresponding to individual phases along with the energy associated are reported [41] elsewhere. The balls depicted as white represent hydrogen atoms, grey – carbon atoms, red – oxygen atoms. The ball and stick diagram of the (DTA+10BAO) is depicted in Fig. 3.2.

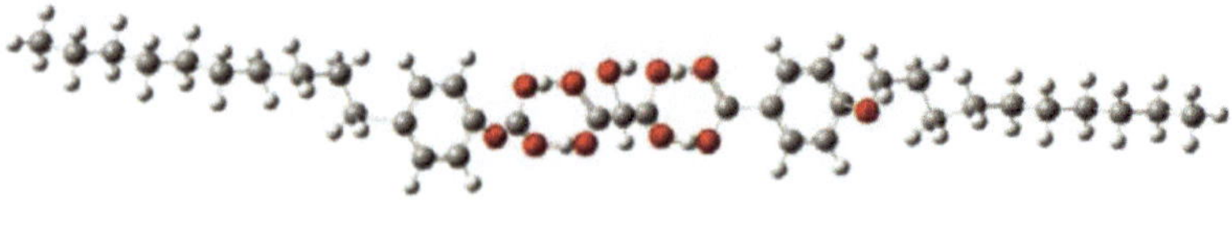

Figure 3.2 Molecular simulate design of DTA+10BAO.

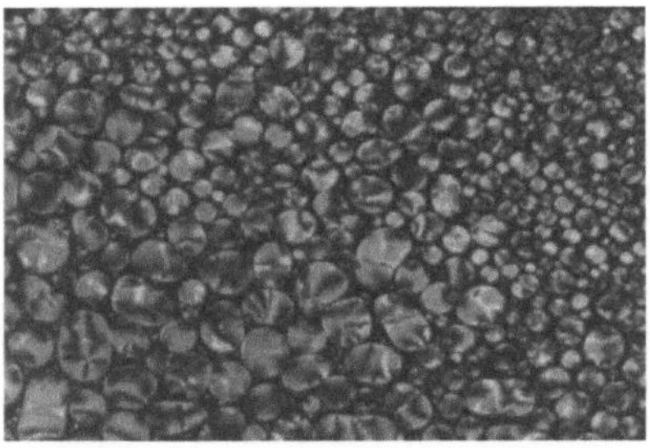

Figure 3.3 Cholesteric droplet texture.

3.3.3.1 Thermo resistance: cholesteric

The sample (DTA+10BAO) is cooled from isotropic to Ch and the inducement of the Ch phase is identified at 143.9°C. The droplet textural observation of this phase is depicted in Fig. 3.3. The temperature is maintained in this phase to attain thermal stability the token of evidence for it is the fully grown Ch texture.

The temperature variation of resistance referred to as a thermo-resistive profile is carried out meticulously in this phase. To the collected data points a theoretical exponential fit obtained from Eq. 3.1 is performed and the thermo-resistance plot is shown in Fig. 3.4. As expected, with the increment of the temperature, the magnitude of the resistance increases exponentially.

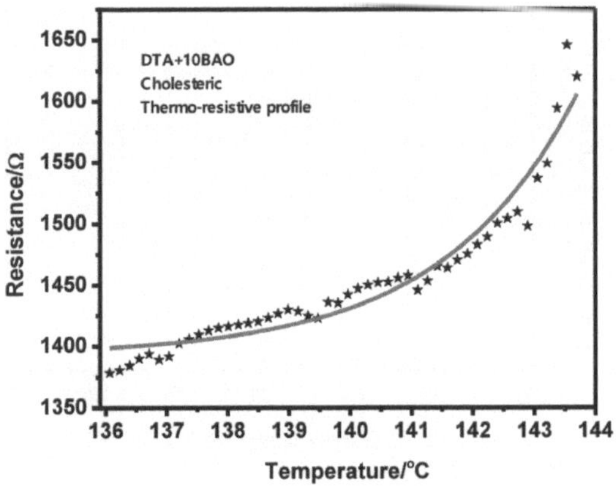

Figure 3.4 Variation of resistance with temperature in cholesteric phase.

This graph is a token of evidence for a positive temperature coefficient. The experimental points and theoretical fit are in concurrence with each other. The magnitude of the material constant β is calculated to be 0.09.

3.3.3.2 Comparison of magnitude of thermistor properties in various phases

In all the liquid crystalline phases of DTA+10BAO namely, Sm C*, Sm X*, re-entrant Sm C* and Sm G* the magnitude of resistance variation per degree is recorded along with the incremental or decremental effect of resistance with temperature. The thermo resistive profiles of all the phases are constructed. From these profiles a plot of magnitudes of resistance in various phases of (DTA+10BAO) is calculated and shown in Fig. 3.5. From Fig. 3.5, it is noticed that Sm C* possesses the highest magnitude of resistance variation per degree while Sm G* has the lowest value.

The above result is not surprising since the origin of helix formation is in Sm C*. The formed helix along with dipoles and conjugation helps the phase Sm C* to attain the highest magnitude of the resistance variation per degree.

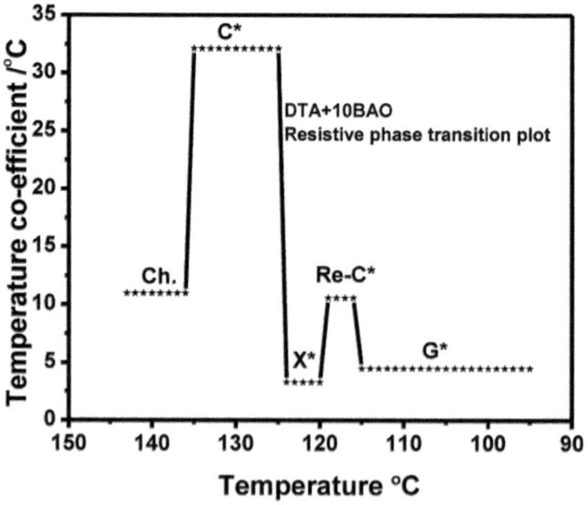

Figure 3.5 Comparison of magnitudes of resistance in various phases of (DTA+10BAO).

Table 3.1 Phase polymorphism and material constant values

Mesogen	Phase	β-value
(DTA+10BAO)	Sm C*	163.41
	Sm X*	57.12
	Re-entrant Sm C*	126.74
RTFLC	Sm C*	0.09

Table 3.1 lists the material constant β values for various phases. It is interesting to note that (DTA+10BAO) in all phases exhibited a positive temperature coefficient except Sm X* phase where a negative temperature coefficient is observed.

3.3.3.3 Molecular modeling

In (DTA+10BAO), as the temperature is decreased Sm C*, Sm X* and re-entrant Sm C* are noticed. Proposed molecular modeling in Sm C*, Sm X* and re-entrant Sm C* is depicted in Fig. 3.6. In Sm C* the molecules orient in a slant manner to the director and the same molecular arrangement is repeated in various Sm layers. In re-entrant Sm C* the molecules are oriented to a director, however, this tilt is 180° to that of molecular tilting in the Sm C* phase.

Sm X* is an intermediate phase amid traditional Sm C* and re-entrant Sm C* with narrow temperature width wherein the molecules are flipped to change their orientation with respect to the director. In this phase, a pair of molecules are in an obtuse angle and this arrangement of flipping of the molecules is repeated

Figure 3.6 Molecular orientations due to various phase difference.

3.3.5 Current versus Voltage Characteristics of Liquid Crystal Thermistors

I-V characteristic profile of any device will precisely give the data for which the device can be exploited. The I-V characteristics of a positive and a negative temperature coefficient (P/NTC) liquid crystal thermistor are discussed. Figure 3.10 illustrates the I-V characteristics of an RTFLC in the Sm C* phase which exhibits an NTC thermistor. The I-V characteristics are experimentally noted in the total thermal span of the Sm C* phase (28°C to 48°C) at eleven different temperatures at a step of 2°C. The profile can be classified as non-linear in the entire region which resembles that of a semiconductor device. As discussed in the previous sections, it not surprising to note that RTFLC exhibits a semiconductor I-V profile as it exhibits NTC.

A keen observation of Fig. 3.10 reveals that the current exponentially varies with voltage and as the temperature is increased the current profile proportionally drifting towards ordinate. This is in accordance with the thermo-resistive profile depicted in Fig. 3.9.

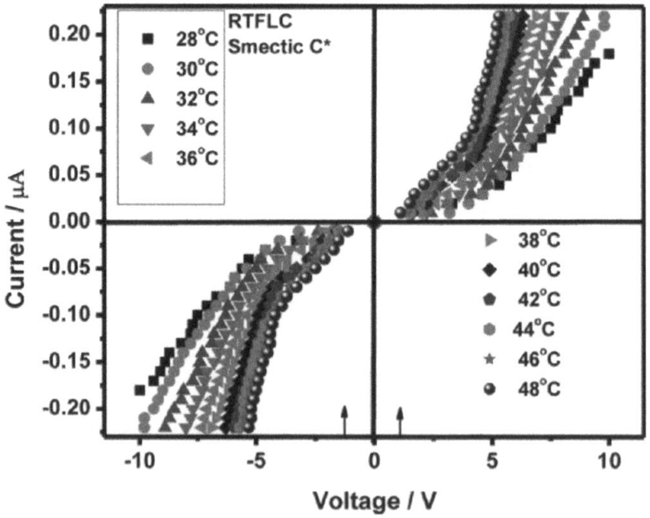

Figure 3.10 I-V characteristics of room temperature ferroelectric liquid crystal.

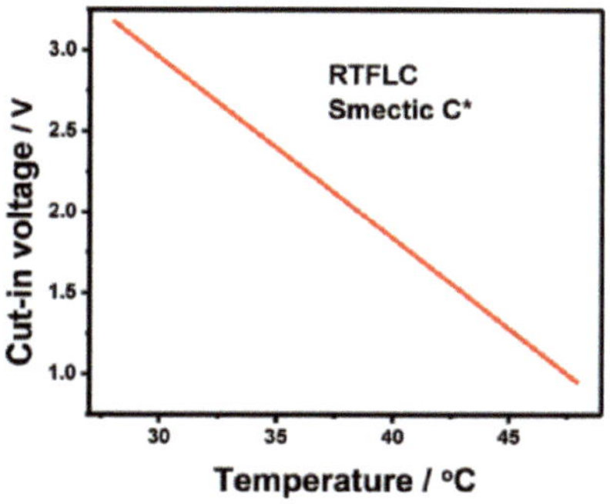

Figure 3.11 Variation of cut-in voltage with temperature.

A further point to be noted is variance the cut in voltage with temperature. Figure 3.11 indicates such a profile where the cut in voltage proportionally decreases with increasing temperature. This experiment severs as another token of evidence for the observed semiconductor behavior discussed in sections 3.3.4.1.

Figure 3.12 illustrates the I-V characteristics of a (DTA+10BAO) in Ch phase which exhibits a PTC thermistor. The I-V characteristics are experimentally noted in the total thermal width of Ch phase (129°C to 137°C) at nine different temperatures at a step of 1°C. From the Fig. 3.12, it can be clearly seen that all the profiles pass

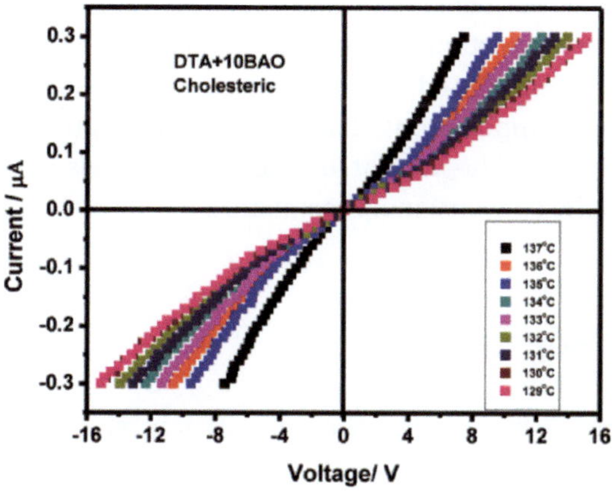

Figure 3.12 I-V characteristics of (DTA+10BAO).

through origin without any cut in voltage. This behavior resembles that of a conductor. At 137°C the I-V characteristics resemble a perfect resistor, this behavior can be correlated to the thermo-resistance graph of (DTA+10BAO) (Fig. 3.4).

3.3.6 Conclusions, Outlook and Challenges

Thermotropic FLC's namely as (DTA+10BAO) and RTFLC are characterized thermally and electrically. The thermo-sensitive resistance property of the above thermotropic ferroelectric liquid crystal materials is exploited commercially as a thermistor for both positive temperature coefficient and negative temperature coefficient. further, it is found that the thermistors obtained through the liquid crystal materials are efficient which is evident through I-V characteristics in the properties like linearity and cut-in voltages hence the thermistor derived from the liquid crystal can be called as a high-performance thermistor. Thus this device can be replaced the conventional thermistors for better-optimized performance.

3.4 Perspectives and Future Scopes

In all the electronic devices this liquid crystal thermistor can be experimented. Performance-wise, chemical stability and operating range are the important improvisationmade. Further, this thermistor can be encapsulated in any desired shape, design to suit the specific device applications. The power consumption of the device is negligibly less, the weight of the device is also comparatively superior to the conventional thermistors.

Acknowledgements

The authors acknowledge the infrastructural support provided by the Bannari Amman Institute of Technology and financial support provided by SERB, New Delhi, India, vide file no.: EMR/2017/001075.

References

1. S. Chandrasekhar, *Liquid Crystals*, Cambridge University Press, Cambridge, United Kingdom, 2010.
2. C. Baristiran Kaynak, et al., High performance thermistor based on $Si_{1-x}Ge_x/Si$ multi quantum wells, *IEEE Electron Device Letters*, 2018, **39**, 753–756.
3. K. Pal, et al., Functionalized graphene oxide dispersedhydrogen bonded liquid crystals efficient electro-optical switching, *Journal of Display Technology*, 2016, **12**(3), 281–287.
4. P. A. Kumar and V. G. K. M. Pisipati, Ambient monocomponent ferroelectric liquid crystals with a wide Smectic-C* range ($<$-20 to $\leq$ 65°C), *Advanced Materials*, 2000, **12**(21), 1617–1619.
5. D.-Y. Kim, K.-U. Jeong, Light responsive liquid crystal soft matters: structures, properties, and applications, *Liquid Crystals Today*, 2019, **28**(2), 34–45.
6. K. L. Woon, et al., Electronic charge transport in extended nematic liquid crystals, *Chemistry of Materials*, 2006, **18**(9), 2311–2317.
7. E. Raynes, *Nematic Liquid Crystals: Applications*, University of Oxford, Oxford, UK, 2016.
8. P. J. Wright, *Encyclopedia of Materials: Science and Technology*, University of Kent, Canterbury, UK, 2001.
9. J. J. Licari and D. W. Swanson, *Adhesives Technology for Electronic Applications*, Elsevier, CA, USA, 2005.
10. N. Madhusudana, Nematic liquid crystals: molecular statistical theories and their applications, in *Reference Materials Science and Materials Engineering*, 2016.
11. D. A. Mosley, Liquid crystal displays: an overview, *Displays*, 1993, **14**(2), 67–73.
12. T. Elze and T. G. Tanner, Temporal properties of liquid crystal displays: implications for vision science experiments, *PLoS One*, 2012, **7**(9), 1–20.
13. T. Elze and T. G. Tanner, Liquid crystal display response time estimation for medical applications, *Medical Physics*, 2009, **36**, 4984–4990.
14. M. E. Becker, LCD response time evaluation in the presence of backlight modulations, *SID Symposium Digest of Technical Papers*, 2008, **39**, 24–27.

15. Badano, Principles of cathode-ray tube and liquid crystal display devices, in *Advances in Digital Radiography Categorical Course in Diagnostic Radiology Physics*, RSNA, Oak Brook, IL, 2003, pp. 91–102.
16. J. Someya and H. Sugiura, Evaluation of liquid-crystal-display motion blur with moving-picture response time and human perception, *Journal of the Society for Information Display*, 2007, **15**(1), 79–86.
17. J.-H. Lee, D. N. Liu and S.-T. Wu, *Introduction to Flat Panel Displays*, Wiley, West Sussex, UK, 2008.
18. J.-K. Song, DCCII: novel method for fast response time in PVA mode, *SID Symposium Digest of Technical Papers*, 2004, **35**, 1344–1347.
19. D. J. R. Cristaldi, S. Pennisi and F. Pulvirenti, *Liquid Crystal Display Drivers*, Springer, Berlin, Germany, 2009.
20. W. den Boer, *Active Matrix Liquid Crystal Displays*, Elsevier, Burlington, 2005.
21. P. Wang and D. Nikolić, An LCD monitor with sufficiently precise timing for research in vision, *Frontiers in Human Neuroscience*, 2011, **5**, 85.
22. M. Schadt, How we made the liquid crystal display, *Nature Electronics*, 2018, **1**, 481–481.
23. G. Brown, Liquid crystals and some of their applications in chemistry, *Analytical Chemistry*, 1969, **41**, 26A–39A.
24. Saupe, Recent results in the field of liquid crystals, *Angewandte Chemie International Edition*, 1968, **7**, 97–112.
25. E. Stepke, Liquid crystals: perspectives, prospects and products, *Electro-Optical Systems Design*, 1972, **4**, 20–31.
26. O. S. Selawry, H. S. Selawry and J. F. Holland, The use of liquid cholesteric crystals for thermographic measurement of skin temperature in man, *Molecular Crystals and Liquid Crystals*, 1969, **1**, 495–502.
27. W. Woodmansee, Aerospace thermal mapping applications of liquid crystals, *Applied Optics*, 1968, **7**, 1721–1728.
28. G. Gottarelli and B. Samorí, Some applications of liquid crystals in organic chemistry, *Inorganica Chimica Acta*, 1980, **40**, X10–X11.
29. N. Kasch, Liquid crystals: applications and industry, *Liquid Crystal Today*, 2013, **22**(3), 70–71.
30. G. H. Brown, Structure, properties, and some applications of liquid crystals, *Journal of the Optical Society of America*, 1973, **63**, 1505–1514.
31. J. L. Fergason, Liquid crystals in nondestructive testing, *Applied Optics*, 1968, **7**(9), 1729–1737.

32. Bregar, M. Štimulak and M. Ravnik, Photonic properties of heliconical liquid crystals, *Optics Express*, 2018, **26**(18), 23265–23277.
33. M. H. Kishor and M. L. N. Madhu Mohan, Realization of memory effect in smectic X* phase, *Journal of Molecular Structure*, 2018, **1168**, 302–308.
34. V. N. Vijayakumar and M. L. N. Madhu Mohan, Double hydrogen bonded liquid crystals: a study of light modulation and field induced transition (FiT), *Molecular Crystals and Liquid Crystals*, 2010, **517**(1), 113–126.
35. M. H. Kishor and M. L. N. Madhu Mohan, An innovative technique to achieve tunable filtering action by ferroelectric material in infrared region, *Journal of Electronic Materials*, 2020, **49**, 2311–2316.
36. H. B. Lu, J. Xing, C. Wei, J. Q. Sha, G. B Zhang, G. Q. Lv, J. Zhu and L. Z. Qiu, Near-infrared light directed reflection in a cholesteric liquid crystal, *Optical Materials Express*, 2017, **7**(11), 4163–4170.
37. J. E. Stockley, G. D. Sharp and K. M. Johnson, Fabry–Perot etalon with polymer cholesteric liquid-crystal mirrors, *Optics Letters*, 1999, **24**(1), 55–57.
38. L. Vicari, *Optical Applications of Liquid Crystals*, CRC Press, Boca Raton, 2019.
39. M. L. N. Madhu Mohan, Light filtering, *Journal of Molecular Liquids*, 2020, **4**(20), 378–385.
40. V. N. Vijayakumar and M. L. N. Madhu Mohan, Experimental evidence of an optical shuttering action in cholesteric phase of a double hydrogen bonded ferroelectric liquid crystal, *Journal of Optoelectronics and Advanced Materials*, 2009, **11**, 1139–1146.
41. M. Hari Kishor and M. L. N. Madhu Mohan, Investigations on smectic X* and re-entrant smectic C* orderings in hydrogen bonded ferroelectric liquid crystals, *Journal of Molecular Liquids*, 2019, **273**, 504–524.
42. Y.-G. Fuh, S.-J. Ho, S.-T. Wu and M.-S. Li, Optical filter with tunable wavelength and bandwidth based on phototunable cholesteric liquid crystals, *Applied Optics*, 2014, **53**, 1658–1662.

Chapter 4

Green Nanostructures Synthesis and Spectroscopic Characterizations

Sreevidya S.,[a] Kirtana Sankara Subramanian,[b] Yokraj Katre,[a] Jai Singh,[c] Ajaya Kumar Singh,[d] Mariya Aleksandrova,[e] and Rabah Khenata[f]

[a]*Department of Chemistry, Kalyan PG College, Bhilai Nagar, Durg, Chhattisgarh 490006, India*
[b]*Department of Food Science, Faculty of Veterinary and Agriculture Science, University of Melbourne, Melbourne, Australia*
[c]*Department of Pure and Applied Physics, Guru Ghasidas University, Bilaspur, Chhattisgarh 495 009, India*
[d]*Department of Chemistry, Govt. V.Y.T. PG. Autonomous College, Durg, Chhattisgarh 491001, India*
[e]*Department of Microelectronics, Technical University of Sofia, 1000 Sofia, Bulgaria*
[f]*Faculty of Sciences and Technology, Mascara University, Mascara, Algeria*
ajayaksingh_au@yahoo.co.in

Crafting or designing and implementing are the prime factors for the nanoengineering of green nanoparticles. The driving force for the formulation of an eco-friendly green nanoparticles are the powerful secondary metabolites present in the plant kingdom. 300,000,000 species of nature's gift are to be looked after for defending nature. The fussy bioactives with voluminous contributions minimize the hazardous potential delivered to nature, in nanomaterial formulation. The crafted nano crores are governed by the bioactives

Nanomaterials for Spectroscopic Applications
Edited by Kaushik Pal

ISBN 978-981-4877-69-5 (Hardcover), 978-1-003-16033-5 (eBook)
www.jennystanford.com

that determine the shape, size and other physicochemical, biological properties of the bioformulated ones. These can be well distinguished and delivered in an operative way according to their features present in them. Technological advancement narrates the valuable sources available for tapping these resourceful properties entrenched in the formulated particles. In the forthcoming sections, a brief note about green nanostructural synthesis and its characterization have been explored.

4.1 Introduction

A little augmentation in research inputs brings massive outputs in technological advances with valuable functions. Technological development is of great assistance to the human race, as it simplifies the lifestyle of the people. However, sometimes it leads to detriments causing a pandemic outbreak like COVID-19 leading to shockwaves globally. Protection of the environment, safety measures, and dangers triggered by irretrievable impairment should be strictly taken care of while developing a new cost-effective technology. This report was brought into the notification at the Rio Declaration of UN (1992) in the Conference on Environment and Development [1].

Technological advances are piled up rapidly with numerous new innovations. Nanotechnology presents itself as one of the strong pillars in the foundation of this new construction. Being adaptable and handy, resourceful with uniqueness, it is attracted not by a small angle, limited to certain fields but with a large degree of expansion in the budding scientific projects. Feynman's concept is treasured at an elevated level in the vast expanding technology. The development of this nanotechnology exponentially increases with the growing thirsty demand of the scientific community. The studies involve maneuvering of the matter present in atomic/molecular states, to deliver particles in nanorange, with structures like the one in the bulk system with incomparable properties to their counterparts [2].

Organic/inorganic/metallic NPs can be formulated by three main routes of biological, chemical and physical [3]. Identification and the physical nature of NPs synthesized can be well interpreted and understood by characterization modules available [4]. Small size, increased energy, large outward surface area and quantum confinement [5] attribute to the unique feature (magnetic, electrical, optical) in the nanorange for the synthesized NPs. The exceptional physio-chemical nature assists NPs to furnish, resourceful utility in diversified fields like optics [6], pharmacology [7], food [8] and agriculture [9], chemical industry and cosmetics [10].

The scientific dreamers are constantly involved in evolving new pathways for developing alternatives for obtaining biodegradable, eco-friendly, non-toxic, simple methods, with high selectivity and stability. Green nanotechnology offers the best solution in an efficient mode with noteworthy characteristics and magnitudes [11–13]. Green nanotechnology (GNT) tries to hire the valuable nanoinventions in physical, material, chemical sciences to craft a sustainable eco-friendly, economical and energy resourceful products. This acceptable green nanoroute furnishes its validity in manufacture, construction and infrastructure industries, transportation, communication, purification, in clean-ups (wastewater, air and soil management), food and agronomy sectors.

Green nanoinfrastructures pressurise to deliver an improved outcome with safety and reliability, to preserve our biosphere, and thereby protect our surroundings [14]. GNT fundamentally leads to the formulation of nanoparticles (NPs) in a cost-effective, non-hazardous and biofriendly way by 'bottom-up' methods. Green called a clean one, evades the use of unsafe chemicals in its biodegradable protocol [15–17]. Anastas and Eghbali [18] narrate some basic principles to be borne in mind while following green nanoroutes for the synthesis of nanostructures. A Green safety triangle with 12 basic ideologies for green nanostructural synthesis is portrayed in Fig. 4.1.

Biosynthesis of nanostructures by GNT utilizes materials from the nature either from a uni/multi-cellular species like fungi [19–22], bacteria [23, 24], viruses [25, 26], yeasts [27, 28], algae [29] plants [30], agro wastes [31], enzyme [32] and protein, find their presence in GNT synthesis. Biological synthesis has captured the scientific forums in choosing better non-toxic alternatives of secondary-metabolites (Sec-Met) as a bioreducer and biostabilizer or biocapping in its eco-friendly itinerary. Utilization of unsafe chemicals (chemical methods) is completely avoided by safer

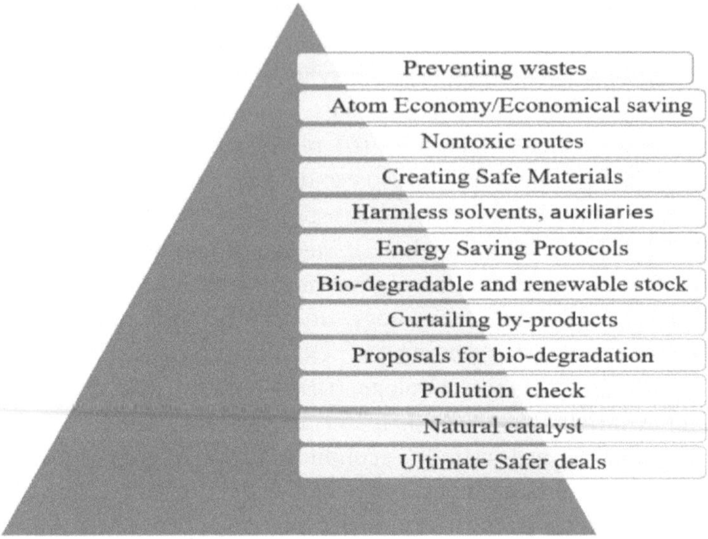

Figure 4.1 A Green safety triangle with 12 basic ideologies for green nanostructural synthesis.

alternatives of plant-green technological attempt (biological method) that are simple, environment-friendly, reliable and economical.

Plant species being non-pathogenic, with its biomolecules actively embedded in it, controls the assembling of the plant-initiated approach of NP formulation. Inadequacies faced by microbial NP formulation are curtailed by GNT-plant-based synthesis [33, 34]. Profound integrated properties of NPs find their utility in a wide range of platforms like catalysis, optics and electronics, biomedical and medical imaging, food and agricultural sectors and pest management [35, 36].

4.2 Sec-Met: An Effective Source for Green Nanostructural Synthesis

300,000,000 biodegradable plant species are presently available in our Mother Earth, with some at the threshold of extermination to be safeguarded for the future engendered (generation) [37]. The biodegradable plant kingdom is acknowledged well as a suitable substitute for physical/chemical techniques in the green nanostructural formulations [38]. The plant kingdom, a royal source of rich secondary metabolites, that serves for defence mechanism can be tapped from its botanical extracts, for an effective utilization in green and clean methods of formulation [39]. GNT effectively employs this potential from plant to engineer NPs [40].

Different plant segments rich in biomolecules (BMs) like (seed, shoot, root, stem, leaf, bud, flower, bark, latex, rhizomes, and fruit) effectively used in the formation of organic/inorganic/metallic NPs by green nanoroutes have been reported by authors [41, 42]. Plant's botanical, extremely rich in BMs like flavonoids, polyphenols, alkaloids, proteins, amino acids, carbohydrates, terpenoids, etc., serve as powerful bioreducer and biostabilizer in green nanostructural synthesis [43]. Plant-mediated, Sec-Met/primary-metabolites, used for biosynthesis of nanostructures are represented in a systematic way in Fig. 4.2.

Many factors like climatic environment, minerals present, nature of the soil, pollution, etc., regulate the concentration and the composition of Sec-Met of the plant [44, 45]. An important keynote linked with variations in morphological size and geometric shape of

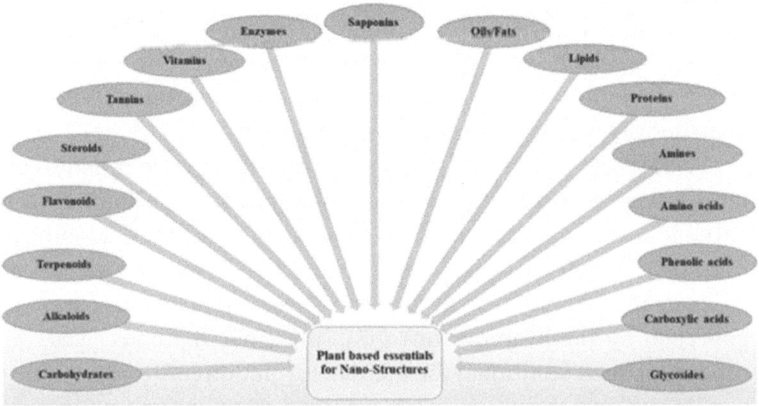

Figure 4.2 Plant Sec-Met/primary-metabolites, used for biosynthesis of nanostructures.

the formulated NPs by green-routes depends upon the structural conformation and molecular concentration of these highly energetic BMs in the plant extract [46].

Utilization of plant Sec-Met in technological advancement is visualized for the green nanostructural fabrication of NPs. These as-synthesized green nanostructures from Sec-Met philosophies find their utility in a wide arena like vegetative food and seafood [47], agricultural [48], pharmaceutical [49], biomedical usages [50], ecological health [51], textile [52], optics [53], and energy [54]. Green nanostructures deliver reliability and sustainability and provide a significant substituted solution for ecological challenges in different venues like waste-water treatment [55], solar-energy garnering [56], H_2 production [57], drug-delivery [58], catalysis [59], and pollution control [60].

4.3 Innovative Focuses on Green Nanostructural Formulation

Scientific ignorance is a great question now. Research innovations are proceeding to face this challenging situation. Nanotechnology is in the first place, duly supported by green-nano routes amidst huge scientific conception with deep-rooted contemplation [61] to solve this difficulty. The three 'E's i.e. 'Energy'/'Environmental'/'Economy' at the marginal level of crunches at present, due to extreme catastrophe are facing challenges to meet sustainability. GNT provides the best solution for these unanswered questions.

The activity of NPs rests upon several aspects like constituent's morphological factors, size dissemination, shape formation, constituent's composition, surface-chemistry, reductant, stabilizer/capping promoters, agglomeration control, dissolution-rate, reactivity, stability, and solubility [62]. These aforesaid features are governed by the prerequisites (T°C, pH, time, methods used, and other reaction conditions) for the NPs formation [63].

GNT utilizes these 3 core constituents, like precursors (organic/inorganic-metallic/non-metallic), reducer, capper/stabilizer for its process of fabrication, with 3 core steps as (1) reduction to

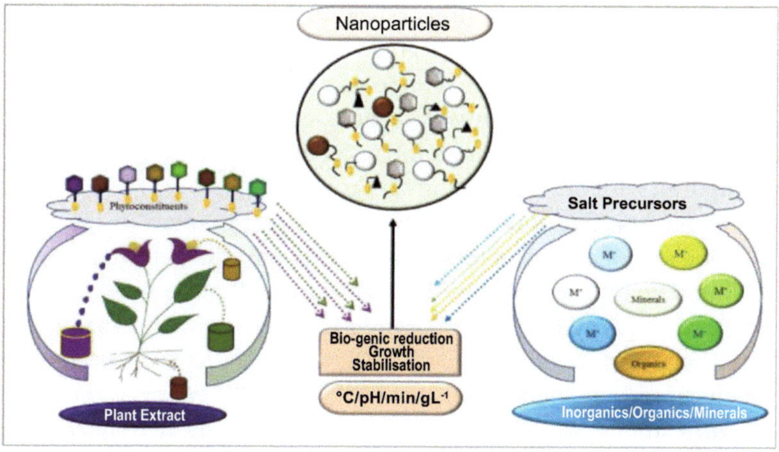

Figure 4.3 Formation of green nanostructured particles from plant's botanicals.

undergo nucleation, (2) growth of nucleation to form particles, (3) stabilization/capping to give stability to the particles formed [64]. The 3 core aspects involved in the biosynthesis of green nanostructures are (1) use of non-toxic biodegradable component, (2) use of no solvent or water solvent system, (3) economical, biological reducer and stabilizer [65]. Figure 4.3, depicts the formation of nanoparticles by green nanosynthesis using plant extracts.

Bio-synthetic methods deliver controlled desired sizes and shapes according to the needs of the application [66]. Plant's bioorganic extracts as reducer and stabilizer deliver the requirement in a distinguished way [67]. The greatest benefit of utilizing plants phyto-molecules is the potential accessibility from a wide range of biological reserves. As a time-saving process, a higher degree of stability and quick solubility in water for the formulated NPs, are the added benefits of this green method [68].

Reducing and stabilizing capacity of the plants depends on the phyto-molecules that participate from various cell tissues within it [69, 70]. Mechanistic exploration being comparatively limited, is yet to be delivered in-depth. Ample literature is available for green synthesis, using various parts of plant species. Authors have exploited inorganic salt precursors for nanoparticle formulation using various approaches of green synthesis. Extraction of plant biomolecules with a large volume of potentials in it, like electron-giving tendency, an important keynote, can be a better eco-friendly, safe option in GNT synthesis for reduction of inorganic ions when compared with other reducing agents.

Many frontline floral species have been exploited in the fields of medicinal, phytochemical, pesticidal and clinical trials. Seeds of *Tectona grandis* i.e. Teak [71], Peel of—Rambutan *Nephelium lappaceum* [72–74], *Citrus reticulata* (Mandarin) [75], banana [76], the shell of *Plukenetia volubilis* [77], *Anacardium occidentale* [78], the bark of *Cinnamon zeylanicum* [79], leaf of *Ziziphora tenuior* [80], *Azadirachta indica* [81], *Plantago asiatica* [82], flowers of *Jacaranda mimosifolia* [83], *Quisqualis indica* [84], *Mari Gold* [85], plant of *Garcinia xanthochymus* [86], resinous gum of *gum karaya* [87], *gum Arabic* [88], *Desmodium gangeticum* [89], and *wheat husk-ash* [90] have been generously used in the formulation of Ag, ZnO, Cu, CuO, Au, Ni, and other NPs. Plant's vital Sec-Met present in the above-said species like sugars, enzymes, polyphenols, flavonoids, terpenoids, fats, alkaloids and others have an essential part in reduction and capping of the salt-precursors into NPs.

Less energy, with a single step approach, is the desired platform in GNT synthesis. Elevated temperatures are an unnecessary component in cost-effective GNT synthesis. Changes in pH, time, temperature, the concentration of extract and salt precursors, modify the structural parameters of the formed NPs. The optimization of various reaction factors leads to the desired outcomes. Table 4.1 gives input on some contribution in green nanosynthesis.

Scientific communities are continuously developing green nanoroutes for the formation of nanoparticles with safe deliveries that limit the negative consequences to humans and the environment. It is important to authenticate the keynotes pertaining to the validation of the nanoproducts formed, its reliability, stability, formation and application operations to ensure a safe endpoint. The pros and cons of nanoparticles have been reported in terms of agriculture and medicine. Nanotoxicity has been commented on but requires clarity in depth.

Table 4.1 Some contributions in green nanosynthesis

Green source-plants	Salt/ nanoparticles	Key notes-phyto-components (PCs)/ groups involved, reaction condition/others	Ref.
Impatiens balsamina, Lantana camara	$AgNO_3/Ag$	Proteins, other Sec-Met, 5 h, 60°C, greyish brown and brownish yellow	[91]
Myristica fragrans	$AgNO_3/Ag$	Essential oil, terpenes and phenyl propenes, 100°C, pH 7, yellow-golden yellow → red	[92]
Berberis vulgaris – leaf, root	$AgNO_3/Ag$	1 h, room temperature, brown color	[93]
Origanum vulgare	$AgNO_3/Ag$	Terpenoids, flavonoids, alkaloids/2 h, 85–90°C, light yellow → dim brown color	[94]
Azadirachta indica	$AgNO_3/Ag$	NH_2, OH – groups, 15 min, room temperature, yellowish → reddish brown	[95]
Eclipta Alba	$HAuCl_4/Au$	Proteins, 10 min to 96 h, room temperature, red coloration	[96]
J. communis, Camellia sinensis,	$HAuCl_4/Au$	Thearubigins, theaflavins, catechins, and other PC, 25°C, 30 min, 50 min, 60 min, pale yellow → purple-red	[97]
Banana peel (Musa paradisiaca)	$HAuCl_4/Au$	80°C, yellow → brown → purplish-pink	[98]
Phyllanthus niruri	$Zn\ (NO_3)_2/ZnO$	OH group, 60°C, calcination 400°C – 2 h, yellow colored suspension to white powder	[99]
Vitex negundo	$Zn\ (NO_3)_2/ZnO$	Alkaloids, phenolic, flavonoid, flavone, isoorientin, 5 h, calcination 450°C.	[100]

(*Contd.*)

Table 4.1 (*Continued*)

Green source-plants	Salt/ nanoparticles	Key notes-phyto-components (PCs)/ groups involved, reaction condition/others	Ref.
Eichhornia crassipes	Zn $(NO_3)_2$/ZnO	6 h, 100°C, calcination 400°C, yellow color → white powder	[101]
Mangnifera indica	ZnO	6 h, 80°C, calcination 450°C	[102]
Aloe vera	Cu $(NO_3)_2$/CuO	–OH, C=C, C=O, 100–120°C, deepblue → colorless → brickred → darkred	[103]
Azadirachta indica	$CuCl_2$/Cu	Terpenoids, polyphenols 85°C, green-yellow-orange-reddish brown-brown-dark brown	[104]
Magnolia kobus	$CuSO_4$/CuO	Alcohols, phenolics, amide – proteins, terpenoids, reducing sugars, 25–95°C	[105]
Camelia sinensis	CuCl/CuO	1 h, 75–85°C, dark green → light green suspension	[106]
Nerium oleander	$CuSO_4$/CuO	20 min, 50°C, light color → dark color	[107]
Enicostemma axillare (Lam.)	$CuSO_4$/CuO	pH 7.0, reddish brown → viscous green	[108]
Lagenaria siceraria	$FeCl_3$/Fe_3O_4	Amino, carboxyl, hydroxyl, 60 min, room temperature, dark → light color	[109]
Physalis angulata	La $(NO_3)_3$/La_2O_3	Alkaloids, saponins, polyphenols, flavonoids 2 h, 50°C, calcined - 2 h/700°C	[110]

4.4 Practices for Characterization of Green-Formulated NPs

Nanoparticles formed varies in composition (biomolecules, synthesized polymers, metal/metal oxides, micelles, etc.). Nanomaterials with unique features are inferred not by a single approach but by different valuable modes. The spectrometric tools offer valuable information about the chemical inputs embedded in these particles. Each spectrometric tool provides unique information specific to the nature and type of its significance and characteristics.

A wide variety of spectrometric tools are used to evaluate the morphological, size, optical, composition and other parameters. Depending upon the nature of the NP (powdered or suspension), one has to decide the type of tool to be chosen for getting adequate characteristic inputs. Figure 4.4, delivers a formative input of characterization analysis based on the nature of NPs. NPs can be inferred by a variable approach of optical (UV-Vis) [111] and functional (FTIR) [112], Raman spectroscopy [113], XPS [114] and XRD [115], SEM [116], TEM [117], EDAX, AFM [118] and solid-state NMR [119]. The examination of a specific size, definite shape, surface-charge, porosity, optical density, and functional groups, stability and composition are some basic fields that give insight to predict the nature of NPs [120]. Nanostructured components encompass a wide segment with micro to nanorange as

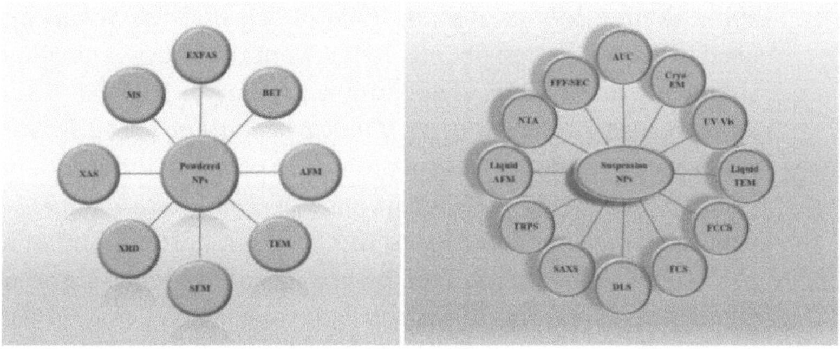

Figure 4.4 A formative input of characterization analysis based on the nature of NPs.

in nanofilms [121], nanotubes [122], nanoflowers [123], nanosheets [124], nanocapsules [125] and many more.

- **Size Distribution**: Size implies the spatial representation of the NP. This is measured by microscopic methods with high-resolutions like XRD where, Pd NPs synthesized using (*Boswellia serrata*), *gum olibanum* showed the presence of the crystal with FCC structure [126], TEM Seeds of *Benincasa hispida* used for Au NP synthesis validated the presence of (10 to 30 nm), 14 nm by TEM [127] if 2D, extracts from the leaves of *Catharanthus roseus* were used for TiO_2 NPs synthesis, inferred that the cubical NP had 25 nm and 110 nm size by SEM 3D [128], DLS reports of [129] infer that S NPs with the size of 41.25 nm were got by using *Acanthe P. Bracteatum* as a reductant, at a single NP sub-level in nm range. The NP size dispersal in an intrinsic quantity that influences the quality of the process and the evaluated value is the average measure that results due to distribution, with each NP having the same mass.
- **Dimensional Shape**: Depending upon the conditions of green synthesis, NPs have varied geometries from regular spherical to irregular matrices [130]. Different characteristic features such as optical/plasmonic effects, surface-binding nature and withhold or give-away various components are basically governed by the shape specification. *Sapindus mukorossi* was used in fabricating MnO_2-needle, Mn_2O_3-spherical, and Mn_3O_4 cubic shaped NPs as seen by TEM [131], *Hibiscus Sabdariffa* used in the formation of CeO_2 NP (3.9 nm) had both crystalline shape and amorphous shape in it according to HRTEM [132], SEM images of S NPs using *Punica granatum* (peel, leaves, fruit) revealed the crystal shapes were spherical, different and sheets respectively [133], fruit juice of *Couroupita guianensis* used in the formation of Fe_3O_4 NPs were oval (10 nm) by AFM [134]. The shape is validated by high-resolution modules that give inputs about the 2D anisotropic concept in the plane and the 3D anisotropic concept of spatial assembly, [135] of the NPs formulated by the green-routes.

- **Surface-Charge**: The surface of the NP contributes a net charge due to the presence of a fringe between the colloidal part and the solid part. The charge is a key factor responsible for the stability of the NPs, else would lead to aggregation, due to electrostatic repulsions [136]. Zeta (ζ) potential is the measure for the colloidal solution's stability, where Au and Ag NPs got by using *Cannabis sativa* had values of −12.3 mV, −20.6 mV and −29.2 mV for fiber, core Au NPs and core-Ag NPs respectively inferring a high colloidal stability [137] and surface charge of NPs. The author used *Catharanthus roseus, Azadirachta indica, Polyalthia longifolia* and *Mangifera indica* to produce S NPs with ζ—potential between −7.12 mV to −34.1 mV [138], which is influenced by ionic strength (solvent) and pH of the system. pH value at zero-point charge deals with the stability of the particles formed [139].
- **Absorbency or Porosity**: Application [60, 61] of porous NPs have expanded exponentially in the field of medicine and in adsorption techniques associated in the field of removing pollutants [140]. An augmentation in the surface: volume ratio delivers the absorbency or porosity of the crystalline NPs formed by green nanoroutes. Gas sorption (BET) [141] analytical technique quantifies the absorbency or porosity associated with nano-objects in the crystalline state. The average pore size of 8.31 nm and surface area of 16.77 $m^2\ g^{-1}$ were seen by BET analytical methods got by using *Sorbaria sorbifolia* (leaves) for synthesizing Carbon spheres [142]. The part of the core and cover of the NPs influenced by the functional moieties fastened to the core surface determine deliverables of the absorbency or porosity of the green synthesized NPs. Basic inputs of various analytical tools used by authors in green nanoway, using plant materials, for acquiring characteristic knowledge of the synthesized NPs are segmented in Table 4.2, and Fig. 4.5 shows a systematic characteristic operation used for different sectors of analysis.

Table 4.2 Analytical tools used for acquiring characteristic knowledge

Characterization trends	Specific response	NPs characterized	Source of plant species	Ref.
UV-Vis, XRD, SEM	λ_{max} 430 nm – Ag, Size 5–35 nm – Ag, Size 5–40 nm – ZnO	Ag, ZnO, Ag/ZnO	*Prosophis fracta, coffee*	[143]
UV-Vis, XRD, FESEM, FETEM, EDX, DLS	Size 12 nm, Spherical	Ag	*Garlic, Clove extracts*	[144]
UV-Vis, XRD, SEM, XPS, TEM, EDX, DLS, HPLC	λ_{max} 420 nm, Size 5–10 nm, clusters	Ag	*Drosera (indica, binata, spatulata, muscipulas)*	[145]
SEM, FTIR, XRD, EDAX, AFM	Size 60–80 nm, Spherical	ZnO	*Cassia alata.*	[146]
FTIR, SEM, XRD, EDX, DLS	λ_{max} 320 nm, ζ −12.3 mV Size 12–35 nm, hexagonal wurtzite	ZnO	*Ceropegia candelabrum*	[147]
FTIR, XRD, SEM, DLS	λ_{max} 316 nm, Size 21 nm/spherical	ZnO	*Ocimum americanum*	[148]
UV-Vis, XRD, TEM, FTIR, EDAX, DLS	λ_{max} 530 nm, Size 26 $\pm$ 11 nm/ spherical hexagonal/triangular	Au	*Couroupita guianensis Aubl.*	[149]
UV-Vis, XRD, HR-TEM, EDX, DLS, ζ potential, FTIR	λ_{max} 534 nm to 551 nm, Size 18.2 nm, FCC	Au	*Eucommia ulmoides*	[150]
UV-Vis, DLS, TEM-EDX, FTIR, TGA, BET	λ_{max} 530 nm, Size 56 nm, Surface area 9.2 m^2/g	Au	*Piper longum*	[151]
UV-Vis, XRD, TEM, EDX, FTIR, TGA, HPLC	λ_{max} 321 nm, 329 nm, Size 1.3–2.6 nm	Pt	*Dates (Ajwa, Barni)*	[152]

UV-Vis, XPS, HR-TEM, FTIR, SEM, EDX, ICP	λ_{max} 477 nm, Size 2–12 nm,	Pt	*Diopyros kaki*	[153]
AAS, FESEM, FETEM, SAED, DLS, TEM, XRD, TGA, FTIR, EPR, ζ potential	Size 47.3 nm, 28.6 nm, 13.2–26.4 nm	Co, Pt, Pt @Co	*Sechium edule*	[154]
UV-Vis, TEM, FTIR, SEM, EDX, DLS, NMR, GC – MS, ζ potential	λ_{max} 460 nm, Size 10–50 nm, (ζ) −22.7 mV	Pd	*Moringa oleifer*	[155]
EDAX, XRD, FTIR, SEM, LSV	LSV −0 387 and −0.5283 V, Size 23 nm	Pd	*Ocimum sanctum*	[156]
UV-Vis, HR-TEM, FTIR, XRD	λ_{max} 421 nm, Spherical Size 50–150 nm	Pd	*Prunus x yedoensis*	[157]
SEM, TEM, DLS, XRD, TGA/DSC, FTIR	Size 6–25 nm	RuO_2	*Acalypha indica*	[158]
UV-Vis, SEM, FTIR, EDX, XRD, fluorescence	λ_{max} 494 nm, λ_{em} 464 nm, Size 25–90 nm	Ru	*Gloriosa superba*	[159]
UV-Vis, TEM, FTIR, EDAX, XRD, photoluminescence	λ_{max} 290 nm, 320 nm, Size 2–8 nm, FCC	CdS QDs	*Asparagus racemosus*	[160]

Systematic characteristic operation used for different sectors of analysis to get valuable inputs.

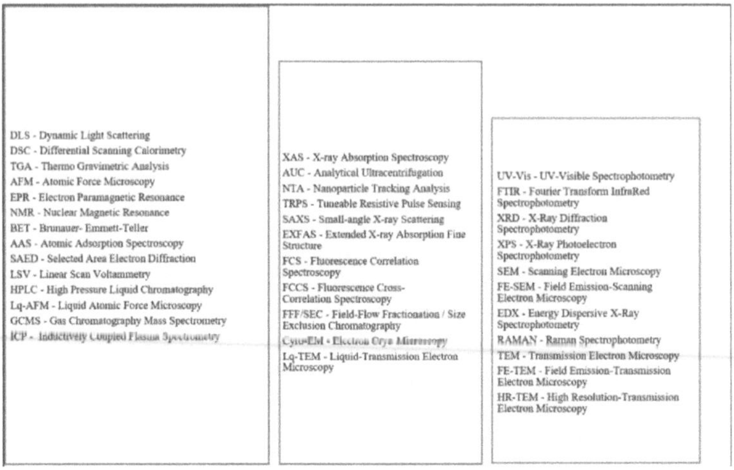

Figure 4.5 A systematic characteristic operation used for different sectors of analysis.

4.5 Optimistic Positivity

The nanostructural formulation by green derivatives provides deliverables that accomplish safety measures and health reflections. The biodegradable employed does not allow the toxic goals to interfere. The bioactives that effectively function as capping or as a bioreducer in nanostructural synthesis, are from the rich reserves of a potential kingdom. Stable and soluble products of the NPs are delivered in a simple and systematic way. This provides a well-sustained platform for an efficient and significant operational module. As an energy and time-saving method, green nanostructural formulation, offers the deliverables in a swift and a speedy way, without any specific downsides.

4.6 Conclusion

A variety of plant sources being utilized, provides a healthy outlook in green nanostructural formation. The Sec-Met present has stimulated the research community to interlace the essence of plants (biological source) with material science, chemistry, and physics to offer a rustic output. Terpenoids, flavonoids, saponins, polysaccharides, proteins, alkaloids, and other natural products effectively provide their functional groups in bioreduction and capping sectors. A simple, cost-efficient, energy and time saving, non-hazardous, products are delivered in a compatible way. The formation of particles basically is regulated by the nature of Sec-Met present. Physio-chemical parameters altered by Sec-Met are vivid when various platforms of characterization tools are employed. This resourceful and meaningful interpretation offers a valuable resource for various modules of the application. The basic limitation is that toxicity, though limited, is to be exploited in depth.

Acknowledgment

This work has been supported by India-Bulgaria joint research program by Department of Science and Technology, India (INT/BLG/P-09/2019).

References

1. Rio Declaration on Environment and Developments: The United Nations Conference on Environment and Development, 12 August 1992, A/CONF.151/26 (Vol. I) Report of The United Nations Conference on Environment and Development (Rio de Janeiro, Brazil, 3–14 June 1992), www.un.org A_CONF.151_26_Vol.I_Declaration.pdf.
2. N. Noah, Chapter 6, Green synthesis: characterization and application of silver and gold nanoparticles, in *Green Synthesis, Characterization and Applications of Nanoparticles*, Micro and Nano Technologies, 2019, pp. 111–135, doi:10.1016/B978-0-08-102579-6.00006-X.
3. S. Iravani, H. Korbekandi, S. V. Mirmohammadi, B. Zolfaghari, Synthesis of silver nanoparticles: chemical, physical and biological methods, *Research in Pharmaceutical Sciences*, 2014, **9**(6), 385–406.
4. C. Dipankar, S. Murugan, The green synthesis, characterization and evaluation of the biological activities of silver nanoparticles synthesized from Iresine herbstii leaf aqueous extracts, *Colloids*

and Surfaces B: Biointerfaces, 2012, **98**, 112–119, doi:10.1016/j.colsurfb.2012.04.006.

5. I. Hussain, N. B. Singh, A. Singh, H. Singh, S. C. Singh, Green synthesis of nanoparticles and its potential application, *Biotechnology Letters*, 2016, **38**(4), 545– 560, doi:10.1007/s10529-015-2026-7.
6. D. R. Raj, S. Prasanth, T. V. Vineeshkumar, C. Sudarsanakumar, Surface plasmon resonance based fiber optic dopamine sensor using green synthesized silver nanoparticles, *Sensors and Actuators B: Chemical*, 2016, **224**, 600–606, doi:10.1016/j.snb.2015.10.106.
7. P. Kuppusamy, M. M. Yusoff, N. Govindan, Biosynthesis of metallic nanoparticles using plant derivatives and their new avenues in pharmacological applications: an updated report, *Saudi Pharmaceutical Journal*, 2016, **24**(4), 473–484, doi:10.1016/j.jsps.2014.11.013.
8. V. S. Ganachari, J. S. Yaradoddi, S. B. Somappa, P. Mogre, R. P. Tapaskar, B. Salimath, A. Venkataraman, V. J. Viswanath, Green nanotechnology for biomedical, food, and agricultural applications, in *Handbook of Ecomaterials*, 2019, pp. 2681–2698, doi:10.1007/978-3-319-68255-6_184.
9. R. Amooaghaie, M. RezaSaeri, M. Aziz, Synthesis, characterization and biocompatibility of silver nanoparticles synthesized from Nigella sativa leaf extract in comparison with chemical silver nanoparticles, *Ecotoxicology and Environmental Safety*, 2015, **120**, 400–408, doi:10.1016/j.ecoenv.2015.06.025.
10. H. Singh, J. Du, P. Singh, T. H. Yi, Role of green silver nanoparticles synthesized from Symphytum ofcinale leaf extract in protection against UVB-induced photoaging, *Journal of Nanostructure in Chemistry*, 2018, **8**, 359–368, doi:10.1007/s40097-018-0281-6.
11. H. S. Salam, R. Sivaraj, R. Venckatesh, Green synthesis and characterization of zinc oxide nanoparticles from Ocimum basilicum L. var. purpurascens Benth-Lamiaceae leaf extract, *Materials Letters*, 2014, **131**(15), 16–18, doi:10.1016/j.matlet.2014.05.033.
12. D. Philip, Honey mediated green synthesis of gold nanoparticles, *Spectrochimica Acta Part A: Molecular and Biomolecular Spectroscopy*, 2009, **73**, 650–653, doi:10.1016/j.saa.2009.03.007.
13. A. G. Ingale, A. N. Chaudhari, Biogenic synthesis of nanoparticles and potential applications: an eco-friendly approach, *Journal of Nanomedicine and Nanotechnology*, 2013, **4**(2), 165, 1–7, doi:10.4172/2157-7439.1000165.

14. S. S. Sanjay, Chapter 2, Safe nano is green nano, in *Green Synthesis, Characterization and Applications of Nanoparticles*, 2019, pp. 27–36, doi:10.1016/B978-0-08-102579-6.00002-2.
15. D. Mandal, M. E. Bolander, D. Mukhopadhyay, G. Sarkar, P. Mukherjee, The use of microorganisms for the formation of metal nanoparticles and their application, *Applied Microbiology and Biotechnology*, 2006, **69**, 485–492, doi:10.1007/s00253-005-0179-3.
16. I. Kumar, M. Mondal, N. Sakthivel, Chapter 3, Green synthesis of phytogenic nanoparticle, in *Green Synthesis, Characterization and Applications of Nanoparticles*, 2019, pp. 37–73, doi:10.1016/B978-0-08-102579-6.00003-4.
17. A. D. Dwivedi, K. Gopal, Biosynthesis of silver and gold nanoparticles using Chenopodium album leaf extract, *Colloids and Surfaces A: Physicochemical and Engineering Aspects*, 2010, **369**, 27–33, doi:10.1016/j.colsurfa.2010.07.020.
18. P. Anastas, N. Eghbali, Green chemistry: principles and practice, *Chemical Society Reviews*, 2010, **39**, 301–312, doi:10.1039/b918763b.
19. G. Subashini, S. Bhuvaneswari, Chapter 39, Nanoparticles from fungi (Myconanoparticles), in *Fungi and their Role in Sustainable Development: Current Perspectives*, 2018, pp. 753–779, doi:10.1007/978-981-13-0393-7_39.
20. S. K. Dorcheh, K. Vahabi, Biosynthesis of nanoparticles by fungi: large-scale production, in *Fungal Metabolites*, 2016, pp. 1–20, doi:10.1007/978-3-319-19456-1_8-1.
21. Z. Molnar, V. Bodai, G. Szakacs, B. Erdelyi, Z. Fogarassy, G. Safran, T. Varga, Z. Konya, E. Toth-Szeles, R. Szucs, I. Lagzi, Green synthesis of gold nanoparticles by thermophilic filamentous fungi, *Scientific Reports*, 2018, **8**, 3943, doi:10.1038/s41598-018-22112-3.
22. G. Krishna, P. Vadapally, M. A. S. Charya, Biogenic synthesis of silver nanoparticles from white rot fungi: their characterization and antibacterial studies, *Open Nano*, 2017, **2**, 64–78, doi:10.1016/j.onano.2017.07.002.
23. K. Kailashwaralal, V. Deepak, S. R. K. Pandian, S. Gurunathan, Biological synthesis of gold nano cubes, from Bacillus licheniformis, *Bioresource Technology*, 2009, **100**(21), 5356–5358, doi:10.1016/j.biortech.2009.05.051.
24. S. Iravani, Bacteria in nanoparticle synthesis: current status and future prospects, *International Scholarly Research Notices*, 2014, **2014**, 1–18, doi:10.1155/2014/359316.

25. M. J. Slocik, R. R. Naik, M. O. Stone, D. W. Wright, Viral templates for gold nanoparticle synthesis, *Journal of Materials Chemistry*, 2005, **15**, 749–753, doi:10.1039/b413074j.
26. A. A. Khan, E. K. Fox, M. L. Gorzny, E. Nikulina, D. F. Brougham, C. Wege, pH Control of the electrostatic binding of gold and iron oxide nanoparticles to tobacco mosaic virus, *Langmuir*, 2013, **29**(7), 2094–2098, doi:10.1021/1a3044126.
27. K. Roy, C. K. Sarkar, C. K. Ghosh, Photocatalytic activity of biogenic silver nanoparticles synthesized using yeast (Saccharomyces cerevisiae) extract, *Applied Nanoscience*, 2015, **5**, 953–959, doi:10.1007/s13204-014-0392-4.
28. A. Mourato, M. Gadanho, A. R. Lino, R. Tenreiro, Biosynthesis of crystalline silver and gold nanoparticles by extremophilic yeasts, *Bioinorganic Chemistry and Applications*, 2011, **2011**, 8, doi:10.1155/2011/546074.
29. R. A. Hamouda, M. H. Hussein, R. A. Aboelmagd, Synthesis and biological characterization of silver nanoparticles derived from the cyanobacterium Oscillatoria limnetica, *Science Reports*, 2019, **9**, 13071, doi:10.1038/s41598-019-49444-y.
30. J. Y. Song, B. S. Kim, Rapid biological synthesis of silver nanoparticles using plant leaf extracts, *Bioprocess Biosynthesis Engineering*, 2009, **32**, 79–84, doi:10.1007/s00449-008-0224-6.
31. G. Bagherzade, M. M. Tavakoli, M. H. Namaei, Green synthesis of silver nanoparticles using aqueous extract of saffron (Crocus sativus L.) wastages and its antibacterial activity against six bacteria, *Asian Pacific Journal of Tropical Biomedicine*, 2017, **7**(3), 227–233, doi:10.1016/j.apjtb.2016.12.014.
32. I. Willner, R. Baron, B. Willner, Growing metal nanoparticles by enzymes, *Advance Materials*, 2006, **18**, 1109–1120, doi:10.1002/adma.200501865.
33. N. Pantidos, L. E. Horsfall, Biological synthesis of metallic nanoparticles by bacteria, fungi and plants, *Journal of Nanomedicine and Nanotechnology*, 2014, **5**(5), doi:10.4172/2157-7439.1000233.
34. S. Ghosh, S. Patil, M. Ahire, R. Rohini Kitture, D. S. Gurav, M. A. Jabgunde, S. Kale, K. Pardesi, V. Shinde, J. Bellare, D. D. Dhavale, B. A. Chopade, Gnidia glauca flower extract mediated synthesis of gold nanoparticles and evaluation of its chemo catalytic potential, *Journal of Nanobiotechnology*, 2012, **10**, 17, doi:10.1186/1477-3155-10-17.

35. V. V. Makarov, A. J. Love, O. V. Sinitsyna, S. S. Makarova, I. V. Yaminsky, M. E. Taliansky, N. O. Kalinina, "Green" nanotechnologies: synthesis of metal nanoparticles using plants, *Acta Naturae*, 2014, **6**(1), 35–44.
36. S. Jafarirad, M. Mehrabi, B. Divband, M. K. Nasa, Biofabrication of zinc oxide nanoparticles using fruit extract of Rosa canina and their toxic potential against bacteria: a mechanistic approach, *Materials Science and Engineering C*, 2016, **59**, 296–302, doi:10.1016/j.msec.2015.09.089.
37. V. De Luca, V. Salim, S. M. Atsumi, F. Yu, Mining the biodiversity of plants: a revolution in the making, *Science*, 2012, **336**(6089), 1658–1661, doi:10.1126/science.1217410.
38. A. Altemimi, N. Lakhssassi, A. Baharlouei, D. G. Watson, D. A. Lightfoo, Phytochemicals: extraction, isolation, and identification of bioactive compounds from plant extracts, *Plants (Basel)*, 2017, **6**(4), 42, doi: 10.3390/plants6040042.
39. J. R. Peasari, S. S. Motamarry, K. S. Varma, P. Anitha, R. B. Potti, Chromatographic analysis of phytochemicals in Costus igneus and computational studies of flavonoids, *Informatics in Medicine Unlocked*, 2018, **13**, 34–40, doi:10.1016/j.imu.2018.10.004.
40. A. K. Mittal, Y. Chisti, U. C. Banerjee, Synthesis of metallic nanoparticles using plant extracts, *Biotechnology Advances*, 2013, **31**, 346–356, doi: 10.1016/j.biotechadv.2013.01.003.
41. R. C. Reddy, G. G. Redhi, Chapter 4, Green synthesis of metal nanoparticles and its reaction mechanisms: synthesis, characterization and their applications, in *Green Metal Nanoparticles*, 2018, pp. 113–139, doi:10.1002/9781119418900.ch4.
42. S. Ahmed, M. Ahmad, B. L. Swami, S. Ikram, A review on plants extract mediated synthesis of silver nanoparticles for antimicrobial applications: a green expertise, *Journal of Advanced Research*, 2016, **7**, 17–28, doi:10.1016/j.jare.2015.02.007.
43. K. S. Sindhura, T. N. V. K. V. Prasad, P. Selvam, O. M. Hussain, Synthesis, characterization and evaluation of effect of phytogenic zinc nanoparticles on soil exo-enzymes, *Applied Nanoscience*, 2014, **4**, 819–827, doi:10.1007/s13204-013-0263-4.
44. L. Jaakola, A. Hohtola, Effect of latitude on flavonoid biosynthesis in plants, *Plant, Cell and Environment*, 2010, **33**, 1239–1247, doi:10.1111/j.1365-3040.2010. 02154.x.

45. R. Akula, G. A. Ravishankar, Influence of abiotic stress signals on secondary metabolites in plants, *Plant Signaling & Behavior*, 2011, **6**(11), 1720–1731, doi:10.4161/psb.6.11.17613.
46. E. C. Njagi, H. Huang, L. Stafford, H. Genuino, H. M. Galindo, J. B. Collins, G. E. Hoag, S. L. Suib, Biosynthesis of iron and silver nanoparticles at room temperature using aqueous sorghum bran extracts, *Langmuir*, 2011, **27**(1), 264–271, doi:10.1021/la103190n.
47. V. P. N. N. Kumar, S. V. N. Pammi, P. Kollu, K. V. V. Satyanarayana, U. Shameem, Green synthesis and characterization of silver nanoparticles using Boerhaavia diffusa plant extract and their antibacterial activity, *Industrial Crops and Products*, 2014, **52**, 562–566, doi:10.1016/j.indcrop.2013.10.050.
48. T. A. Shalaby, Y. Bayoumi, N. Abdalla, H. Taha, T. Alshaal, S. Shehata, M. Amer, É. Domokos-Szabolcsy, H. El-Ramady, Chapter 10, Nanoparticles, soils, plants and sustainable agriculture, in *Nanoscience in Food and Agriculture 1, Sustainable Agriculture*, 2016, **20**, pp. 283–312, doi:10.1007/978-3-319-39303-2_10.
49. H. Korbekandi, M. R. Chitsazi, G. Asghari, R. B. Najafi, A. Badii, S. Iravani, Green biosynthesis of silver nanoparticles using Quercus brantii (oak) leaves hydroalcoholic extract, *Pharmaceutical Biology*, 2015, **53**(6), 807–812, doi:10.3109/13880209.2014.942868.
50. V. N. Kalpana, V. D. Rajeswari, A review on green synthesis, biomedical applications, and toxicity studies of ZnO NPs, *Bioinorganic Chemistry and Applications*, 2018, **2018**, 12, doi:10.1155/2018/3569758.
51. M. Maksimoviu, O. E. MikliSSanin, Towards green nanotechnology: maximizing benefits and minimizing harm, in *CMBEBIH 2017, IFMBE Proceedings*, 2017, **62**, 164–170, doi:10.1007/978-981-10-4166-2_26.
52. P. S. Vankar, D. Shukla, Biosynthesis of silver nanoparticles using lemon leaves extract and its application for antimicrobial finish on fabric, *Applied Nanoscience*, 2012, **2**, 163–168, doi:10.1007/s13204-011-0051-y.
53. S. Pugazhendhi, E. Kirubha, P. K. Palanisamy, R. Gopalakrishnan, Synthesis and characterization of silver nanoparticles from Alpinia calcarata by Green approach and its applications in bactericidal and nonlinear optics, *Applied Surface Science*, 2015, **357B**, 1801–1808, doi:10.1016/j.apsusc.2015.09.237.
54. P. K. Rai, V. Kumar, S. S. Lee, N. Raza, H. Kim, Y. S. Ok, D. C. W. Tsang, Nanoparticle-plant interaction: implications in energy,

environment, and agriculture, *Environment International*, 2018, **119**, 1–19, doi:10.1016/j.envint.2018.06.012.

55. B. Devi, M. Ahmaruzzaman, Bio-inspired facile and green fabrication of Au@Ag@AgCl Core–double shells nanoparticles and their potential applications for elimination of toxic emerging pollutants: a green and efficient approach for wastewater treatment, *Chemical Engineering Journal*, 2017, **317**, 726–741, doi:10.1016/j.cej.2017.02.082.
56. K. W. Guo, Green nanotechnology of trends in future energy: a review, *International Journal of Energy Research*, 2012, **36**, 1–17, doi:10.1002/er.1928.
57. D. Nath, A. K. Manhar, K. Gupta, D. Saikia, S. K. Das, M. Mandal, Phytosynthesized iron nanoparticles: effects on fermentative hydrogen production by Enterobacter cloacae DH-89, *Bulletin of Material Science*, 2015, **38**(6), 1533–1538, doi:10.1007/s12034-015-0974-0.
58. M. Ganeshkumar, M. Sathishkumar, T. Ponrasu, M. G. Dinesh, L. Suguna, Spontaneous ultra fast synthesis of gold nanoparticles using Punica granatum for cancer targeted drug delivery, *Colloids and Surfaces B: Biointerfaces*, 2013, **106**, 208–216, doi:10.1016/j.colsurfb.2013.01.035.
59. R. C. Fierascu, A. Ortan, S. M. Avramescu, I. Fierascu, Phyto-nanocatalysts: green synthesis, characterization, and applications, *Molecules*, 2019, **24**, 3418, doi:10.3390/molecules2419341.
60. S. U. Ganaie, T. Abbasi, J. Anuradha, S. A. Abbasi, Biomimetic synthesis of silver nanoparticles using the amphibious weed ipomoea and their application in pollution control, *Journal of King Saud University–Science*, 2014, **26**(3), 222–229, doi:10.1016/j.jksus.2014.02.004.
61. S. Palit, C. M. Hussain, Chapter 1, Recent advances in green nanotechnology and the vision for the future, in *Part I - Future Vision of Green Nanotechnology, Green Metal Nanoparticle*, 2018, pp. 1–21, doi:10.1002/9781119418900.ch1.
62. M. S. Akhtar, J. Panwar, Y. S. Yun, Biogenic synthesis of metallic nanoparticles by plant extracts, *ACS Sustainable Chemistry and Engineering*, 2013, **1**(6), 591–602, doi:10.1021/sc300118u.
63. P. Malik, R. Shankar, V. Malik, N. Sharma, T. K. Mukherjee, Green chemistry based benign routes for nanoparticle synthesis, *Journal of Nanoparticles*, 2014, **2014**, 14, doi:10.1155/2014/302429.
64. X. F. Zhang, Z. G. Liu, W. Shen, S. Gurunathan, Silver nanoparticles: synthesis, characterization, properties, applications, and therapeutic

approaches, *International Journal of Molecular Sciences*, 2016, **17**, 1534, doi:10.3390/ijms17091534.

65. K. Ramanujam, M. Sundrarajan, Antibacterial effects of biosynthesized MgO nanoparticles using ethanolic fruit extract of Emblica officinalis, *Journal of Photochemistry and Photobiology B: Biology*, 2014, **141**, 296–300, doi:10.1016/j.jphotobiol.2014.09.011.
66. S. Pirtarighat, M. Ghannadnia, S. Baghshahi, Green synthesis of silver nanoparticles using the plant extract of Salvia spinosa grown in vitro and their antibacterial activity assessment, *Journal of Nanostructure in Chemistry*, 2019, **9**(1), 1–9, doi:10.1007/s40097-018-0291-4.
67. P. Dauthal, M. Mukhopadhyay, Noble metal nanoparticles: plant mediated synthesis, mechanistic aspects of synthesis and applications, *Industrial and Engineering Chemical Research*, 2016, **55**(36), 9557–9577, doi:10.1021/acs.iecr.6b00861.
68. Z. Haghparasti, M. M. Shahri, Green synthesis of water-soluble nontoxic inorganic polymer nanocomposites containing silver nanoparticles using white tea extract and assessment of their in vitro antioxidant and cytotoxicity activities, *Materials Science and Engineering C*, 2018, **87**, 139–148, doi:10.1016/j.msec.2018.02.026.
69. V. Kattumuri, K. Katti, S. Bhaskaran, E. J. Boote, S. W. Casteel, G. M. Fent, D. J. Robertson, M. Chandrasekhar, R. Kannan, K. V. Katti, Gum Arabic as a phytochemical construct for the stabilization of gold nanoparticles: in vivo pharmacokinetics and X-ray-contrast-imaging studies, *Small*, 2007, **3**(2), 333–341, doi:10.1002/smll.200600427.
70. J. Fowsiya, I. V. Asharani, S. Mohapatra, A. Eshapula, P. Mohi, N. Thakar, S. Monad, G. Madhumitha, Aegle marmelos phytochemical stabilized synthesis and characterization of ZnO nanoparticles and their role against agriculture and food pathogen, *Green Process Synthesis*, 2019, **8**(1), 488–495, doi:10.1515/gps-2019-0017.
71. A. Rautela, J. Rani, M. D. Das, Green synthesis of silver nanoparticles from Tectona grandis seeds extract: characterization and mechanism of antimicrobial action on different microorganisms, *Journal of Analytical Science and Technology*, 2019, **10**, 5, doi:10.1186/s40543-018-0163-z.
72. B. Kumar, K. Smita, L. Cumbal, Y. Angulo, Fabrication of silver nanoplates using Nephelium lappaceum (Rambutan) peel: a sustainable approach, *Journal of Molecular Liquids*, 2015, **211**, 476–480, doi: 10.1016/j.molliq.2015.07.067.

73. P. Lestari, I. Pratiwi, A. Juliani, Green synthesis of silver nanoparticle using rambutan (Nephelium lappaceum L.) peel extract and its antibacterial activity against Salmonella parathypi A, *MATEC Web of Conferences*, 2018, **154**, 01024, doi:10.1051/matecconf/201815401024.

74. R. Yuvakkumar, J. Suresh, A. J. Nathanael, M. Sundrarajan, S. I. Hong, Novel green synthetic strategy to prepare ZnO nanocrystals using rambutan (Nephelium lappaceum L.) peel extract and its antibacterial applications, *Materials Science and Engineering C*, 2014, **41**, 17–27, doi:10.1016/j.msec.2014.04.025.

75. N. T. H. Trang, P. D. Dung, L. S. Ngoc, P. T. H. Van, P. Q. Thang, Green synthesis of copper nanoparticles using mandarin (Citrus reticulata) peel extract and antifungal study, *Asian Journal of Biotechnology and Bioresource Technology*, 2018, **3**(3), 1–9, doi:10.9734/AJB2T/2018/39126.

76. M. Aminuzzamana, L. M., Kei, W. H. Liang, Green synthesis of copper oxide (CuO) nanoparticles using banana peel extract and their photocatalytic activities, green and sustainable technology, *AIP Conference and Proceedings*, 2017, **1828**, 020016-1–020016-5, doi:10.1063/1.4979387.

77. B. Kumar, K. Smita, L. Cumbal, A. Debut, Sacha inchi (Plukenetia volubilis L.) shell biomass for synthesis of silver nanocatalyst, *Journal of Saudi Chemical Society*, 2014, **21**(1), S293–S298, doi:10.1016/j.jscs.2014.03.005.

78. P. Velmurugan, S. H., Jang, S. C. Hong, P. I. Yi, E. S. Jung, J. S. Park, S. Sivakumar, Green crystallization and characterization of copper oxide (CuO) nanoparticles using Anacardium occidentale shell liquid and their biomedical applications, *Journal of Nano Research*, 2016, **40**, 167–173, doi:10.4028/www.scientific.net/JNanoR.40.167.

79. M. Sathishkumar, K. Sneha, S. W. Won, C. W. Cho, S. Kim, Y. S. Yun, Cinnamon zeylanicum bark extract and powder mediated green synthesis of nano-crystalline silver particles, *Colloids and Surfaces B: Biointerfaces*, 2009, **73**(2), 332–338. doi:10.1016/j.colsurfb.2009.06.005.

80. B. Sadeghi, F. Gholamhoseinpoor, A study on the stability and green synthesis of silver nanoparticles using Ziziphora tenuior (Zt) extract at room temperature, *Spectrochimica Acta Part A: Molecular and Biomolecular Spectroscopy*, 2015, **134**, 310–315, doi:10.1016/j.saa.2014.06.046.

81. T. Bhuyan, K. Mishra, M. Khanuja, R. Prasad, A. Varma, Biosynthesis of zinc oxide nanoparticles from Azadirachta indica for antibacterial

and photocatalytic applications, *Materials Science in Semiconductor Processing*, 2015, **32**, 55–61, doi:10.1016/j.mssp.2014.12.053.

82. M. Nasrollahzadeh, S. S. Momeni, M. Sajadi, Green synthesis of copper nanoparticles using Plantago asiatica leaf extract and their application for the cyanation of aldehydes using $K_4Fe(CN)_6$, *Journal of Colloid and Interface Science*, 2017, **506**, 471–477, doi:10.1016/j.jcis.2017.07.072.
83. D. Sharma, M. I. Sabela, S. Kanchi, P. S. Mdluli, G. Singh, T. A. Stenstrom, K. Bisetty, Biosynthesis of ZnO nanoparticles using Jacaranda mimosifolia flowers extract: synergistic antibacterial activity and molecular simulated facet specific adsorption studies, *Journal of Photochemistry and Photobiology B: Biology*, 2016, **162**, 199–207, doi:10.1016/j.jphotobiol.2016.06.04.
84. R. Mukhopadhyay, J. Kazi, M. C. Debnath, Synthesis and characterization of copper nanoparticles stabilized with Quisqualis indica extract: evaluation of its cytotoxicity and apoptosis in B16F10 melanoma cells, *Biomedicine and Pharmacotherapy*, 2018, **97**, 1373–1385, doi:10.1016/j.biopha.2017.10.167.
85. H. Padalia, P. Moteriya, S. Chanda, Green synthesis of silver nanoparticles from marigold flower and its synergistic antimicrobial potential, *Arabian Journal of Chemistry*, 2015, **8**(5), 732–741, doi:10.1016/j.arabjc.2014.11.015.
86. P. C. Nethravathi, G. S. Shruthi, D. Suresh, Udayabhanu, H. Nagabhushana, S. C. Sharma, *Garcinia xanthochymus* mediated green synthesis of ZnO nanoparticles: photoluminescence, photocatalytic and antioxidant activity studies, *Ceramics International*, 2015, **41**(7), 8680–8687, doi:10.1016/j.ceramint.2015.03.084.
87. V. V. T. Padil, M. Cerník, Green synthesis of copper oxide nanoparticles using gum karaya as a biotemplate and their antibacterial application, *International Journal of Nanomedicine*, 2013, **8**, 889–898, doi:10.2147/IJN.S40599.
88. C. Wu, D. Chen, Facile green synthesis of gold nanoparticles with gum arabic as a stabilizing agent and reducing agent, *Gold Bulletin*, 2010, **43**, 234–240, doi:10.1007/BF03214993.
89. S. Sudhasree, S. A. Shakila Banu, P. Brindha, G. A. Kurian, Synthesis of nickel nanoparticles by chemical and green route and their comparison in respect to biological effect and toxicity, *Toxicological and Environmental Chemistry*, 2014, **96**(5), 743–754, doi:10.1080/02772248.2014.923148.

90. F. Buazar, Impact of biocompatible nano silica on green stabilization of subgrade soil, *Scientific Reports*, 2019, **9**, 15147, doi:10.1038/s41598-019-51663-2.
91. H. F. Aritonang, H. Koleangan, A. D. Wuntu, Synthesis of silver nanoparticles using aqueous extract of medicinal plants' (Impatiens balsamina and Lantana camara) fresh leaves and analysis of anti-microbial activity, *International Journal of Microbiology*, 2019, **2019**, 8, doi:10.1155/2019/8642303.
92. V. Vilas, D. Philip, J. Mathew, Catalytically and biologically active silver nanoparticles synthesized using essential oil, *Spectrochimica Acta Part A: Molecular and Biomolecular Spectroscopy*, 2014, **132**, 743–750, doi:10.1016/j.saa.2014.05.046.
93. M. Behravan, A. H. Panahi, A. Naghizadeh, M. Ziaee, R. Mahdavi, A. Mirzapour, Facile green synthesis of silver nanoparticles using Berberis vulgaris leaf and root aqueous extract and its antibacterial activity, *International Journal of Biological Macromolecules*, 2019, **124**, 148–154, doi:10.1016/j.ijbiomac.2018.11.101.
94. M. R. Shaik, M. Khan, M. Kuniyil, A. Warthan, H. Z. Alkhathlan, M. R. H. Siddiqui, J. P. Shaik, A. Ahamed, A. Mahmood, M. Khan, S. F. Adil, Plant-extract-assisted green synthesis of silver nanoparticles using Origanum vulgare L. extract and their microbicidal activities, *Sustainability*, 2018, **10**, 913, doi:10.3390/su10040913.
95. S. Ahmed, Saifullah, M. Ahmad, B. L. Swami, S. Ikram, Green synthesis of silver nanoparticles using *Azadirachta indica* aqueous leaf extract, *Journal of Radiation Research and Applied Sciences*, 2016, **9**, 1–7, doi:10.1016/j.jrras.2015.06.006.
96. S. Mukherjee, V. Sushma, S. Patra, A. K. Barui, M. P. Bhadra, B. Sreedhar, C. R. Patra, Green chemistry approach for the synthesis and stabilization of biocompatible gold nanoparticles and their potential applications in cancer therapy, *Nanotechnology*, 2012, **23**, 455103, 1–13, doi:10.1088/0957-4484/23/45/455103.
97. A. N. Geraldes, A. A. da Silva, J. Leal, G. M. Estrada-Villegas, N. Lincopan, K. V. Katti, A. B. Lugao, Green nanotechnology from plant extracts: synthesis and characterization of gold nanoparticles, *Advances in Nanoparticles*, 2016, **5**, 176–185, doi:10.4236/anp.2016.53019.
98. A. Bankar, B. Joshi, A. R. Kumar, S. Zinjarde, Banana peel extract mediated synthesis of gold nanoparticles, *Colloids and Surfaces B: Biointerfaces*, 2010, **80**, 45–50, doi:10.1016/j.colsurfb.2010.05.029.

99. M. Anbuvannan, M. Ramesh, G. Viruthagiri, Synthesis, characterization and photocatalytic activity of ZnO nanoparticles prepared by biological method, *Spectrochimica Acta Part A: Molecular and Biomolecular Spectroscopy*, 2015, **143**, 304–308, doi:10.1016/j.saa.2015.01.124.
100. S. Ambika, M. Sundrarajan, Antibacterial behaviour of Vitex negundo extract assisted ZnO nanoparticles against pathogenic bacteria, *Journal of Photochemistry and Photobiology B: Biology*, 2015, **146**, 52–57, doi:10.1016/j.jphotobiol.2015.02.020.
101. P. Vanathi, P. Rajiv, S. Narendhran, S. Rajeshwari, P. K. S. M. Rahman, R. Venckatesh, Biosynthesis and characterization of phyto mediated zinc oxide nanoparticles: a green chemistry approach, *Materials Letters*, 2014, **134**, 13–15, doi:10.1016/j.matlet.2014.07.029.
102. S. Rajeshkumar, S. V. Kumar, A. Ramaiah, H. Agarwal, T. Lakshmi, S. M. Roopan, Biosynthesis of zinc oxide nanoparticles using Mangifera indica leaves and evaluation of their antioxidant and cytotoxic properties in lung cancer (A549) cells, *Enzyme and Microbial Technology*, 2018, **117**, 91–95, doi:10.1016/j.enzmictec.2018.06.009.
103. P. P. N. V. Kumar, U. Shameem, P. Kollu, R. L. Kalyani, S. V. N. Pammi, Green synthesis of copper oxide nanoparticles using Aloe vera leaf extract and its antibacterial activity against fish bacterial pathogens, *BioNanoScience*, 2015, **5**, 135–139, doi:10.1007/s12668-015-0171-z.
104. N. Nagar, V. Devra, Green synthesis and characterization of copper nanoparticles using Azadirachta indica leaves, *Materials Chemistry and Physics*, 2018, **213**, 44–51, doi:10.1016/j. matchemphys.2018.04.007.
105. H. J. Lee, J. Y. Song, B. S. Kim, Biological synthesis of copper nanoparticles using Magnoliakobus leaf extract and their antibacterial activity, 2013, (Wiley online library.com), doi:10.1002/jctb.4052.
106. S. Ibrahim, N. Z. J. Zakaria, S. Rozali, N. N. N. Ghazali, M. S. A. Karim, M. F. M. Sabri, Biosynthesis of copper oxide nanoparticles using *Camellia Sinensis* plant powder, in *Advances in Material Sciences and Engineering*, Lecture Notes in Mechanical Engineering, 2020, pp. 233–238, doi:10.1007/978-981-13-8297-0_26.
107. M. Gopinath, R. Subbaiya, M. M. Selvam, D. Suresh, Synthesis of copper nanoparticles from Nerium oleander leaf aqueous extract and its antibacterial activity, *International Journal of Current Microbiology and Applied Sciences*, 2014, **3**(9), 814–818, http://www.ijcmas.com.
108. S. C. Mali, R. Shani, R. Trivedi, Biosynthesis of copper oxide nanoparticles using Enicostemma axillare (Lam.) leaf extract, *Biochemistry*

and Biophysics Reports, 2019, **20**, 100699, doi:10.1016/j.bbrep.2019.100699.

109. S. Kanagasubbulakshmi, K. Kadirvelu, Green synthesis of Iron oxide nanoparticles using Lagenariasiceraria and evaluation of its Antimicrobial activity, *Defence Life Science Journal*, 2017, **2**(4), 422–427, doi:10.14429/dlsj.2.12277.
110. N. Sulaiman, Y. Yulizar, D. O. B. Apriandanu, Eco-friendly method for synthesis of La_2O_3 nanoparticles using Physalis angulata leaf extract, *AIP Conference Proceedings*, 2017, **2023**, 020105-1–020105-5, doi:10.1063/1.5064102.
111. N. Nazar, I. Bibi, S. Kamal, M. Iqbal, S. Nouren, K. Jilani, M. Umair, S. Ata, Cu nanoparticles synthesis using biological molecule of P. granatum seeds extract as reducing and capping agent: growth mechanism and photo-catalytic activity, *International Journal of Biological Macromolecules*, 2018, **106**, 1203–1210, doi:10.1016/j.ijbiomac.2017.08.126.
112. M. Jayandran, M. M. Haneefa, V. Balasubramanian, Green synthesis and characterization of manganese nanoparticles using natural plant extracts and its evaluation of antimicrobial activity, *Journal of Applied Pharmaceutical Science*, 2015, **5**(12), 105–110, doi:10.7324/JAPS.2015.501218.
113. F. T. Thema, E. Manikandan, M. S. Dhlamini, M. Maaza, Green synthesis of ZnO nanoparticles via Agathosma betulina natural extract, *Materials Letters*, 2015, **161**, 124–127, doi:10.1016/j.matlet.2015.08.052.
114. R. A. Raj, M. S. AlSalhi, S. Devanesan, Microwave-assisted synthesis of nickel oxide nanoparticles using Coriandrum sativum leaf extract and their structural-magnetic catalytic properties, *Materials*, 2017, **10**(5), 460, doi:10.3390/ma10050460.
115. K. Karthik, S. Dhanuskodi, C. Gobinath, S. Prabukumar, S. Sivaramakrishnan, Andrographis paniculata extract mediated green synthesis of CdO nanoparticles and its electrochemical and antibacterial studies, *The Journal of Materials Science: Materials in Electronics*, 2017, **28**, 7991–8001, doi:10.1007/s10854-017-6503-8.
116. R. Dobrucka, Synthesis of MgO nanoparticles using Artemisia abrotanum herba extract and their antioxidant and photocatalytic properties, *Iranian Journal of Science and Technology, Transactions of Electrical Engineering*, 2018, **42**, 547–555 doi:10.1007/s40995-016-0076-x.

117. R. A. Solano, A. P. Herrera, D. Maestre, A. Cremades, Fe-TiO_2 nanoparticles synthesized by green chemistry for potential application in waste water photocatalytic treatment, *Journal of Nanotechnology*, 2019, **2019**, 11, doi: 10.1155/2019/4571848.

118. P. Yugandhar, N. Savithramma, Biosynthesis, characterization and antimicrobial studies of green synthesized silver nanoparticles from fruit extract of Syzygium alternifolium (Wt.) Walp. an endemic, endangered medicinal tree taxon, *Applied Nanoscience*, 2016, **6**, 223–233, doi:10.1007/s13204-015-0428-4.

119. R. K. Das, N Gogoi, U. Bora, Green synthesis of gold nanoparticles using Nyctanthes arbortristis flower extract, *Bioprocess and Biosystems Engineering*, 2011, **34**, 615–619, doi:10.1007/s00449-010-0510-y.

120. J. K. Patra, K. H. Baek, Green nanobiotechnology: factors affecting synthesis and characterization techniques, *Journal of Nanomaterials*, 2014, **2014**, 12, doi: 10.1155/2014/417305.

121. N. M. Shinde, A. C. Lokhande, C. D. Lokhande, A green synthesis method for large area silver thin film containing nanoparticles, *Journal of Photochemistry and Photobiology B: Biology*, 2014, **136**, 19–25, doi:10.1016/j.jphotobiol.2014.04.011.

122. J. Qu, C. Luo, X. Yuan, Synthesis of hybrid carbon nanotubes using Brassica juncea L. application to photodegradation of bisphenol A, *Environmental and Science Pollution Research*, 2013, **20**, 3688–3695, doi:10.1017/s11356-012-1289-y.

123. D. Borah, M. Hazarika, P. Tailor, Starch-templated bio-synthesis of gold nanoflowers for in vitro antimicrobial and anticancer activities, *Applied Nanoscience*, 2018, **8**, 241–253, doi:10.1007/s13204-0180793-x.

124. S. B. Maddinedi, B. K. Mandal, R. Vankayala, P. Kalluru, S. R. Pamanji, Bioinspired reduced graphene oxide nanosheets using Terminalia chebula seeds extract, *Spectrochimica Acta Part A: Molecular and Biomolecular Spectroscopy*, 2015, **145**, 117–124, doi:10.1016/j.saa.2015.02.037.

125. A. Esmaeili, M. Ebrahimzadeh, Polymer-based of extract-loaded nanocapsules Aloe vera L. delivery, *Synthesis and Reactivity in Inorganic, Metal-Organic, and Nano-Metal Chemistry*, 2015, **45**(1), 40–47, doi:10.1080/15533174.2013.818027.

126. A. J. Kora, L. Rastogi, Catalytic degradation of anthropogenic dye pollutants using palladium nanoparticles synthesized by gum olibanum, a glucuronoarabinogalactan biopolymer, *Industrial Crops and Products*, 2016, **81**, 1–10, doi:10.1016/j.indcrop.2015.11.055.

127. S. Aromal, D. Philip, Benincasa hispida seed mediated green synthesis of gold nanoparticles and its optical nonlinearity, *Physica E: Low-dimensional Systems and Nanostructures*, 2012, **44**, 1329–1334, doi: 10.1016/j.physe.2012.02.013.

128. K. Velayutham, A. A. Rahuman, G. Rajakumar, T. Santhoshkumar, S. Marimuthu, C. Jayaseelan, A. Bagavan, A. V. Kirthi, C. Kamaraj, A. A. Zahir, G. Elango, Evaluation of Catharanthus roseus leaf extract-mediated biosynthesis of titanium dioxide nanoparticles against Hippobosca maculata and Bovicola ovis, *Parasitology Research*, 2012, **111**, 2329–2337, doi:10.1007/s00436-011-2676-x.

129. V. J. Kouzegaran, K. Farhadi, Green synthesis of sulphur nanoparticles assisted by a herbal surfactant in aqueous solutions, *Micro and Nano Letters*, 2017, **12**(5), 329–334, doi:10.1049/mnl.2016.0567.

130. S. C. C. Arruda, A. L. D. Silva, R. M. Galazzi, R. A. Azevedo, M. A. Z. Arruda, Nanoparticles applied to plant science: a review, *Talanta*, 2015, **131**, 693–705, doi:10.1016/j.talanta.2014.08.050.

131. V. Jassal, U. Shanker, S. Gahlot, B. S. Kaith, Kamaluddin, Md. A. Iqubal, P. Samuel, Sapindus mukorossi mediated green synthesis of some manganese oxide nanoparticles interaction with aromatic amines, *Applied Physics A*, 2016, **122**, 271, doi:10.1007/s00339-016-9777-4.

132. N. Thovhogi, A. Diallo, A. G. Fakim, M. Maaza, Nanoparticles green synthesis by Hibiscus Sabdariffa flower extract: main physical properties, *Journal of Alloys and Compounds*, 2015, **647**, 392–396, doi:10.1016/j.jallcom.2015.06.076.

133. N. M. Salem, S. L. Albanna, A. M. Awwad, Green synthesis of sulfur nanoparticles using Punica granatum peels and the effects on the growth of tomato by foliar spray applications, *Environmental Nanotechnology, Monitoring and Management*, 2016, **6**, 83–87, doi:10.1016/j.enmm.2016.06.006.

134. G. Sathishkumar, C. Rajkuberan, K. Manikandan, S. Prabukumar, J. D. John, S. Sivaramakrishnan, Facile biosynthesis of antimicrobial zinc oxide (ZnO) nanoflakes using leaf extract of Couroupita guianensis, Aubl, *Materials Letters*, 2017, **188**, 383–386, doi:10.1016/j.matlet.2016.11.100.

135. G. A. Molina, R. Esparza, J. L. L. Miranda, A. R. H. Martínez, B. L. E. Sanchez, E. A. E. Pena, M. Estevez, Green synthesis of Ag nanoflowers using Kalanchoe Daigremontiana extract for enhanced photocatalytic and antibacterial activities, *Colloids and Surfaces B: Biointerfaces*, 2019, **180**, 141–149, doi:10.1016/j.colsurfb.2019.04.044.

136. H. Muthukumar, M. Manickam, Amaranthus spinosus leaf extract mediated FeO nanoparticles: physicochemical traits, photocatalytic and antioxidant activity, *ACS Sustainable Chemistry and Engineering*, 2015, **3**(12), 3149–3156, doi:10.1021/acssuschemeng.5b00722.
137. P. Singh, S. Pandit, J. Garnaes, S. Tunjic, V. Mokkapati, A. Sultan, A. Thygesen, A. Mackevica, R. V. Mateiu, A. E. Daugaard, A. Baun, I. Mijakovic, Green synthesis of gold and silver nanoparticles from Cannabis sativa (industrial hemp) and their capacity for biofilm inhibition, *International Journal of Nanomedicine*, 2018, **13**, 3571–3591, doi:10.2147/ IJN.S157958.
138. P. Paralikar, M. Rai, Bio-inspired synthesis of sulphur nanoparticles using leaf extract of four medicinal plants with special reference to their antibacterial activity, *IET Nanobiotechnology*, 2017, 1–7, doi:10.1049/iet-nbt.2017.0079.
139. E. S. Onal, T. Yatkin, T. Aslanov, M. Ergut, A. Ozer, Biosynthesis and characterization of iron nanoparticles for effective adsorption of Cr(VI), *International Journal of Chemical Engineering*, 2019, **2019**, 13, doi:10.1155/2019/2716423.
140. S. P. Goutama, G. Saxena, V. Singh, A. K. Yadav, R. N. Bharagava, K. B. Thapa, Green synthesis of TiO_2 nanoparticles using leaf extract of Jatropha curcas L. for photocatalytic degradation of tannery wastewater, *Chemical Engineering Journal*, 2018, **336**, 386–396, doi:10.1016/j.cej.2017.12.029.
141. J. O. Tijani, O. Ugochukwu, L. A. Fadipe, One-step green synthesis of WO_3 nanoparticles using Spondias mombin aqueous extract: effect of solution pH and calcination temperature, *Applied Physics A*, 2019, **125**, 162, doi:10.1007/s00339-019-2450-y.
142. S. P. Dubey, A. D. Dwivedi, M. Sillanpaa, Y. N. Kwon, C. Lee, Single-step green synthesis of imine-functionalized carbon spheres and their application in uranium removal from aqueous solution, *RSC Advances*, 2014, **4**, 46114–46121, doi:10.1039/c4ra06890d.
143. M. Khatami, R. S. Varma, N. Z. Zafarnia, H. Yaghoobi, M. Sarani, V. G. Kumar, Applications of green synthesized Ag, ZnO and Ag/ZnO nanoparticles for making clinical antimicrobial wound-healing bandages, *Sustainable Chemistry and Pharmacy*, 2018, **10**, 9–15, doi:10.1016/j.scp.2018.08.001.
144. M. Ahamed, M. A. M. Khan, M. K. J. Siddiqui, M. S. AlSalhi, S. A. Alrokayan, Green synthesis, characterization and evaluation of biocompatibility

of silver nanoparticles, *Physica E: Low-dimensional Systems and Nanostructures*, 2011, **43**(6), 1266–1271, doi:10.1016/j.physe.2011.02.014.

145. R. Banasiuk, M. Krychowiak, D. Swigon, W. Tomaszewicz, A. Michalak, A. Chylewska, M. Ziabka, M. Lapinski, B. Koscielska, M. Narajczyk, A. Krolicka, Carnivorous plants used for green synthesis of silver nanoparticles with broad-spectrum antimicrobial activity, *Arabian Journal of Chemistry*, 2020, **13**(1), 1415–1428, doi:10.1016/j.arabjc.2017.11.013.
146. A. Happy, M. Soumya, S. V. Kumar, S. Rajeshkumar, R. D. Sheba, T. Lakshmi, V. D. Nallaswamy, Phyto-assisted synthesis of zinc oxide nanoparticles using Cassia alata and its antibacterial activity against Escherichia coli, *Biochemistry and Biophysics Reports*, 2019, **17**, 208–211, doi:10.1016/j.bbrep.2019.01.002.
147. M. Murali, C. Mahendra, Nagabhushan, N. Rajashekar, M. S. Sudarshana, K. A. Raveesha, K. N. Amruthesh, Antibacterial and antioxidant properties of biosynthesized zinc oxide nanoparticles from Ceropegia candelabrum L.- an endemic species, *Spectrochimica Acta Part A: Molecular and Biomolecular Spectroscopy*, 2017, **179**, 104–109, doi:10.1016/j.saa.2017.02.027.
148. H. K. N. Kumar, C. N. Mohana, B. R. Nuthan, Phyto-mediated synthesis of zinc oxide nanoparticles using aqueous plant extract of Ocimum americanum and evaluation of its bioactivity, *SN Applied Science*, 2019, **1**, 651, doi:10.1007/s42452-019-0671-5.
149. G. Sathishkumar, P. K. Jha, V. Vignesh, R. Kuberan, M. Jeyaraj, M. Selvakumar, R. Jhaf, S. Sivaramakrishnan, Cannonball fruit (Couroupita guianensis, Aubl.) extract mediated synthesis of gold nanoparticles and evaluation of its antioxidant activity, *Journal of Molecular Liquids*, 2016, **215**, 229–236, doi:10.1016/j.molliq.2015.12.043.
150. M. Guo, L. Wei, F. Yang, H. Liu, Controllable biosynthesis of gold nanoparticles from a Eucommia ulmoides bark aqueous extract, *Spectrochimica Acta Part A: Molecular and Biomolecular Spectroscopy*, 2015, **142**, 73–79, doi:10.1016/j.saa.2015.01.109.
151. J. R. Nakkala, R. Mata, S. R. Sadras, The antioxidant and catalytic activities of green synthesized gold nanoparticles from Piper longum fruit extract, *Process Safety and Environmental Protection*, 2016, **100**, 288–294, doi:10.1016/j.psep.2016.02.007.
152. N. S. Al-Radadi, Green synthesis of platinum nanoparticles using Saudi's Dates extract and their usage on the cancer cell treatment,

Arabian Journal of Chemistry, 2019, **12**(3), 330–349, doi:10.1016/j.arabjc.2018.05.008.

153. J. Y. Song, E. Y. Kwon, B. S. Kim, Biological synthesis of platinum nanoparticles using Diopyros kaki leaf extract, *Bioprocess and Biosystems Engineering*, 2010, **33**, 159–164, doi: 10.1007/s00449-009-0373-2.
154. V. R. Chelli, A. K. Golder, One pot green synthesis of Pt, Co and Pt@Co core-shell nanoparticles using Sechium edule, *Journal of Chemical Technology and Biotechnology*, 2019, **94**(3), 911–918, doi:10.1002/jctb.5838.
155. K. Anand, C. Tiloke, A. Phulukdaree, B. Ranjan, A. Chuturgoon, S. Singh, R. M. Gengan, Biosynthesis of palladium nanoparticles by using Moringa oleifera flower extract and their catalytic and biological properties, *Journal of Photochemistry and Photobiology B: Biology*, 2016, **165**, 87–95, doi:10.1016/j.jphotobiol.2016.09.03.
156. C. Soundarrajan, A. Sankari, P. Dhandapani, S. Maruthamuthu, S. Ravichandran, G. Sozhan, N. Palaniswamy, Rapid biological synthesis of platinum nanoparticles using Ocimum sanctum for water electrolysis applications, *Bioprocess and Biosystem Engineering*, 2012, **35**, 827–833, doi:10.1007/s00449-011-0666-0.
157. V. Manikandan, P. Velmurugan, J. H. Park, N. Lovanh, S. K. Seo, P. Jayanthi, Y. J. Park, M. Cho, B. T. Oh, Synthesis and antimicrobial activity of palladium nanoparticles from Prunus x yedoensis leaf extract, *Materials Letters*, 2016, **185**, 335–338, doi:10.1016/j.matlet.2016.08.120.
158. S. K. Kannan, M. Sundrarajan, Green synthesis of ruthenium oxide nanoparticles: characterization and its antibacterial activity, *Advanced Powder Technology*, 2015, **26**(6), 1505–1511, doi:10.1016/j.apt.2015.08.009.
159. K. Gopinath, V. Karthika, S. Gowri, Antibacterial activity of ruthenium nanoparticles synthesized using Gloriosa superba L. leaf extract, *Journal of Nanostructures in Chemistry*, 2014, **4**, 83, doi:10.1007/s40097-014-0083-4.
160. K. S. Prasad, T. Amin, S. Katuva, M. Kumari, K. Selvaraj, Synthesis of water soluble CdS nanoparticles and study of their DNA damage activity, *Arabian Journal of Chemistry*, 2017, **10**(2), S3929–S3935, doi:10.1016/j.arabjc.2014.05.033.

Chapter 5

Synthesis, Designing and Challenges of Functionalized Polymeric Nanomaterials and Their Spectroscopic Applications

Jitha S. Jayan,[a] Sethulekshmi A. S.,[a] Gopika Venu,[a] Appukuttan Saritha,[a] and Kuruvilla Joseph[b]

[a] *Department of Chemistry, School of Arts and Sciences, Amrita Vishwa Vidyapeetham, Amritapuri, Kollam 690525, Kerala, India*

[b] *Department of Chemistry, Indian Institute of Space Science and Technology, Valiyamala PO, Thiruvananthapuram 695547, Kerala, India*

sarithatvla@gmail.com

Polymeric nanoparticles are sub-micron level colloidal particles in the form of nanospheres or nanocapsules. These nanoparticles can be formed by either modifying nanoparticles using a polymer or by the sequential arrangement of long polymeric chains in the form of micro or nanostructures. Dendrimers, block copolymers, peptides, etc. can be arranged in such a way as to get well-defined nanostructures and can also be used for the modification nanoparticles. These polymeric nanomaterials have wide applications in the field of drug delivery, bioimaging, storage, photocatalysis, biosensing, chemosensing, semiconduction, water purification, composite fabrication, etc. This chapter tries to bring to light the various fabrication and characterization methods of polymeric

Nanomaterials for Spectroscopic Applications
Edited by Kaushik Pal

ISBN 978-981-4877-69-5 (Hardcover), 978-1-003-16033-5 (eBook)
www.jennystanford.com

nanoparticles. Applications of these polymeric nanoparticles are also discussed in detail with special emphasis on spectroscopic investigations.

5.1 Introduction

Polymeric nanoparticles are materials consisting of natural or synthetic polymers that are in the nano regime. They can be either solids or in colloidal form and are most commonly used for drug loading applications and cancer therapy [1–3]. Polymeric nanoparticles are also finding potential applications in the field of spectroscopy and sensing [4, 5]. Generally, polymeric nanoparticles are considered as drug-loaded polymers. But Polymeric nanoparticles can be obtained either by the sequential arrangement of the polymer itself in nano form or by grafting polymers onto nanomaterials [6, 7]. These are called self-assembled polymer nanoparticles [8, 9]. These polymeric nanoparticles can be used effectively in various applications due to the tunable properties attained by their orderly arrangement [10, 11]. The classification of polymeric nanostructures is depicted in Fig. 5.1. The synthesis and modification of these materials along with a systematic analysis of their sequential arrangement has not been much probed into. Hence the attempt of this chapter is to provide a summary of such factors which have not been much discussed in the reviews published so far.

Self-assembly of polymers is an important behavior since it has tremendous applications in various fields due to the formation of hierarchical structures in different organic solvents [12–14]. Self-assembly is a process by which un-ordered components are getting ordered due to the internal interactions without any external physical force. Self-assembly is a very important parameter as we are familiar with the self-assembly of molecules, lipids, polymers and colloids that occur in nature [15–17]. This process often leads to the formation of functional structures and the easiest and simple example of self-assembly is the cleansing action of soap. The arrangement of soap molecules in water to form a micellar structure by the internal interactions of hydrophilic and phobic ends of soap molecules with water [18, 19] is similar to the self-assembly of

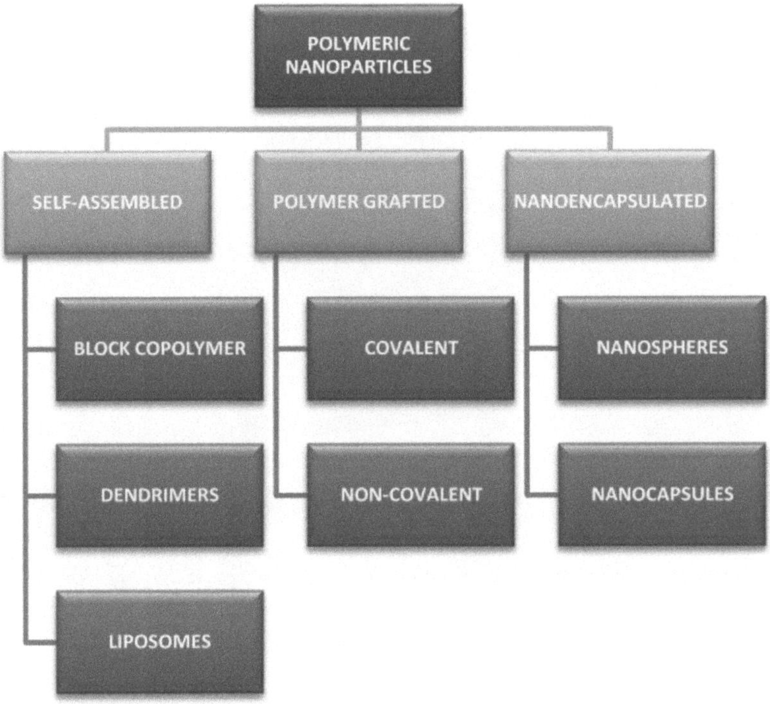

Figure 5.1 Schematic representation of a classification of polymeric nanoparticles.

block copolymers and macromolecules like lipids and dendrimers. In short, self-assembly is the conversion of large molecules to nano levels [20, 21].

5.1.1 Self-Assembly to Form Polymeric Nanoparticles

5.1.1.1 Block copolymers

Block copolymers are a class of copolymers in which different homopolymer units are linked together [22]. There are different factors determining the nature of the self-assembly like copolymer concentration, mobility of chains, Flory-Huggins parameter, nature of solvents, presence of additives and other thermodynamical and kinetical aspects [23]. Ludwigs et al. [24] found that ABC type

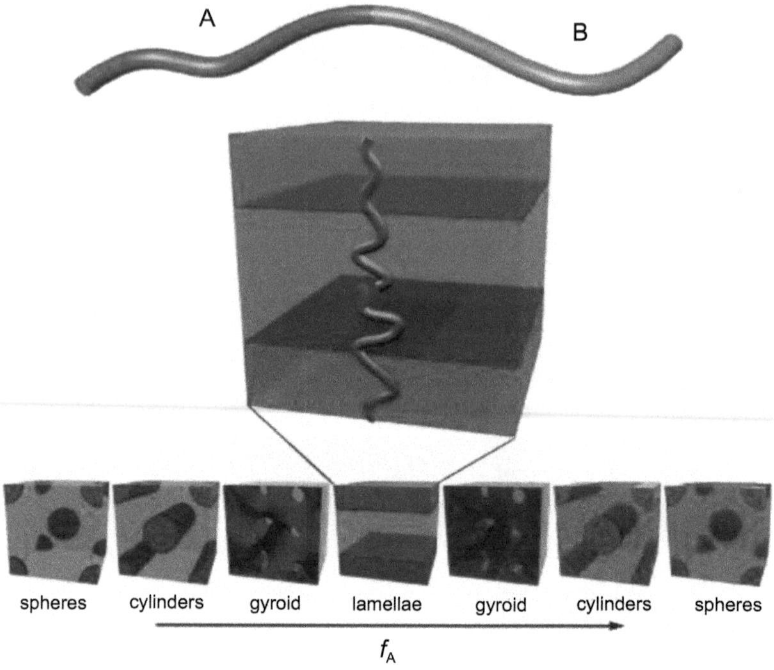

Figure 5.2 Schematic representation of nanostructure developed due to the self-assembly of the block copolymer as a function of f_A [25]. Reproduced with permission from Elsevier.

triblock copolymers like polystyrene-block-poly (2-vinylpyridine)-block-poly (t-butyl methacrylate) also capable of forming a core like structure. Darling et al. [25] observed that the thermodynamic incompatibility between the polymer chains is responsible for the microphase separation and thus it leads to the formation of different nanostructures. Depending on the concentration of one of the blocks (f_A) in the block copolymer the structure of the self-assembled micelle varies as shown in Fig. 5.2.

The arrangement of a linear block copolymer into nanostructure having a high domain size is demonstrated in Fig. 5.3 [26, 27]. The amphiphilic nature of block copolymers leads to the formation of different nanostructures like spherical, wormlike, vesicles, etc. ... [28, 29]. Hence the polarity of the end chains are very important in determining the structure of the self-assembled micelle [30, 31].

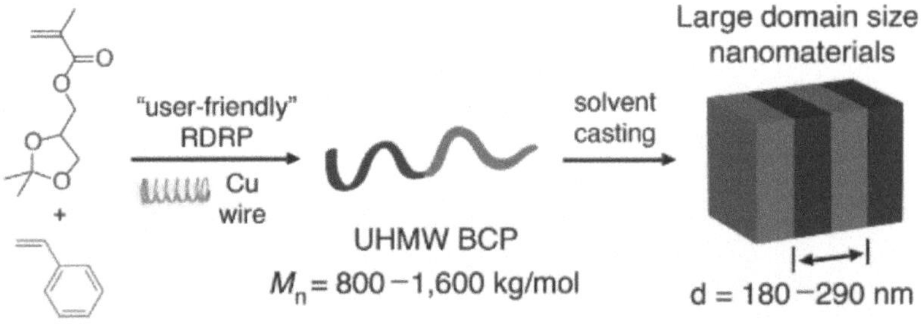

Figure 5.3 Schematic representation of the method of fabrication of PGMA-PHPMA block copolymeric nanoparticles by radical polymerization [27]. Reproduced with permission from American Chemical Society.

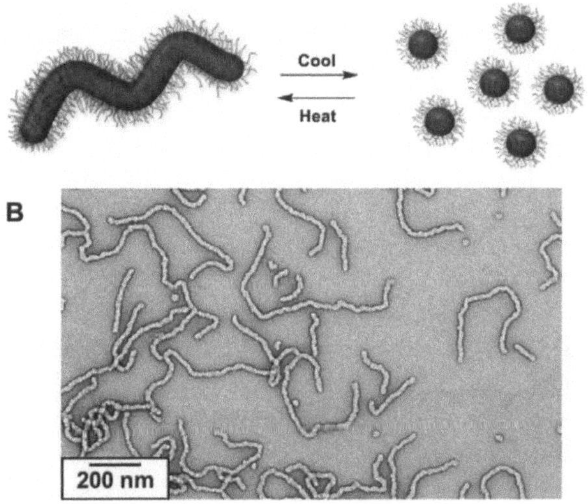

Figure 5.4 TEM image of the wormlike arrangement of PGMA-PHPMA diblock copolymer [31]. Reproduced with permission from Wiley.

The microscopic image of a worm-like micellization of a linear amphiphilic diblock copolymer with different polar ends is shown in Fig. 5.4.

5.1.1.2 Dendrimers

Dendrimers are fascinating polymeric architectures that are currently gaining momentum in various emerging applications like

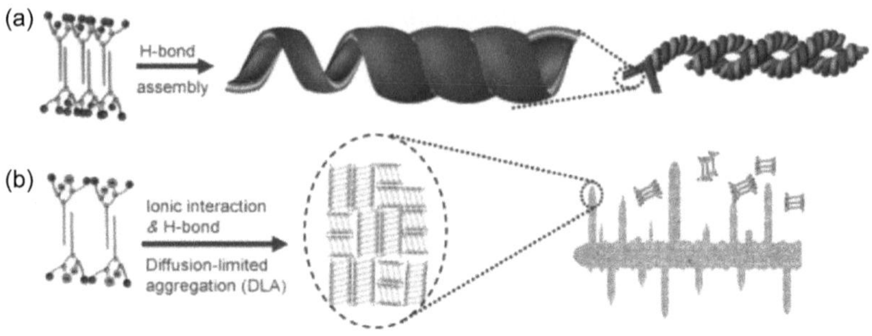

Figure 5.5 A representative scheme of self-assembly of amphiphilic dendron containing L-glutamic acids [36]. Reproduced with permission from Wiley.

drug delivery and sensing. However, the high manufacturing cost as well as the absence of versatile preparative techniques limit their utility in various applications. Dendrimers are highly branched macromolecules capable of self-assembling into nanoscale dimensions [32, 33]. There are different studies showing the self-assembly of dendrimers by simple modification using thiophene, alkyl groups, glutamic acids, etc.... [34–36]. Moreover varying the conditions like pH, concentration, polarity, of the medium one can tune the self-assembly of the supramolecular dendrimers. Figure 5.5 shows the self-assembly induced by the alkyl modification to dendron.

Setaro et al. [37] used the self-assembling ability of dendrimers to make nanoparticles of biohybrid protein. They have synthesized protein-dendrimer biohybrid nanoparticles that could be used as a potential therapeutic delivery agent due to the photosensitization effect of phthalocyanine. They also found that the number of primary branches, the density of the radical branches and distribution of chemical moieties are the factors which affect the arrangements of the dendrimers. Percec et al. [38] developed a dendrimer called Janus by chemically linking two dendritic building blocks and observed the spherical symmetry. Xu et al. [39] observed the self-assembly of poly (L-lysine) dendrimers and linear poly (L-leucine) into a capsid-like biomimetic nanoarchitecture. Yue et al. [40] observed the effect of DNA based dendrimers into nanostructures.

5.1.1.3 Lipids

Like dendrimers and block copolymers, the amphiphilic nature of lipids helps in the formation of different nanostructures [41]. Another emerging area is the study of lipid—polymer hybrid nanoparticles which exhibit the synergistic effect of both the systems in various applications. This enabled the study of lipids through surface-enhanced Raman spectroscopy by means of integrating liposomes with nanoparticles. García-Díaz et al. [42] utilized the self-assembly of lipids to enhance the drug loading in polymer nanoparticles. Lee et al. [43] studied the self-assembly of lipids and single-walled nanotubes. These self-assembled lipids were also capable of showing adsorption on SWCNT. John et al. [44] studied the helical morphology of cardanyl glycoside. For this study, they have treated cardanol with D-glucose and the components obtained by the fractionation of cardanyl glycoside were separated and the self-assembly studies were carried out. The obtained nanostructures were of high axial ratio. Suzuki et al. [45] utilized the effect of lipid self-assembly for the synthesis of DNA origami nanostructures into 2D lattices using zwitterionic lipid 1,2-dioleoyl-sn-glycero-3-phosphocholine (DOPC). The templated synthesis is shown in Fig. 5.6. In another study, Tsai et al. [46] observed the self-assembly of peptide lipids into a dimensional nanostructure with superhelical twisting. Hence the possibility of arrangement block copolymers, dendrimers and lipids in the form of nanostructures can be utilized for spectroscopic applications.

5.1.2 Surface Engineered Polymeric Nanoparticles

Surface engineering is a unique phenomenon by which the nanoparticles or polymers can be tailored to meet the specifications for a particular application. By means of this technique, a vast array of polymeric nanoparticles could be formulated so that the shortcomings of virgin nanofillers could be overcome. The polymer can be attached to the nanoparticles by means of covalent and non-covalent approaches [47]. As we reported recently, the uniform grafting of macromolecules onto the nanoparticles can also offer self-assembly [48, 49]. The covalent methods are of two types,

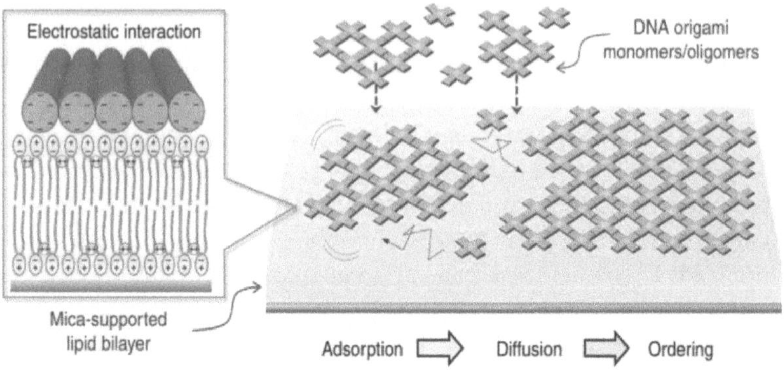

Figure 5.6 Schematic illustration of the method of lipid-bilayer-assisted self-assembly [45]. Reproduced with permission from Nature Communications.

'grafting to' and 'grafting from' approaches. In 'grafting to' approach, the pre-synthesized polymer chains are attached to the surface of nanomaterials or functionalized particles by means of simple chemical reactions. In grafting from, the polymerization is starting from the surface of nanoparticles itself [49, 50]. Recently, our group [51] grafted polyethylene glycol (PEG) onto the surface of graphene oxide (GO) by means of 'grafting to' technique. Akcora et al. [52] by the theoretical study observed the possibility of self-assembly of polymer grafted nanoparticles into different extended nanostructures. Jayaraman et al. [53] also confirmed the self-assembly of the polymer grafted hybrid nanofillers by means of models and theory. It was also observed that the solvent-free polymer grafted nanoparticles can form 2D films. If there is a dislike between core nanoparticle and polymer chains, the chance of self-assembly higher. The possibility of self-assembly is shown in Fig. 5.7 [54].

Zhao et al. [55] compared the self-assembly of PS grafted and PS poly(2-vinylpyridine) (P2VP) grafted silica particles in the PS matrix. Depending on the graft density and graft to matrix ratio, it is possible to tune the dispersion. Gao et al. [56] made hydrophilic plasmonic films by grafting hydrophilic polymer onto silver nanocubes. These hydrophilic nano polymers can self-assemble in hydrophobic solvents or other polymeric matrices.

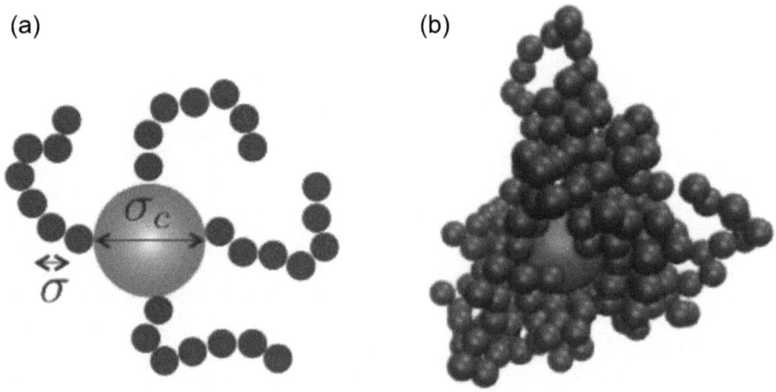

Figure 5.7 2D and 3D representation of polymer grafted nanoparticle obtained from the coarse-grained model [54]. Reproduced with permission from Royal Society of Chemistry.

Ye et al. [57] grafted PS onto the gold nanocrystals by means of ligand exchange and demonstrated that it is possible to create different 2D and 3D structures by adjusting the Molecular weight of the attached polymer chain and the size of the core of the grafted filler. Divandari et al. [58] grafted cyclic polymer poly(2-ethyl 2-oxazoline) (PEOXA) brushes, onto the surface of nano titania. These polymers are capable of forming self-assembled structures over nanomaterial as shown in Fig. 5.8.

In 'grafting from' technique, all the possible polymerization techniques are followed for grafting the polymer onto the surface of nanoparticle [59]. Generally, controlled polymerization techniques are followed for this purpose as polymerization starts over the surface. Either the nanoparticle itself or the macroinitiator over it can bring about the polymerization.

Chevigny et al. [60] and Liu et al. [61] grafted polystyrene onto the surface of silica nanoparticles by means of nitroxide mediated polymerization method (NMP) radical addition-fragmentation chain transfer (RAFT) mechanisms respectively. In another study, Bartholome et al. [59] followed the NMP method using a triethoxysilyl-terminated alkoxy amine-based initiator. Using atom transfer radical polymerization (ATRP) Shen et al. [62] grafted poly(glycidyl methacrylate) onto magnetic nanoparticle using the

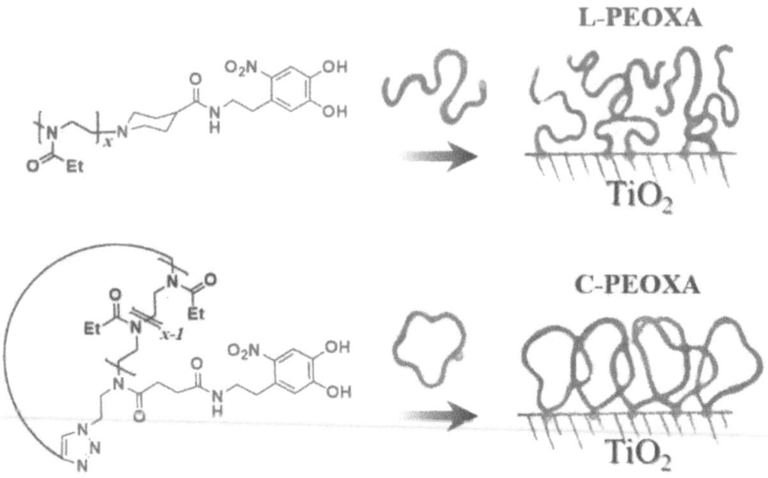

Figure 5.8 Schematic representation of grafting of PEOXA brushes onto TiO_2 surfaces from aqueous solutions [58]. Reproduced with permission from Elsevier.

glycidyl methacrylate monomer. Sunday et al. [63] grafted PS onto the silica nanoparticles by means of ATRP and RAFT and analyzed its self-assembly in the PS matrix. They observed a phase separation which confirmed the self-assembly of the grafted nanoparticle. Li et al. [64] followed the method of ATRP, and grafted poly(N-isopropylacrylamide) and poly(methoxy-oligo(ethylene glycol) methacrylate) onto the gold nanoparticles. The grafted polymeric nanoparticles were capable of showing thermosensitive nanostructures due to the self-assembly. The polymeric nanoparticles were showing a core–shell structure and the structure was responsive to temperature with two critical points. Hong et al. [65] used the RAFT method for the polymerization of mesoporous silica nanoparticles and made particles having core–shell nanostructures as shown in Fig. 5.9. Gupta et al. [66] studied the self-assembly of diblock copolymer grafted by means of theoretical modeling by varying factors like a fraction of block copolymer, loading of nanoparticle, number and size of grafted chain and confinement degree. Possible structures like a ring, disk, helical, and concentric lamellar phases were identified. Barandiaran et al. [67] made

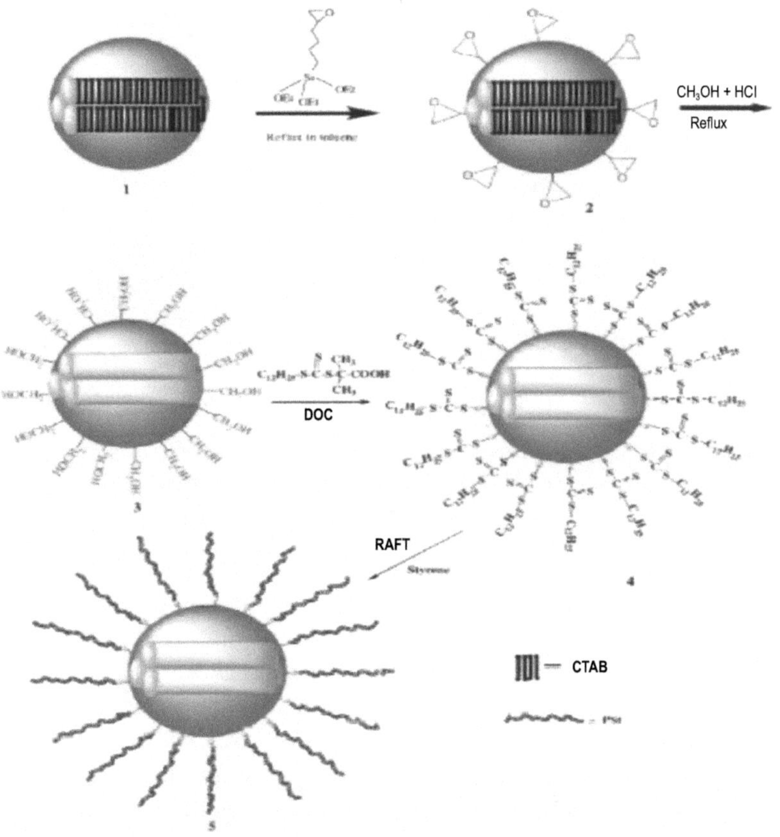

Figure 5.9 Schematic representation of grafting PS shell onto MSN via surface RAFT polymerization [65]. Reproduced with permission from Elsevier.

lamellar shaped poly(styrene-b-methylmethacrylate) (PS-b-PMMA) block copolymer by grafting it onto magnetic Fe_3O_4 nanoparticles.

5.1.3 Nano Encapsulated Polymer

Encapsulation of nanoparticles in polymeric materials is yet another way of making polymer nanomaterials [68]. Generally, they are two types called nanospheres and nanocapsules. These materials are generally referred to as drug-loaded polymers [69].

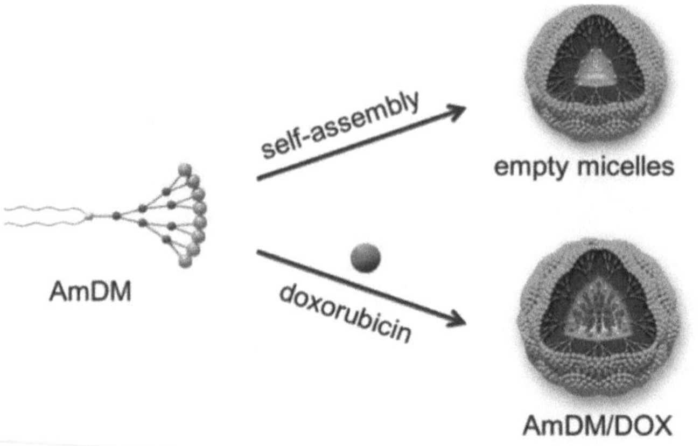

Figure 5.10 Schematic representation of the formation of drug-loaded nanomicelles [72]. Reproduced with permission from PNAS.

When the nanoparticles are dispersed in polymers it is called nanospheres, whereas, when polymer membranes are surrounded by nanoparticles it is termed as nanocapsules [70]. Jang et al. [71] prepared polypyrrole and carbon-based core/shell structured nanocapsules by following the nano emulsification process. With the help of surfactants and pyrrole monomers, a core/shell structure is attained which is carbonized to form a carbon-based core/shell nanostructure. Wei et al. [72] used an amphiphilic dendrimer for loading the anti-cancerous drug doxorubicin. The ability of dendrimers to self-assemble into nanostructures is utilized in the study and it is schematically demonstrated in Fig. 5.10. We have utilized the oil-water emulsification of polyethylene glycol-b-polyethylene oxide-b-polyethylene glycol block copolymers in loading the drug Amoxicillin and made nanocapsules out of it [73]. Kandirmaz et al. [74] studied the lavender oil encapsulation efficiency of the same block copolymer and its usage in the paper industry.

Li et al. [75] fabricated magnetic nanoparticles by encapsulating PEGylated poly (lactic-co-glycolic acid) block copolymer. The core/shell structure of the nanocapsules was confirmed with the help of HRTEM and AFM images and the particle distribution was also analyzed in detail. These nanocapsules were as payloads for

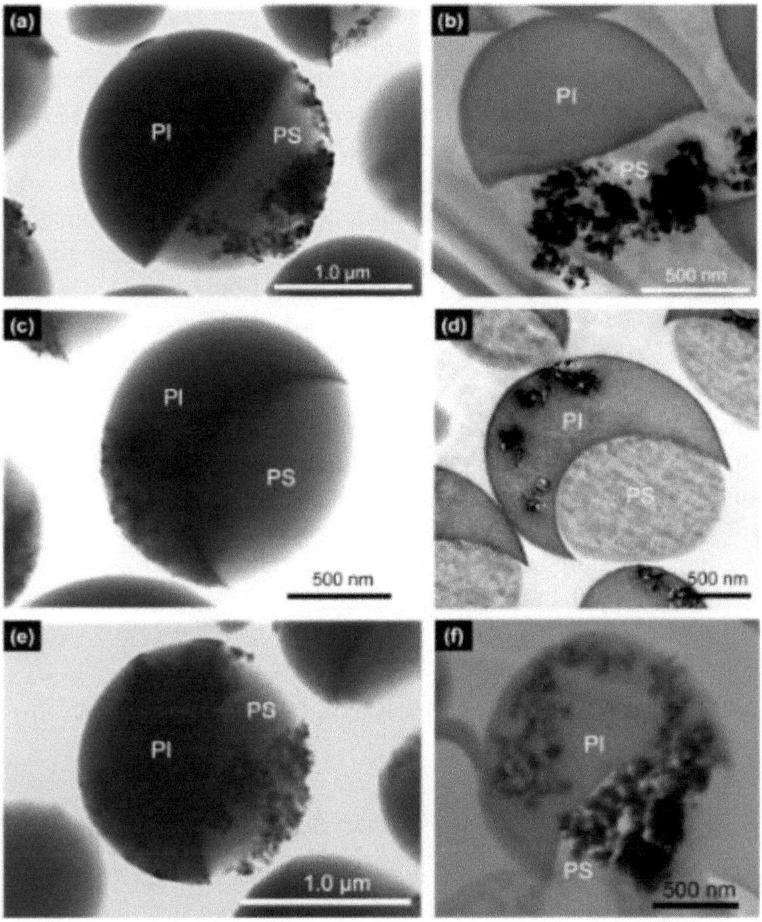

Figure 5.11 STEM and TEM images of Janus particles containing PDMA-r-PS-encapsulated Fe_3O_3 nanoparticles [76]. Reproduced with permission from American Chemical Society.

anticancer drugs. Yabu et al. [76] encapsulated titania and γ-Fe_2O_3 nanoparticles into polydimethylacrylamide based amphiphilic block copolymers like PDMA-r-PS and PDMA-r-PDA. The nanostructures were confirmed with the help of HRTEM and STEM images as shown in Fig. 5.11. Hence the encapsulation of inorganic nanoparticles in block copolymers is a challenging fact and they developed a universal strategy of double phase functionalization.

Mable et al. [77] loaded silica nanoparticles into the vesicles of poly(glycerol monomethacrylate)-poly(2-hydroxypropyl methacrylate) diblock copolymer by utilizing the encapsulation property. The encapsulation is carried out by following the polymer induced self-assembly. Later, Mable et al. [78] studied the thermally triggered release rate of silica from the vesicle and the kinetics in detail. Ning et al. [79] developed silica loaded tetrablock copolymer vesicles by the occlusion of silica into poly (glycerol monomethacrylate)-poly (2-hydroxypropyl methacrylate)-poly (ethylene glycol di-methacrylate)-poly (methacrylic acid) tetrablock copolymer by following the RAFT mechanism of block copolymer synthesis. Thus the polymerization induced self-assembly is utilized for the encapsulation and the optimal chain length associated with the inorganic nanoparticles was calculated. Lglesias et al. [80] studied the confinement of polystyrene stabilized gold nanoparticles in mesoporous silica by solvent-induced self-assembly. Mei et al. [81] studied the catalytic activity of palladium in polyelectrolyte brushes and in core–shell particles.

5.2 Spectroscopic Applications of Polymeric Nanoparticles

5.2.1 Surface Enhanced Raman Spectroscopy (SERS)

SERS materials have potential applications in the field of biological labeling, medical diagnosis and chemical sensing. It is widely used for chemical applications, environmental monitoring, food safety and medical applications [82]. The polymer modifications to nanomaterials are an effective way of enhancing the Raman intensity. Mainly gold and silver nanoparticles are used for this purpose as the SERS materials have wide applications in the environmental fields [83]. Generally, the polymer and protein parts are helpful in quenching the aggregation of nanoparticles by encapsulating them and thereby acting as platform for SERS analysis [84]. Braun et al. [85] developed a new class of SERS substrate that is a kind of cluster material by following a linking protocol. They have used the SERS signal as a parameter for optimization purposes.

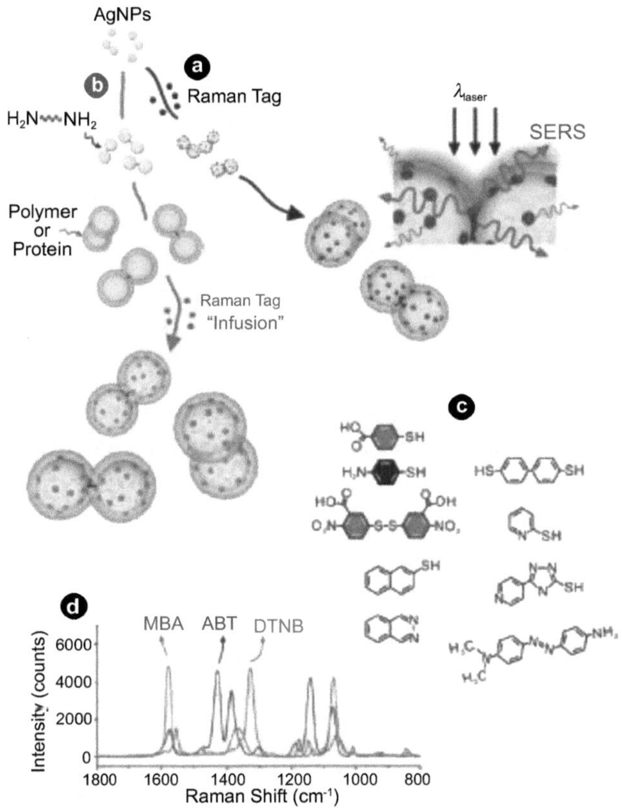

Figure 5.12 Schematic representation and SERS analysis of Raman active nanocapsule [85]. Reproduced with permission from American Chemical Society.

The linking is followed by the quenching step in order to stabilize the nano aggregates. They have used Raman active linkers to tag the material in order to act as a hot spot. The detailed synthesis method and the SERS spectra are shown in Fig. 5.12.

SERS immunosensors developed by the polymeric magnetic nanoparticles can be used for finding pathogens like S. typhimurium .[86] as it ensures rapid detection and easy separation in a cost-effective way. Dey et al. [87] analyzed the SERS performance of self-assembled hyperbranched Au nanoparticles. The polymer

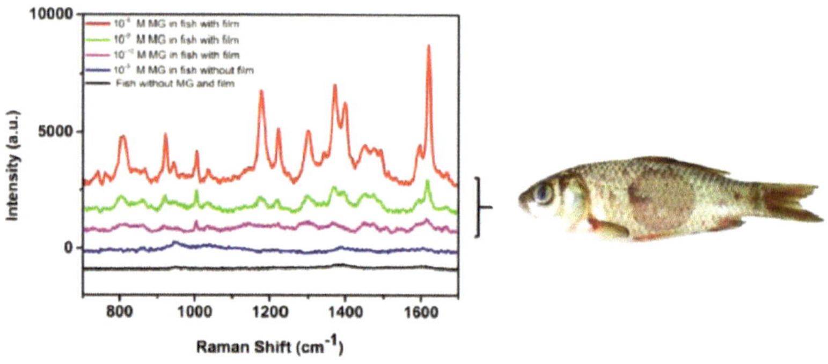

Figure 5.16 Detection of MG on the fish surface under different conditions using SERS analysis [94]. Reproduced with permission from American Chemical Society.

should not show extravascular leakage and rapid renal excretion are [95] and hence the iodinated polymeric nanoparticles, and their emulsions have tremendous application in the X-ray imaging application as iodine is an important contrast agent [96]. The in vitro and in vivo contrasting properties enable the polymeric nanoparticle as a good diagnostic agent. Emulsification polymerization is an effective way of fabricating radiopaque polymeric nanoparticles as it avoids aggregation of synthesized particles [97]. As shown in Fig. 5.17 the intravenous injection offers a better understanding of the interior organs by providing highly contrasted images.

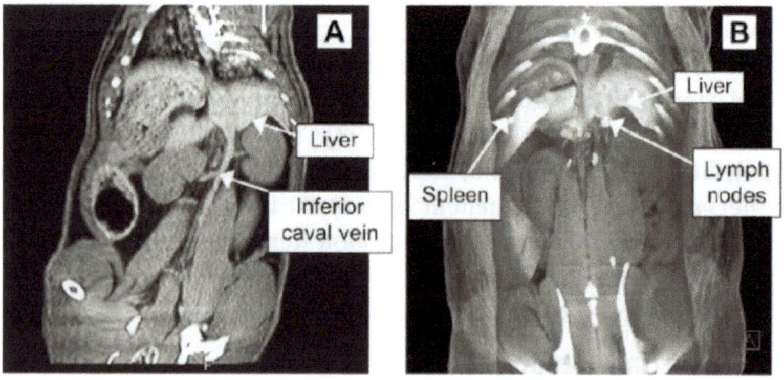

Figure 5.17 CT scanning images of RAT after using the radiopaque polymeric nanoparticles [97]. Reproduced with permission from Elsevier.

But the iodinated contrast agents have several drawbacks like renal toxicity, short imaging times as a result of the rapid renal clearance, and vascular permeation. In order to avoid this Kim et al. [98] developed new polymeric nanoparticles using gold-covered PEG and used it as an effective contrast agent in CT analysis. They have combined the anti-biofouling effect of PEG and the X-ray absorption capability of gold nanoparticles. The polymer nanoparticles were active as blood pool imaging and also as a detection agent. Mieszawska et al. [99] analyzed the effect of a lipid-coated polylactic-co-glycolic acid (PLGA) nanoparticles for CT imaging. The Au nanoparticle incorporated core–shell system was capable of showing high X-ray attenuation superior to iodine-based contrast agents. Wang et al. [100] made Fe_3O_4 and iodine-containing rare earth compounds containing trifunctional polymer-based composites for CT analysis applications. The study showed that the trifunctional polymer nanoparticles with iodine are efficient in providing high contrast CT images. Wallyn et al. [101] utilized the method of nanoprecipitation of the PEGylated polymer synthesized from the monomer 2-methacryloyloxyethyl (2,3,5-triiodobenzoate) monomer (MAOTIB) by RAFT polymerization. The in-vivo micro CT scans show (Fig. 5.18) the excellent efficiency of the polymeric nanoparticles as X-ray contrast agents. The study shows that the X-ray attenuation offered by these particles is high and the accumulations occurred after short circulation time. These contrast agents were non-toxic and hence more suitable for small animals as well as for preclinical imaging. Yin et al. [102] also analyzed the efficiency of poly(iohexol) nanoparticles as contrast agents.

5.2.3 Fluorescence Imaging

The incorporation of polymeric matrixes and surface functionalization are other suitable methods for developing fluorescence probes. Li et al. [103] utilized the efficiency of conjugated polymers as a fluorescent probe. Due to high fluorescence, non-toxicity and photostability, conjugated polymers (CP) can be utilized for the fluorescence probe. Delocalized Π electrons present on the backbone of the CP are responsible for the emission spectra. The nanostructured PEG-based polymeric nanoparticles are able to

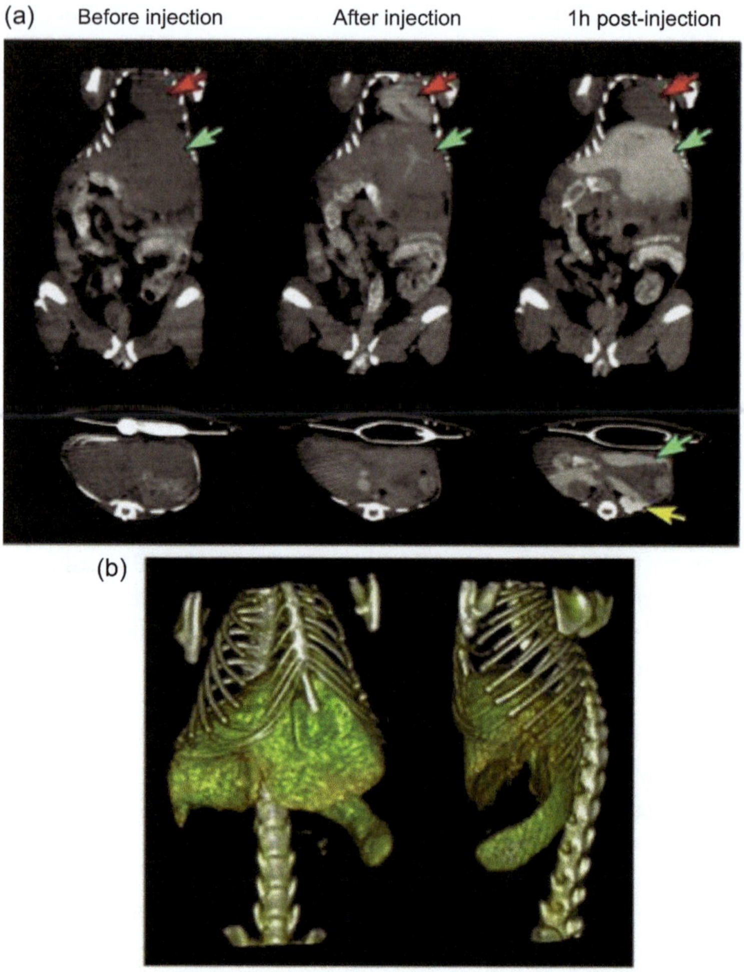

Figure 5.18 (A) CT scan images of heart, liver and spleen are respectively indicated by red, green and yellow arrows. (B) Liver and spleen 3D volume rendering imaging [101]. Reproduced with permission from Elsevier.

show fluorescence imaging applications in the near-infrared region by incorporating a fluorescent dye indocyanine green [104]. These nanoparticles were found effective for the imaging of ovarian cancer cells. The CP nanoparticles were utilized for imaging the brain of zebrafish larvae as shown in Fig. 5.19 [105].

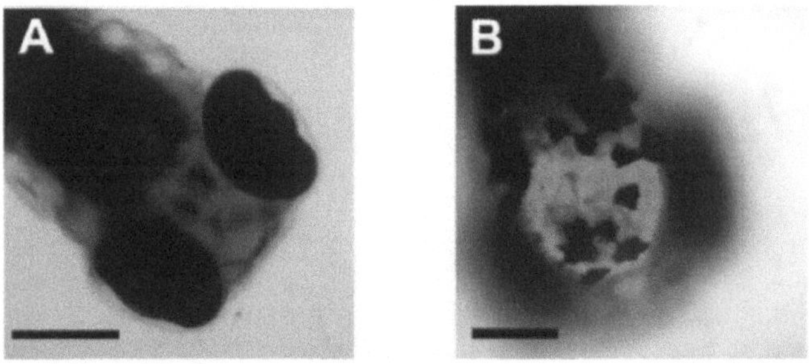

Figure 5.19 Live imaging of zebrafish brain by confocal laser scanning microscopy after microinjecting fluorescent PFPBA nanoparticles into the brain ventricle [105]. Reproduced with permission from American Chemical Society.

Utilizing the efficiency of copolymers to self-assemble into fluorescent polymeric nanoparticles different block copolymers are utilized for the purpose of making a CT contrast agent [106]. RAFT polymerization was generally followed for the synthesis and thus prepared polymers were able to show aggregation induced emission (AIE). The fluorescence spectra confirm the fluorescence stability, which is an important parameter of imaging. Zhang et al. [107] prepared gold nanoparticle and block copolymer-based nanoparticles by the method of re-precipitation for the Fluorescence imaging applications as shown in Fig. 5.20.

The spherical shape and core–shell structured polymeric nanoparticles are found effective in fluorescence imaging as the infrared dyes and photosensitizer can be incorporated easily into it [108] and thus prepared materials were able to show fluorescence in visible, near and short IR region. By following simple organic reactions like Mannich reaction and utilizing the click-chemistry methods, AIE fluorescence contracting polymeric nanoparticles can be prepared as these particles are capable of forming different micellar structures with structural integrity and stability [109, 110].

5.2.4 Other Applications

By exploiting the properties of polymeric nanoparticles, they can be used for wider spectroscopy-based applications like magnetic

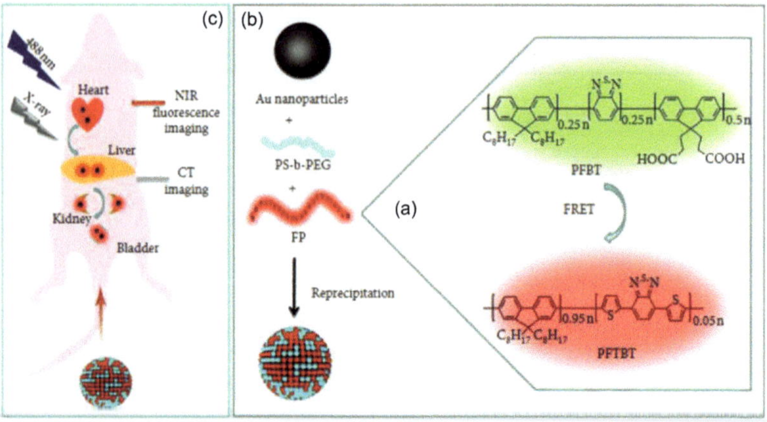

Figure 5.20 Schematic illustration of the fabrication procedure of Au@FP nanoparticles and their application for dual-modal imaging [107]. Reproduced with permission from Hindawi.

resonance imaging, optical sensing, photon sensing, etc. It is widely used for magnetic resonance imaging which is a kind of magnetic resonance spectroscopic technique. Commonly known as nuclear magnetic resonance, it is a non-invasive technique used in medical diagnosis. Magnetic resonance imaging (MRI) is another application offered by the polymeric nanoparticles. The presence of transition metal ions and lanthanide ions in the polymeric nanoparticles makes active signals and can be utilized for making active MRI probes [110–114]. Wang et al. [100] used polymeric nanoparticles for optical imaging applications. Optical images of liver and spleen were taken using a confocal laser scanning microscope and the efficiency of optical imaging application of the synthesized polymeric nanoparticle was demonstrated. Shen et al. [115] doped a photosensitizer into the conjugated polymer nanoparticles for a photon imaging application. Due to the 2 photon emission provided by the conjugated polymer, it can be used for an imaging application. Cannizo et al. [116] also utilized polymeric nanoparticles for the development of optical nanosensors. They have synthesized nanolatex by the method of microemulsion polymerization method. The grafting of phenylboronic acid maintains the selectivity for the detection of diols. Behbahani et al. [117] made a colorimetric sensor out of iron imprinted polymeric nanoparticles for the detection of uranyl ions.

5.3 Conclusions and Future Scope

In summary, the chapter deals with the significant fabrication of polymeric nanoparticle spectroscopic investigations and novelty of innovations. The methods of synthesis of polymeric nanoparticles like self-assembly, functionalization and encapsulation are portrayed in a detailed manner. The utilization of properties of different polymeric nanoparticles for various spectroscopic applications like SERS, fluorescence imaging, MR imaging and optical sensing are helpful in fabricating novel devices. Highly conjugated polymeric structure and its self-assembled micellar nanostructures can offer ERS, Fluorescence and X-ray imaging activity. The electrical hotspots, radiopaque properties are highly beneficial and can be utilized further for better enhancement in the specified application. We have synthesized fluorescent block copolymer grafted carbon dots which could function as SERS substrates. We have also confirmed the formation of graphene oxide grafted polymeric nanoparticles that exist as vesicle and spherical shaped micelle. The spectroscopic applications of these materials are still in the infancy stage and a lot of work has to be done to unveil their vast possibility in this field. The vast array of nanofillers as well as the various sophisticated fabrication techniques might enable the synthesis of novel polymeric nanoparticles in the future so that specific systems could be tailored for the definite application.

References

1. Krishnaswamy, K., and Orsat, V. (2017). Sustainable delivery systems through green nanotechnology, in *Nano- and Microscale Drug Delivery Systems*, Elsevier, pp. 17–32.
2. Pund, S., and Joshi, A. (2017). Nanoarchitectures for neglected tropical protozoal diseases: challenges and state of the art, in *Nano- and Microscale Drug Delivery Systems*, Elsevier, pp. 439–480.

3. Yadav, H. K. S., Almokdad, A. A., Sumia, I. M., and Debe, M. S. (2019). Polymer-based nanomaterials for drug-delivery carriers, in *Nanocarriers for Drug Delivery*, Elsevier, pp. 531–556.
4. Tessier, P. M., Velev, O. D., Kalambur, A. T., Lenhoff, A. M., Rabolt, J. F., and Kaler, E. W. (2001). Structured metallic films for optical and spectroscopic applications via colloidal crystal templating. *Advanced Materials*, **13**(6), 396–400.
5. Tan, H., Liu, B., and Chen, Y. (2012). Lanthanide coordination polymer nanoparticles for sensing of mercury (II) by photoinduced electron transfer. *ACS Nano*, **6**(12), 10505–10511.
6. Kang, Y., and Taton, T. A. (2005). Core/shell gold nanoparticles by self-assembly and crosslinking of micellar, block-copolymer shells. *Angewandte Chemie International Edition*, **44**(3), 409–412.
7. Mai, Y., and Eisenberg, A. (2012). Self-assembly of block copolymers. *Chemical Society Reviews*, **41**(18), 5969–5985.
8. Rong, M. Z., Ji, Q. L., Zhang, M. Q., and Friedrich, K. (2002). Graft polymerization of vinyl monomers onto nanosized alumina particles. *European Polymer Journal*, **38**(8), 1573–1582.
9. Paris, J. L., Cabañas, M. V., Manzano, M., and Vallet-Regí, M. (2015). Polymer-grafted mesoporous silica nanoparticles as ultrasound-responsive drug carriers. *ACS Nano*, **9**(11), 11023–11033.
10. Sahiner, N., and Ilgin, P. (2010). Synthesis and characterization of soft polymeric nanoparticles and composites with tunable properties. *Journal of Polymer Science Part A: Polymer Chemistry*, **48**(22), 5239–5246.
11. Chen, J., Zhang, P., Fang, G., Yi, P., Zeng, F., and Wu, S. (2012). Design and synthesis of FRET-mediated multicolor and photoswitchable fluorescent polymer nanoparticles with tunable emission properties. *Journal of Physical Chemistry B*, **116**(14), 4354–4362.
12. De Volder, M., and Hart, A. J. (2013). Engineering hierarchical nanostructures by elastocapillary self-assembly. *Angewandte Chemie International Edition*, **52**(9), 2412–2425.
13. Son, J. G., Hannon, A. F., Gotrik, K. W., Alexander-Katz, A., and Ross, C. A. (2011). Hierarchical nanostructures by sequential self-assembly of styrene-dimethylsiloxane block copolymers of different periods. *Advanced Materials*, **23**(5), 634–639.
14. Hirai, T., Leolukman, M., Hayakawa, T., Kakimoto, M., and Gopalan, P. (2008). Hierarchical nanostructures of organosilicate nanosheets

within self-organized block copolymer films. *Macromolecules*, **41**(13), 4558–4560.

15. Barry, E., and Dogic, Z. (2010). Entropy driven self-assembly of nonamphiphilic colloidal membranes. *Proceedings of the National Academy of Sciences of the United States of America*, **107**(23), 10348–10353.
16. Kostarelos, K., Emfietzoglou, D., and Tadros, T. F. (2005). Light-sensitive fusion between polymer-coated liposomes following physical anchoring of polymerisable polymers onto lipid bilayers by self-assembly. *Faraday Discussions*, **128**, 379–388.
17. Rasch, M. R., Rossinyol, E., Hueso, J. L., Goodfellow, B. W., Arbiol, J., and Korgel, B. A. (2010). Hydrophobic gold nanoparticle self-assembly with phosphatidylcholine lipid: membrane-loaded and janus vesicles. *Nano Letters*, **10**(9), 3733–3739.
18. Jonas, U., and Krüger, C. (2002). The effect of polar, nonpolar, and electrostatic interactions and wetting behavior on the particle assembly at patterned surfaces. *Journal of Supramolecular Chemistry*, **2**(1–3), 255–270.
19. Harkins, W. D. (1950). Soap solutions: salt, alcohol, micelles, rubber. *The Scientific Monthly*, **70**(4), 220–228.
20. Wang, D., and Wang, X. (2013). Amphiphilic azo polymers: molecular engineering, self-assembly and photoresponsive properties. *Progress in Polymer Science*, **38**(2), 271–301.
21. Elemans, J. A. A. W., Rowan, A. E., and Nolte, R. J. M. (2003). Mastering molecular matter. Supramolecular architectures by hierarchical self-assembly. *Journal of Materials Chemistry*, **13**(11), 2661–2670.
22. Alexandridis, P., and Lindman, B. (2000). *Amphiphilic Block Copolymers: Self-Assembly and Applications*, Elsevier.
23. Karayianni, M., and Pispas, S. (2016). Self-assembly of amphiphilic block copolymers in selective solvents, in *Fluorescence Studies of Polymer Containing Systems*, Springer, pp. 27–63.
24. Ludwigs, S., Böker, A., Voronov, A., Rehse, N., Magerle, R., and Krausch, G. (2003). Self-assembly of functional nanostructures from ABC triblock copolymers. *Nature Materials*, **2**(11), 744–747.
25. Darling, S. B. (2007). Directing the self-assembly of block copolymers. *Progress in Polymer Science*, **32**(10), 1152–1204.
26. Bolton, J., and Rzayev, J. (2012). Tandem RAFT-ATRP synthesis of polystyrene–poly (methyl methacrylate) bottlebrush block copoly-

mer architecture in hybrid assembly formation and SERS activity. *Langmuir*, **30**(8), 2249–2258.

89. Song, J., Duan, B., Wang, C., Zhou, J., Pu, L., Fang, Z., Wang, P., Lim, T. T., and Duan, H. (2014). SERS-encoded nanogapped plasmonic nanoparticles: growth of metallic nanoshell by templating redox-active polymer brushes. *Journal of the American Chemical Society*, **136**(19), 6838–6841.

90. Zhang, J., Gao, Y., Alvarez-Puebla, R. A., Buriak, J. M., and Fenniri, H. (2006). Synthesis and SERS properties of nanocrystalline gold octahedra generated from thermal decomposition of $HAuCl_4$ in block copolymers. *Advanced Materials*, **18**(24), 3233–3237.

91. Tokareva, I., Minko, S., Fendler, J. H., and Hutter, E. (2004). Nanosensors based on responsive polymer brushes and gold nanoparticle enhanced transmission surface plasmon resonance spectroscopy. *Journal of the American Chemical Society*, **126**(49), 15950–15951.

92. Xu, P., Jeon, S.-H., Mack, N. H., Doorn, S. K., Williams, D. J., Han, X., and Wang, H.-L. (2010). Field-assisted synthesis of SERS-active silver nanoparticles using conducting polymers. *Nanoscale*, **2**(8), 1436–1440.

93. Yang, M., Chen, T., Lau, W. S., Wang, Y., Tang, Q., Yang, Y., and Chen, H. (2009). Development of polymer-encapsulated metal nanoparticles as surface-enhanced Raman scattering probes. *Small*, **5**(2), 198–202.

94. Yang, N., You, T.-T., Gao, Y.-K., Zhang, C.-M., and Yin, P.-G. (2018). Fabrication of a flexible gold nanorod polymer metafilm via a phase transfer method as a SERS substrate for detecting food contaminants. *Journal of Agricultural and Food Chemistry*, **66**(26), 6889–6896.

95. Galperin, A., and Margel, S. (2006). Synthesis and characterization of new radiopaque microspheres by the dispersion polymerization of an iodinated acrylate monomer for X-ray imaging applications. *Journal of Polymer Science Part A: Polymer Chemistry*, **44**(12), 3859–3868.

96. Galperin, A., Margel, D., Baniel, J., Dank, G., Biton, H., and Margel, S. (2007). Radiopaque iodinated polymeric nanoparticles for X-ray imaging applications. *Biomaterials*, **28**(30), 4461–4468.

97. Aviv, H., Bartling, S., Kieslling, F., and Margel, S. (2009). Radiopaque iodinated copolymeric nanoparticles for X-ray imaging applications. *Biomaterials*, **30**(29), 5610–5616.

98. Kim, D., Park, S., Lee, J. H., Jeong, Y. Y., and Jon, S. (2007). Antibiofouling polymer-coated gold nanoparticles as a contrast agent for in vivo

X-ray computed tomography imaging. *Journal of the American Chemical Society*, **129**(24), 7661–7665.

99. Mieszawska, A. J., Gianella, A., Cormode, D. P., Zhao, Y., Meijerink, A., Langer, R., Farokhzad, O. C., Fayad, Z. A., and Mulder, W. J. M. (2012). Engineering of lipid-coated PLGA nanoparticles with a tunable payload of diagnostically active nanocrystals for medical imaging. *Chemical Communications*, **48**(47), 5835–5837.
100. Wang, X., Tu, M., Yan, K., Li, P., Pang, L., Gong, Y., Li, Q., Liu, R., Xu, Z., and Xu, H. (2015). Trifunctional polymeric nanocomposites incorporated with Fe_3O_4/iodine-containing rare earth complex for computed X-ray tomography, magnetic resonance, and optical imaging. *ACS Applied Materials & Interfaces*, **7**(44), 24523–24532.
101. Wallyn, J., Anton, N., Serra, C. A., Bouquey, M., Collot, M., Anton, H., Weickert, J.-L., Messaddeq, N., and Vandamme, T. F. (2018). A new formulation of poly (MAOTIB) nanoparticles as an efficient contrast agent for in vivo X-ray imaging. *Acta Biomaterialia*, **66**, 200–212.
102. Yin, Q., Yap, F. Y., Yin, L., Ma, L., Zhou, Q., Dobrucki, L. W., Fan, T. M., Gaba, R. C., and Cheng, J. (2013). Poly (iohexol) nanoparticles as contrast agents for in vivo X-ray computed tomography imaging. *Journal of the American Chemical Society*, **135**(37), 13620–13623.
103. Li, K., and Liu, B. (2012). Polymer encapsulated conjugated polymer nanoparticles for fluorescence bioimaging. *Journal of Materials Chemistry*, **22**(4), 1257–1264.
104. Bahmani, B., Guerrero, Y., Bacon, D., Kundra, V., Vullev, V. I., and Anvari, B. (2014). Functionalized polymeric nanoparticles loaded with indocyanine green as theranostic materials for targeted molecular near infrared fluorescence imaging and photothermal destruction of ovarian cancer cells. *Lasers in Surgery and Medicine*, **46**(7), 582–592.
105. Qian, C.-G., Zhu, S., Feng, P.-J., Chen, Y.-L., Yu, J.-C., Tang, X., Liu, Y., and Shen, Q.-D. (2015). Conjugated polymer nanoparticles for fluorescence imaging and sensing of neurotransmitter dopamine in living cells and the brains of zebrafish larvae. *ACS Applied Materials & Interfaces*, **7**(33), 18581–18589.
106. Dong, J., Liu, M., Jiang, R., Huang, H., Wan, Q., Wen, Y., Tian, J., Dai, Y., Zhang, X., and Wei, Y. (2018). Synthesis and biological imaging of cross-linked fluorescent polymeric nanoparticles with aggregation-induced emission characteristics based on the combination of RAFT polymerization and the Biginelli reaction. *Journal of Colloid and Interface Science*, **528**, 192–199.

107. Zhang, M., Liu, J., and Wang, G. (2019). Highly biocompatible nanoparticles of Au@ fluorescent polymers as novel contrast agent for in vivo bimodality NIR fluorescence/CT imaging. *Contrast Media & Molecular Imaging*, **2019**, 8085039.
108. Chepurna, O. M., Yakovliev, A., Ziniuk, R., Nikolaeva, O. A., Levchenko, S. M., Xu, H., Losytskyy, M. Y., Bricks, J. L., Slominskii, Y. L., and Vretik, L. O. (2020). Core–shell polymeric nanoparticles co-loaded with photosensitizer and organic dye for photodynamic therapy guided by fluorescence imaging in near and short-wave infrared spectral regions. *Journal of Nanbiotechnology*, **18**(1), 19.
109. Jiang, R., Liu, H., Liu, M., Tian, J., Huang, Q., Huang, H., Wen, Y., Cao, Q., Zhang, X., and Wei, Y. (2017). A facile one-pot Mannich reaction for the construction of fluorescent polymeric nanoparticles with aggregation-induced emission feature and their biological imaging. *Materials Science and Engineering C*, **81**, 416–421.
110. Hu, J., Liu, T., Zhang, G., Jin, F., and Liu, S. (2013). Synergistically enhance magnetic resonance/fluorescence imaging performance of responsive polymeric nanoparticles under mildly acidic biological milieu. *Macromolecular Rapid Communications*, **34**(9), 749–758.
111. Cheung, E. N. M., Alvares, R. D. A., Oakden, W., Chaudhary, R., Hill, M. L., Pichaandi, J., Mo, G. C. H., Yip, C., Macdonald, P. M., and Stanisz, G. J. (2010). Polymer-stabilized lanthanide fluoride nanoparticle aggregates as contrast agents for magnetic resonance imaging and computed tomography. *Chemistry of Materials*, **22**(16), 4728–4739.
112. Hu, D., Chen, L., Qu, Y., Peng, J., Chu, B., Shi, K., Hao, Y., Zhong, L., Wang, M., and Qian, Z. (2018). Oxygen-generating hybrid polymeric nanoparticles with encapsulated doxorubicin and chlorin e6 for trimodal imaging-guided combined chemo-photodynamic therapy. *Theranostics*, **8**(6), 1558.
113. Lee, H., Lee, E., Kim, D. K., Jang, N. K., Jeong, Y. Y., and Jon, S. (2006). Antibiofouling polymer-coated superparamagnetic iron oxide nanoparticles as potential magnetic resonance contrast agents for in vivo cancer imaging. *Journal of the American Chemical Society*, **128**(22), 7383–7389.
114. Xu, H., Cheng, L., Wang, C., Ma, X., Li, Y., and Liu, Z. (2011). Polymer encapsulated upconversion nanoparticle/iron oxide nanocomposites for multimodal imaging and magnetic targeted drug delivery. *Biomaterials*, **32**(35), 9364–9373.

115. Shen, X., Li, L., Wu, H., Yao, S. Q., and Xu, Q.-H. (2011). Photosensitizer-doped conjugated polymer nanoparticles for simultaneous two-photon imaging and two-photon photodynamic therapy in living cells. *Nanoscale*, **3**(12), 5140–5146.

116. Cannizzo, C., Amigoni-Gerbier, S., and Larpent, C. (2005). Boronic acid-functionalized nanoparticles: synthesis by microemulsion polymerization and application as a re-usable optical nanosensor for carbohydrates. *Polymer (Guildf).*, **46**(4), 1269–1276.

117. Behbahani, M., Salimi, S., Abandansari, H.S., Omidi, F., Salarian, M., and Esrafili, A. (2015). Application of a tailor-made polymer as a selective and sensitive colorimetric sensor for reliable detection of trace levels of uranyl ions in complex matrices. *RSC Advances*, **5**(74), 59912–59920.

Chapter 6

Fabrication of Ultra-Pure Anisotropic Nanoparticles, Spectroscopic Studies and Biological Applications

Tean Zaheer
Department of Parasitology, University of Agriculture, Faisalabad, Pakistan
teanzaheer942@gmail.com

Anisotropic nanoparticles (ANPs) constitute a fascinating world of nanomaterials. The unique shape and size properties make them more adaptable for throughput downstream applications in biosciences. A number of synthesis techniques have been trialed for ANPs, employing either top-down or bottom-up approaches. Interesting shapes of ANPs viz. nanotubes, nanowires, nano-triangles, nanostars, nanoprisms, nanorods, nanocubes, nanorings, nanoplatelets, nanowalls, nanocages, nanothorns, nanoplates and nanoflowers have been widely fabricated. Moreover, different analytical tools for quantitation of ANPs have revolutionized to bring forward nanoparticles possessing better suited biological applications. ANPs have wide range of applications in medical/life sciences, agricultural sciences, waste management, bioremediation, food industry, paint and textile industry, robotics, energy and systems engineering, aviation and aeronautics and many others. The

Nanomaterials for Spectroscopic Applications
Edited by Kaushik Pal

ISBN 978-981-4877-69-5 (Hardcover), 978-1-003-16033-5 (eBook)
www.jennystanford.com

chapter presents an overview of techniques for synthesis, spectroscopic quantitation and biological domains employing anisotropic nanoparticles.

6.1 Introduction

Nanotechnology is a dynamic science dealing with nanoscale (1–100 nm) materials (Mahmoud et al., 2020). This field has attracted multi-trillion-dollar investment, engaging more than 5,00,000 scientists, producing thousands of commercially important products. Nanoscience has wide-ranging applications to almost every basic science and has reshaped the approach to mankind's well-being. Primarily, the fabrication of nanoparticles follows either bottom-up (smaller materials combine to form nanoparticles) or top-down (larger particles are broken down to nanoparticles) approach. The fabrication of nanoparticles is possible via various routes, leading to the production of either isotropic or anisotropic nanoparticles. Anisotropic nanoparticles are the nanomaterials that have variation in shape/size/conformation/particle characteristics, contrary to those of isotropic nanoparticles.

Table 6.1 Overview of influence of various factors on nanoparticle properties

Factor	Effect on synthesis
pH	higher pH, lower size nps
Technique of synthesis	Physical, chemical and biological = most widely accepted due to eco-friendly
Temperature	higher for physical, medium for biological, lower for chemical methods
Stabilizing agent	thin layer may cause aggregation
Pressure	reduction of ions increased at ambient pressure
Particle surface area	has direct relation to physical properties, e.g. melting point decreases sharply when size is of atomic scale
Particle proximity effect	Physicochemical properties altered by reduced particle size

Several factors influence the properties of nanoparticles. pH has an inverse effect on the size of nanoparticles. Similarly, the technique adopted for fabrication dictates the temperature range and pressure requirement. A brief account of major factors influencing nanoparticle characteristics has been given in Table 6.1. Among others, biomediated synthesis or green synthesis has gained wide acceptance due to eco-friendly nature and facile synthesis.

This chapter presents an overview of the fabrication and spectroscopic quantitation of anisotropic nanoparticles. Further, it highlights key areas utilizing ANPs for research and commercial purposes. Tremendous research in biological sciences has helped to prevent, diagnosing and treating superbugs of the 21st century. Bio-assimilation, toxicity and hazard analysis of nanoparticles are major concerns and direction of future research in nanotechnology.

6.2 Fabrication Approaches for Synthesis

Anisotropic nanoparticles possessing different dimensions and shapes can be fabricated using various approaches, are briefly discussed in this chapter. The nanotubes, nanowires, nanotriangles, nanostars, nanoprisms, nanorods, nanocubes, nanorings, nanoplatelets, nanowalls, nanocages, nanothorns, nanoplates and nanoflowers are among most commonly synthesized type of anisotropic nanoparticles (Fig. 6.1).

6.2.1 Polyol Synthesis

Originally proposed by Fievet et al. and Viau et al. in 1989, the currently used method has been modified by the one developed by Sun and Wiley and co-workers (Wiley et al., 2005). Briefly, the method involves the reduction of salt ions (inorganic) using any of the polyol compounds. Stabilizers like polyvinyl pyrrolidone (PVP), polyacrylamide (PAM), poly-(diallyldimethylammonium) chloride (PDDA) are used to prevent agglomeration during reduction by polyol. The reactivity of seeds depends on the morphology

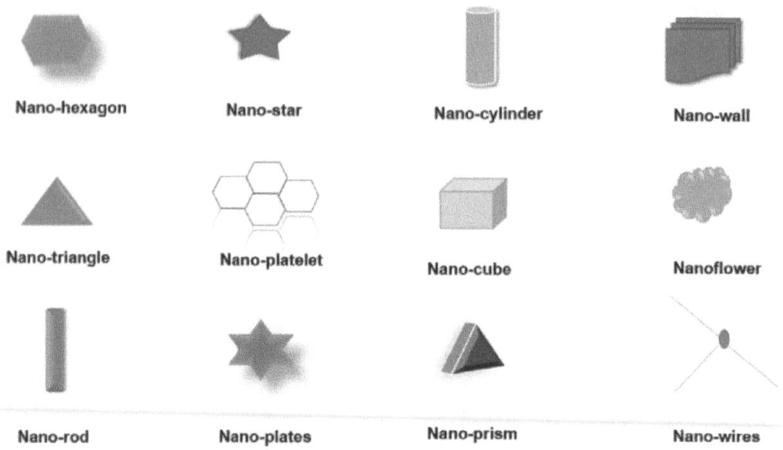

Figure 6.1 Possible shapes of anisotropic nanoparticles.

(singly or multiply twinned, plates having faulty stacking, single-crystal seeds having different folds symmetry). It is a facile and controllable synthesis method for octahedral gold nanoparticles (Li et al., 2008).

6.2.2 Biological Synthesis

Nanoparticles having various shapes and sizes could be synthesized using biological systems of plants, bacteria, fungi, viruses and algae. It may also be termed as green nano-synthesis or nano-bio synthesis. The biological system acts as a reducing and capping agent, yielding nanoparticles. To this end, plants have been most widely used to synthesize ANPs possessing diverse physicochemical characteristics. Nanowires, nanotriangles, nanocubes, nanoprisms, nanoplates, etc. have been fabricated using biological synthesis. It is by far the most facile, least toxigenic and environment-friendly approach to synthesize nanoparticles. The reaction usually takes place from room temperature to 100°C. The green synthesis route needs further optimization to yield more nanoparticles of uniform characteristics.

6.2.3 Seed Mediated Synthesis

One of the most widely used methods having good yields of desired nanoshapes. It is basically a modified version of Zsigmondy's 'nuclear' method, involving nanoparticle synthesis in 2 steps (Zsigmondy, 1917). Briefly, the first step involves the fabrication of seed nanoparticles which arise by simple reduction. Secondly, the desired shapes can be formed using seed nanoparticles. The use of counter anions from surfactants (e.g. cetyltrimethylammonium bromide CTAB) can further transform the shapes of nanoparticles. Factors like temperature, electric field, use of ionic surfactants can affect the shapes and properties of ANPs (Millstone et al., 2006). Core-shell, star, trigonal, pentagonal, hexagonal, cube, branched and tetrapod nanoparticles having various metals compounds have been synthesized using this approach (Samal et al., 2010).

6.2.4 Template Mediated Synthesis

ANPs of single dimension and controllable physical characteristics could be fabricated using templates like alumina or polycarbonate of nanoporous nature. This approach for synthesis follows the principle of electrodeposition at atmospheric conditions. The pore size of the basic template dictates the diameter of newly synthesized nanoparticles. Template dependent approach is easier, cost-effective and more adaptable with different materials.

6.2.5 Hydrothermal Approach

Nanoparticles are synthesized under high temperature and pressure, on a substrate, in the presence of water. Sometimes polar or non-polar solvents may be added to broaden the applicability of synthetic hydrothermal process, also known as solvothermal method. ANPs of fascinating morphologies like nanoleaves, nanocabbages, dandelion-like nanoparticles, nanoflowers, nanoloops and dendritic have been fabricated using hydrothermal reduction (Chang et al., 2008). This approach is relatively facile, eco-friendly, economical (could be up-scaled), template-free, having a range of morphological yields (up to 3D) utilizing a single substrate for the entire reaction (Tian et al., 2008).

6.2.6 Electrochemical

Nanoparticles of uniform size, lower cost, good quality and reduced reaction temperature can be procured using the electrochemical route of ANP synthesis. Two electrodes (anode and cathode) are set, spaced by any non-conducting material (e.g. teflon) and electrolyte surfactant (e.g. CTAB) to act as a stabilizer as well as micelle that may control size and uniformity of nanoparticles. The de-atoms are formed by reduction at electrodes, which are then stabilized and transformed into desired nanostructures. However, nano-geometry may be controlled merely by an electric potential applied where no surfactant and template are added (Gu and Zhang, 2008).

6.2.7 Photochemical

This method employs photochemical or radiolytic reduction of metallic salts at a known rate of reaction. Additionally, ANPs have higher yields, more stable optical properties, with a minimum requirement of reagents/equipment for fabrication. The properties of nanoparticles depend on the duration, type and wavelength of a light source used. Shapes like nanoprisms, nanowires, nanorods, crown-shaped nanoparticles have been developed using a photochemical reduction approach.

6.3 Spectroscopic Analysis of Fabricated Nanoparticles

The post-fabrication analysis helps to characterize and further validate nanoparticles. There are several spectroscopic techniques that have a common basic principle of evaluating the particles based on their ability to absorb/transmit light waves. The techniques, however, may differ depending upon the approach and determination of specific properties of nanoparticles. An analysis is a significant component that aids in the optimization of protocol, reaction conditions and possible applications. Table 6.2 presents a brief comparison and properties of commonly used techniques.

Table 6.2 Summary of spectroscopic techniques for ANP analysis

Technique	**Principle**	**Information obtained**	**Reference**
X-ray diffraction (XRD)	Position and intensity peaks are generated from dried samples using Scherrer equation	Grain size, structure and composition of nanocrystals	Mourdikoudis et al., 2018
X-ray absorption spectroscopy (XAS)	Ability of nanoparticles to absorb X-rays to determine structural and kinetic properties	Debye–Waller factor, distances between atoms, partial coordination number	Srabionyan et al., 2016
Visible spectroscopy (UV-Vis)	Measures intensity of reflected light by nanoparticles compared to reference material	Size, purity, optical property, degree of agglomeration	Ray et al., 2015
Dynamic light scattering (DLS)	Nanoparticle diameter is detected by virtue of Brownian motion in colloidal solution, exhibited as a scattering of light	Size (hydrodynamic), presence of agglomeration	Kestens et al., 2016
Nuclear magnetic resonance (NMR)	Differences in spin up and spin down states are quantified by applying strong magnetic fields	Atomic conformation, electronic core assembly, ligand properties and their effect on ANPs	Marbella and Millstone, 2015
Raman scattering (RS)	Based on elastic or inelastic scattering, it determines the vibrational energy of nanoparticles when subjected to photons	Electronic and vibrational characteristics, conformation of metals and conjugates, size (hydrodynamic)	Mahmoud et al., 2020
Mass spectroscopy (MS)	Electron ionization results in production of various ions that are separated, analyzed and detected based on mass to charge ratio	Size, composition, concentration, surface characteristics, bioconjugation quantitation	Costa-Fernández et al., 2016

(*Contd.*)

Table 6.2 (*Continued*)

Technique	**Principle**	**Information obtained**	**Reference**
Fourier transform infrared spectroscopy (FTIR)	The dipole motion of nanoparticle absorbing infra-red wavelength is recorded on a spectrum	Composition (functional groups, bond nature and strength)	Buso-Rogero et al., 2016
Circular dichroism (CD)	Nanoparticles are subject to photo-thermal imaging based on light polarization in various directions	Conformational changes and thermal properties	Spaeth et al., 2019
Scanning tunneling spectroscopy (STS)	Measures electrons as a function of energy on a spatial resolution	Electron density and band gap	Zandvliet et al., 2009
Inductively coupled plasma optical emission spectrometry (ICP-OES)	Plasma induced high frequency current bring electrons to excitation state followed by emission of photons, whose intensity is recorded	Composition, size	Cui et al., 2014
Graphite furnace atomic absorption spectroscopy (GFAAS)	Nanoparticles are coated on pyroelectric carbon, vaporized and atomized by heat	Concentration	Leopold et al., 2017
Photoluminescence spectroscopy (PL)	Fluorescent nanoparticles emit light on absorbing photons	Optical and related conformational properties	Andersen et al., 2015
Mössbauer spectroscopy	The excited state of atom absorbs energy when bombarded with γ-radiation and transmits it, which is analyzed	Anisotropic energy, differentiation of iron oxides, symmetry, state of oxidation	Ushakov et al., 2020
Nanoparticle tracking analysis (NTA)	Nanoparticles suspended in liquid are subject to laser beam that causes light scattering based on Brownian motion of particles	Size (more sensitive to lower sizes than DLS), refractive indices, capping efficiency of stabilizing agents	van der Pol et al., 2014

6.4 Biological Applications of Anisotropic Nanoparticles

Nanoparticles have revolutionized the scope of research in biomedical and agricultural sciences. Industrial applications of ANPs have attracted more commercial interest towards this field. Infectious and genetic diseases of high significance could be detected at early stages, difficult to treat clinical infections/injuries/amputations and genetic manipulations of sub-cellular level could be made possible using nanoparticles. Some potential areas of biological research and commercialization have been highlighted in Fig. 6.2.

6.4.1 Toxicity of Pathogens

Inhibition of activity and mortality of several pathogens of concern has been validated at both in vitro and in vivo scales, using ANPs. Moreover, important life cycle stages of vectors (living organisms transmitting microbes) of diseases have been controlled using ANPs. Nanoparticles against mosquitoes, ticks, flies, lice, fleas, bugs and

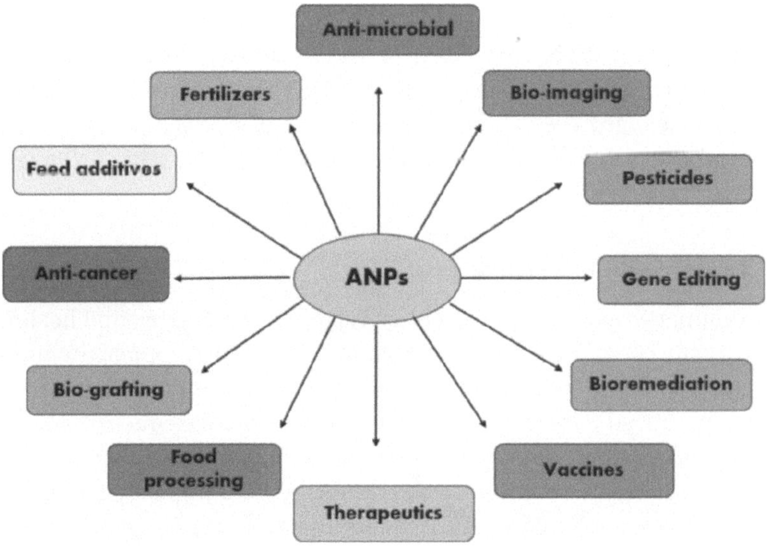

Figure 6.2 Biological applications of anisotropic nanoparticles.

snails have remarkably lower LC_{50} and LC_{99}, compared to those of conventional chemical therapeutics. The exact mode of action for toxicity is not completely understood, however, it is proposed that nanoparticles may accelerate or inhibit cellular pathways of death.

6.4.2 Medical

Various ANPs have been functionalized with biological entities to prevent, diagnose and treat metabolic, hereditary, infectious and idiopathic ailments. Silver and gold ANPs, followed by zinc nanoparticles have been most widely employed for biomedical applications.

6.4.2.1 Bio-imaging and biosensing

Significant diseases like cancer, malaria, trypanosomiasis, influenza, could be diagnosed at early or pre-clinical stages using ANPs. ANPs serving as a substrate, conjugating with target pathogen, have been visualized via surface-enhanced Raman spectroscopy. Also, use of ANPs has been combined with clinical tools like X-rays, ultrasonography, magnetic resonance imaging, positron emission tomography, etc. (Kohout et al., 2018). This property of nanoparticles has helped in detection of drug resistant bacteria, viruses, parasites, protozoa, cancer, etc. (Kang et al., 2017). Nano biosensors mediators (enzymes, growth factors, proteins, etc.) of cellular pathways are mapped with the help of conjugation/interaction with ANPs (He et al., 2018).

6.4.2.2 Wound healing

Nanoparticles can serve as either vector (mediating wound healing capacity) or sole treatment of a wound. Silver, gold, copper, titanium, zinc and chitosan nanoparticles of various shapes and origins, have been utilized to decrease microbial burden, increase recruitment of pro-inflammatory mediators, enhance wound healing. Nanoparticle impregnated to secondary vehicles including ointments, bandages, membranes, scaffolds, gels, anti-septics has shown large promise in treating wounds with high microbial loads (Lu et al., 2018). One of the most significant application of nano-wound healing is for diabetic and complicated wounds, which present a great deal of clinical challenges.

6.4.2.3 Gene editing and gene delivery

Double-stranded nucleotides have been delivered and expressed using ANPs having various benefits among other materials. To this end, chitosan nanoparticles have been used to silence gene expression of undesirable traits in mosquito (Dhandapani et al., 2019). Also, the efficient particle size and properties of nanoparticles have made them suitable vectors for gene delivery in agricultural and livestock models. Inorganic nano-vectors such as gold ANPs, dendrimers, quantum dots and magnetic nanoparticles can be utilized as efficient cargo for nucleotide delivery (Sanzari et al., 2019).

6.4.2.4 Therapeutics

Treatment of bacterial, viral, carcinogenic, parasitological, fungal, anti-microbial resistant, multi-drug resistant pathogens could be done using ANPs. Extensive work has been carried out to cure the diseases of epidemic or pandemic nature. Cancer radiotherapy has been made more selective by the use of gold ANPs which can induce reactive oxygen species by virtue of secondary electrons, possessing shorter ranges (Goel et al., 2017).

6.4.2.5 Vaccinology

Polymers, self-assembling, ISCOMs, virus-like particles, inorganic nanoparticles possessing high biocompatibility such as polystyrene, chitosan, peptides and proteins, hyaluronic acid and many vaccines adjuvant delivery systems have offered protection against many infectious diseases (Al-Halifa et al., 2019). High stability and safety, biocompatibility, efficient particle characteristics and specific binding domains are key features of nanoparticles, offering protective immunity (Espinosa et al., 2018). Chimeric nano-vaccines have provided an insight to target multi-pronged immunity towards infectious diseases.

6.4.2.6 Bio-grafting

Medical implants (bones, teeth, skin) may cause adverse reactions by the recipient's body. Nanoparticles derived from biological sources like tri-calcium phosphate from poultry eggshell and chitosan from crustacean shells have offered more biocompatibility, better tensile strength and relatively non-toxic alternative to conventional biografts (Demirel et al., 2016).

6.4.2.7 Cosmetics

Nano-cosmetics and dermatological preparations have shown large promise and consumer readiness to use them. Major types of nanoparticle-based products are skin protectants (sunscreens, sun-blocks), routine skincare (soaps, moisturizers, shampoos, deodorant, shaving and body creams, anti-aging, cleansers, nail polish, eyeliner) scar lighteners and anti-microbial treatment of fungal/bacterial infections, etc. (Niska et al., 2018).

6.4.2.8 Disinfectants

History of nano-powder as decontaminant is as old as World War I where wounds of soldiers were treated with silver powder to prevent microbial infection. ANPs have shown efficacy in reducing bacterial loads from the de-salinized water samples. Silver nanoparticles have wide applications as surface, aerosol, water disinfectants at industrial/commercial/farm/healthcare settings. The safety profile of nanoparticle disinfectant is higher, making application at food processing units, animal farms and hospitals possible (Deshmuk et al., 2019).

6.4.3 Food Industry

Food safety and security are major concerns of the food industry globally. ANPs could be efficiently commercialized to enhance the quality of packaged food. The shelf life of perishable food items could be increased by treatment with anti-microbial nanoparticles.

6.4.3.1 Feed additives

The nutraceutical industry has recently introduced the use of minerals/vitamins/enzymes for human use and overall well-being. Weight gain and optimum production of beef and dairy animals are critical parameters of animal farming. Nutrients crucial to an animal's overall production and performance can be used as nano-coated feed additives (nano-minerals, nano-vitamins, nano toxin binders, etc.). Improved quality and quantity of eggs, meat and milk have been achieved using nano-minerals (Cr, Cu, Se, Zn), nano-vitamins (vitamin E), nano-enzymes, nano-fatty acids supplementation in animal feed (Konkol et al., 2018). Similarly, the use of some nano-probiotics for improvising animal's gut health and digestive function has exhibited promising results.

6.4.3.2 Food packaging and processing

Increasing shelf life while maintaining food quality is the mainstay of quality assurance. Microbial loads (of bacteria, fungi, yeasts), toxins, other factors (pH, temperature, humidity, etc.) may cast adverse effects on food (Arfat et al., 2017). Detection of contaminants/ heavy metals/gases/microbes in food could be done by utilizing nanoparticles. Also, silver ANPs have been widely utilized during the processing and packaging of various food materials.

6.4.4 Bioremediation

The ever-growing rise in the human population has cast adverse effects on the natural eco-biology of indigenous flora and fauna. Increased deforestation and urbanization have led to devastating aftermath in the shape of change in climate, globe wide. Also, Waste-water management and recycling has attracted a lot of research focus over time. Pollutants in the form of microbes (bacteria, fungi, algae, viruses, protozoa, etc.), dyes, effluents, toxic chemicals, heavy metals, etc. could be eliminated using ANPs. This approach has been used to generate potable quality drinking water, green alternatives to hazardous procedures, a cost-effective and eco-friendly approach to recycle wastewater. The emission of nitrogenous gases due to fossil fuel combustion can also be mitigated using nanotechnology.

6.4.5 Agricultural

Nanoparticles have been widely used to enhance the growth (nano-supplementation through sustained release), genetically modify/transform, eliminate crop pests (nano-pesticides) and fertilize the soils (nano-fertilizers) for increased agricultural production and disease prevention. ANPs can help quantify the levels of bacteria, viruses, fungi and mycotoxins in plants. Carbon nanotubes-based platforms for gene/nucleotide delivery can help the expression of desirable genes. Also, the use of quantum dots for phytoplasma analysis and DNA probing for molecular analysis of genetics. Nanoparticle-based biosensors could be used to monitor mineral/contaminants profile in soils (Sanzari et al., 2019).

6.5 Conclusion and Future Outlook

Nanoscience is a multi-dimensional field, integrating many sciences to promote the well-being of humans, animal and the environment. It has brought forward intelligent solutions towards prophylaxis, theranostics, diagnostics and therapeutics of many significant diseases. Moreover, ANPs have further accelerated the progress of this field due to unique properties and better-suited applications. Manipulation of nano-scale materials inside biological systems, however, may raise some valid concerns. There's a dearth of data regarding how biological systems assimilate and distribute nano-sized materials, without casting carryover effects on the subject and associated ecologies.

References

Al-Halifa, S., Gauthier, L., Arpin, D., Bourgault, S., and Archambault, D. (2019). Nanoparticle-based vaccines against respiratory viruses. *Frontiers in Immunology*, **10**, 22.

Andersen, S. K. H., Pors, A., and Bozhevolnyi, S. I. (2015). Gold photoluminescence wavelength and polarization engineering. *ACS Photonics*, **2**, 432.

Arfat, Y. A., Ejaz, M., Jacob, H., and Ahmed, J. (2017). Deciphering the potential of guar gum/Ag-Cu nanocomposite films as an active food packaging material, *Carbohydrate Polymers*, **157**, 65–71.

Branda, F., Silvestri, B., Costantini, A., and Luciani, G. (2015). Effect of exposure to growth media on size and surface charge of silica based Stöber nanoparticles: a DLS and ζ-potential study. *Journal of Sol-Gel Science and Technology*, **73**, 54–61.

Buso-Rogero, C., Brimaud, S., Solla-Gullon, J., Vidal-Iglesias, F. J., Herrero, E., Behm, R. J., and Feliu, J. M. (2016). Ethanol oxidation on shape-controlled platinum nanoparticles at different pHs: a combined in situ IR spectroscopy and online mass spectrometry study. *Journal of Electroanalytical Chemistry*, **763**, 116.

Chang, C.-C., Wu, H.-L., Kuo, C.-H., and Huang, M. H. (2008). Hydrothermal synthesis of monodispersed octahedral gold nanocrystals with five different size ranges and their self-assembled structures. *Chemistry of Materials*, **20**, 7570–7574.

Costa-Fernández, J. M., Menéndez-Miranda, M., Bouzas-Ramos, D., Encinar, J. R., and Sanz-Medel, A. (2016). Mass spectrometry for the characterization and quantification of engineered inorganic nanoparticles. *Trends in Analytical Chemistry*, **84**, 139–148.

Cui, C., He, M., Chen, B., and Hu, B. (2014). Chitosan modified magnetic nanoparticles based solid phase extraction combined with ICP-OES for the speciation of Cr(III) and Cr(VI). *Analytical Methods*, **6**, 8577–8583.

Demirel, M., and Aksakal, B. (2016). The synthesis of eggshell-derived nano- and microscale hydroxyapatite bioceramic bone grafts. *Journal of Sol Gel Science and Technology*, **78**, 126–134.

Deshmukh, S. P., Patil, S. M., Mullani, S. B., and Delekar, S. D. (2019). Silver nanoparticles as an effective disinfectant: a review. *Materials Science & Engineering C*, **97**, 954–965.

Dhandapani, R. K., Gurusamy, D., Howell, J. L., et al. (2019). Development of CS-TPP-dsRNA nanoparticles to enhance RNAi efficiency in the yellow fever mosquito, Aedes aegypti. *Scientific Reports*, **9**, 8775.

Espinosa, D., Mendy, J., Manayani, D., Vang, L., Wang, C., Richard, T., et al. (2018). Passive transfer of immune sera induced by a Zika virus-like particle vaccine protects AG129 mice against lethal Zika virus challenge. *EBioMedicine*, **27**, 61–70.

Goel, S., Ni, D., and Cai, W. (2017). Harnessing the power of nanotechnology for enhanced radiation therapy. *ACS Nano*, **11**, 5233–5237.

Gu, C., and Zhang, T.-Y. (2008). Electrochemical synthesis of silver polyhedrons and dendritic films with superhydrophobic surfaces. *Langmuir*, **24**, 12010–12016.

He, S., Kyaw, Y. M. E., Tan, E. K. M., Bekale, L., Kang, M. W. C., Kim, S. S. Y., Tan, I., Lam, K. P., and Kah, J. C. Y. (2018). Quantitative and label-free detection of protein kinase A activity based on surface-enhanced Raman spectroscopy with gold nanostars. *Analytical Chemistry*, **90**, 6071–6080.

Kang, W. C. M., Wong, Y. L., Piotrowski, M., Teo, W. P. J., He, S., and Kah, J. C. (2017). A gold nanostar-based SERS platform for point-of-care diagnostics of carbapenemase-producing enterobacteriacae. *Proceedings of SPIE*, **10340**, 1034011.

Kestens, V., Roebben, G., Herrmann, J., Jamting, A., Coleman, V., Minelli, C., Clifford, C., de Temmerman, P.-J., Mast, J., Junjie, L., Babick, F., Colfen, H., and Emons, H. (2016). Challenges in the size analysis of a silica nanoparticle mixture as candidate certified reference material. *Journal of Nanoparticle Research*, **18**, 171.

Kohout, C., Santi, C., and Polito, L. (2018). Anisotropic gold nanoparticles in biomedical applications. *International Journal of Molecular Sciences*, **19**, 3385.

Konkol, D., and Wojnarowski, K. (2018). The use of nanominerals in animal nutrition as a way to improve the composition and quality of animal products. *Journal of Chemistry*, **2018**, 5927058.

Leopold, K., Brandta, A., and Tarren, H. (2017). Sizing gold nanoparticles using graphite furnace atomic absorption spectrometry. *Journal of Analytical Atomic Spectrometry*. **32**, 723–730.

Li, C., Shuford, K. L., Chen, M., Lee, E. J., and Cho, S. O. (2008). A facile polyol route to uniform gold octahedra with tailorable size and their optical properties. *ACS Nano*, **2**, 1760–1769.

Li, W., Zamani, R., Rivera Gil, P., Pelaz, B., Ibanez, M., Cadavid, D., Shavel, A., Alvarez-Puebla, R. A., Parak, W. J., Arbiol, J., and Cabot, A. (2013). CuTe nanocrystals: shape and size control, plasmonic properties, and use as SERS probes and photothermal agents. *Journal of the American Chemical Society*, **135**, 7098–7101.

Lu, M. M., Bai, J., Shao, D., Qiu, J., Li, M., Zheng, X., Xiao, Y., Wang, Z., Chang, Z. M., Chen, L., Dong, W. F., and Tang, C. B. (2018). Antibacterial and biodegradable tissue nano-adhesives for rapid wound closure. *International Journal of Nanomedicine*, **13**, 5849–5863.

Mahmoud, A. Y. F., Rusin, C. J., and McDermott, M. T. (2020). Gold nanostars as a colloidal substrate for in-solution SERS measurements using a handheld Raman spectrometer. *The Analyst*, **145**(4), 1396–1407.

Marbella, L. E., and Millstone, J. E. (2015). NMR techniques for noble metal nanoparticles. *Chemistry of Materials*, **27**, 2721–2739.

Millstone, J. E., Métraux, G. S., and Mirkin, C. A. (2006). Controlling the edge length of gold nanoprisms via a seed-mediated approach. *Advanced Functional Materials*, **16**, 1209–1214.

Mourdikoudis, S., Pallares, R. M., and Thanh, N. T. K. (2018). Characterization techniques for nanoparticles: comparison and complementarity upon studying nanoparticle properties. *Nanoscale*, **10**, 12871–12934.

Niska, K., Zielinska, E., Radomski, M. W., and Inkielewicz-Stepniak, I. (2018). Metal nanoparticles in dermatology and cosmetology: interactions with human skin cells. *Chemico-Biological Interactions*, **295**, 38–51.

Paramasivam, G., Kayambu, N., Rabel, A. M., Sundramoorthy, A. K., and Sundaramurthy, A. (2017). Anisotropic noble metal nanoparticles: synthesis, surface functionalization and applications in biosensing, bioimaging, drug delivery and theranostics. *Acta Biomaterialia*, **49**, 45–65.

Ray, T. R., Lettiere, B., de Rutte, J., and Pennathur, S. (2015). Quantitative characterization of the colloidal stability of metallic nanoparticles using UV–vis absorbance spectroscopy. *Langmuir*, **31**, 3577–3586.

Samal, A., Sreeprasad, T., and Pradeep, T. (2010). Investigation of the role of $NaBH_4$ in the chemical synthesis of gold nanorods. *Journal of Nanoparticle Research*, **12**, 1777–1786.

Sanzari, I., Leone, A., and Ambrosone, A. (2019). Nanotechnology in plant science: to make a long story short. *Frontiers in Bioengineering and Biotechnology*, **7**, 120.

Spaeth, P., Adhikari, S., Le, L., Jollans, T., Pud, S., Albrecht, W., Bauer, T., Caldarola, M., Kuipers, L., and Orrit, M. (2019). Circular dichroism measurement of single metal nanoparticles using photothermal imaging. *Nano Letters*, **19**(12), 8934–8940.

Srabionyan, V. V., Pryadchenko, V. V., Kurzin, A. A., Belenov, S. V., Avakyan, L. A., Guterman, V. E., and Bugaev, L. A. (2016). Atomic structure of PtCu nanoparticles in PtCu/C catalysts from EXAFS spectroscopy data. *Physics of the Solid State*, **58**, 752–762.

Tian, L., Tan, H. Y., and Vittal, J. J. (2008). Morphology-controlled synthesis of Bi_2S_3 nanomaterials via single- and multiple-source approaches. *Crystal Growth and Design*, **8**, 734–738.

Upadhyay, S., Parekh, K., and Pandey, B. (2016). Influence of crystallite size on the magnetic properties of Fe_3O_4 nanoparticles. *Journal of Alloys and Compounds*, **678**, 478–485.

Ushakov, M. V., Sousa, M. H., Morais, P. C., Kuzmann, E., Semionkin, V. A., and Oshtrakh, M. I. (2020). Effect of iron oxide nanoparticles functionalization by citrate analyzed using Mössbauer spectroscopy. *Hyperfine Interact*, **241**, 27.

van der Pol, E., Coumans, F. A. W., Sturk, A., Nieuwland, R., and van Leeuwen, T. G. (2014). Refractive index determination of nanoparticles in suspension using nanoparticle tracking analysis. *Nano Letters*, **14**, 6195.

Wiley, B., Sun, Y., Mayers, B., and Xia, Y. (2005). Shape-controlled synthesis of metal nanostructures: the case of silver. *Chemistry A European Journal*, **11**, 454–463.

Zandvliet, Harold J.W., and van Houselt, A. (2009). Scanning tunneling spectroscopy. *Annual Review of Analytical Chemistry*, **2**(1), 37–55.

Zsigmondy, R. (1917). *The Chemistry of Colloids*. John Wiley, New York.

Chapter 7

Bioinspired Nanomaterials for Improving Sensing and Imaging Spectroscopy

Janti Qar,[a] Alaa A. A. Aljabali,[b] Tasnim Al-Zanati,[c] Mazhar S. Al Zoubi,[d] Khalid M. Al-Batanyeh,[a] Poonam Negi,[e] Gaurav Gupta,[f] Dinesh M. Pardhi,[g] Kamal Dua,[h] and Murtaza M. Tambuwala[i]

[a]*Department of Biological Sciences, Yarmouk University, Irbid 21163, Jordan*
[b]*Department of Pharmaceutics and Pharmaceutical Technology, Faculty of Pharmacy, Yarmouk University, Irbid 21163, Jordan*
[c]*Health Informatics, International Medical Corps, Amman, Jordan*
[d]*Department of Basic Medical Studies, Yarmouk University, Irbid 21163, Jordan*
[e]*School of Pharmaceutical Sciences, Shoolini University of Biotechnology and Management Sciences, Solan, 173229, India*
[f]*School of Pharmacy, Suresh Gyan Vihar University, Jagatpura, Jaipur 302017, India*
[g]*School of Pharmacy, University of Eastern Finland, FI-70211 Kuopio, Finland*
[h]*Discipline of Pharmacy, Graduate School of Health, University of Technology, Sydney, NSW 2007, Australia*
[i]*SAAD Centre for Pharmacy and Diabetes, School of Pharmacy and Pharmaceutical Science, Ulster University, Coleraine BT52 1SA, UK*
alaaj@yu.edu.jo

Inorganic nanomaterials with outstanding physical and chemical properties can be used as containers for viral-based nanoparticles. The integration of nanomaterials into the internal cavity of viruses has opened several prospects in the field of biology and medicine for imaging applications. Inorganic nanoparticles (NPs)

Nanomaterials for Spectroscopic Applications
Edited by Kaushik Pal

ISBN 978-981-4877-69-5 (Hardcover), 978-1-003-16033-5 (eBook)
www.jennystanford.com

encapsulation in virus particles can achieve nanoprobe-labeling of viruses and preserve the first exterior surface properties of viral capsid simultaneously. Besides, viruses are distinguished from conventional nanocarriers used for drug delivery by their strong durability, their quickly changed surface, and immense variety in shape and size. Besides, several herbal and bacterial viruses (e.g., phages) were studied and added as drug carriers. The summary of this chapter describes the latest developments and applications of bionanomaterials and the processes of drug developments and drug delivery mechanisms and discusses, along with the identified difficulties and advantages, the present state of bionanomaterials of clinical science. Multifunctional, nanoscaled materials for developments in fields of drug delivery, diagnosis, biosensing, and bioengineering have been reviewed in this relation.

7.1 Introduction

For many reasons, the use of protein cages, plant viruses, and peptides as natural nanoparticles (NPs) is beneficial; they are not infectious and non-toxic to humans and deemed safe for intravital imaging and drug delivery. Protein cages consist of repetitive units that self-assemble to form different geometrical arrangements with three surfaces, the interior, the exterior, and the interunit, resulted from the self-assembling of the protein subunits. In the interior of nanocages, therapeutic and diagnostic molecules can be loaded to enhance the selectivity and the targeting potential. At the same time, their outer surfaces were designed to improve biocompatibility and to target diseased cells [1]. In the design and implementation of synthetic nanocarriers, high toxicity and low drug distribution efficiency were limited. For example, cationic fat, micelle, copolymer block, carbohydrates, dendrimers, inorganic NPs. Natural nanocarriers are ideal solutions because they satisfy biocompatibility, biodegradability, water solubility, and cell toxicity [2]. For example, natural nanostructures contain protein cages, such as viruses, ferritin, heat-shock proteins, vaults, and many more that comprise of a cage-like structural arrangement. Monodispersity of naturally occurring nanosystems is not matched by their synthetic counterparts and may be altered with chemical and genetic approaches to impart the desired functionalities [3]. Protein related biological structures such as viruses, and countless natural inspired nanocarriers, have undergone an extensive analysis over the past two decades as they are essential for the penetration of cell membranes. The viral arrangement best reflects the concepts of the natural assembly of proteins. Structural analyzes of viruses reveal a hollow protein architecture containing a specific number of asymmetric subunits that cover the genome and protect it [4–14]. Naturally, viral capsids are designed to target host cells and cell entry. Viruses are stable structures that can resist environmental conditions (internal pH, temperatures, lytic enzymes, etc.) but are sensitive enough to detect signaling triggers, thereby releasing nuclear acids within the targeted cells within the targeted cellular microenvironment. They are the means of intermediation between the various chemical environments [2]. However, protein nanocages pose various challenges. Protein-based NPs lack an inherent natural capacity for transmitting therapeutics substances, lack of specificity toward selective cells, and a low efficiency in cellular uptake, short and limited time of circulation, the ability to induce an immune response. They lack tunable and controllable drug release properties [7, 10, 12]. Besides, due to the structure's low stability and cellular permeability, clinical applications of nanocage are minimal [5].

The temporal modulation of functional groups and protein-binding ligands differentiate between nanocage-based structures and other inorganic NPs. Intelligent nanocarriers with huge potential were created by the combining of bioconjugation and genetic manipulation. For particular covalent chemical modifications, the use of natural amino acids outside or within the capsid is helpful, such as lysine, glutamic acid, aspartic acid, and cysteine in protein nanocages [3–6]. Bionanomaterials provide a fascinating platform technology, which has been enriched by detailed structural, functional studies that have established design guidelines for the concept of the most suitable bionano material for various applications. Once aimed at the molecular imaging endothelium, these organic nanomaterials display advantageous marginality properties while preventing immune clearance.

7.2 Types of Bionanomaterials

7.2.1 Peptides

A peptide is a short sequence of amino acids linked by a peptide bond. The uncomplicated peptides are dipeptides, followed by tripeptides, tetrapeptides. A polypeptide is an extensive, continuous, unbranched sequence of peptides. A peptide consisting of new building blocks, for example, sensors, catalysis, magnetic structures, medicine, tissue processing, ion canals, and diagnostic imaging, with specific uses [14]. Peptides' chemical and physical properties, their capability to form a three-dimensional structure, biocompatible, self-assembly in vitro, and in vivo into various nanoscale structures [2–4, 6]. Self-assembly property of peptides has been studied extensively over the previous few decades, observed that molecules

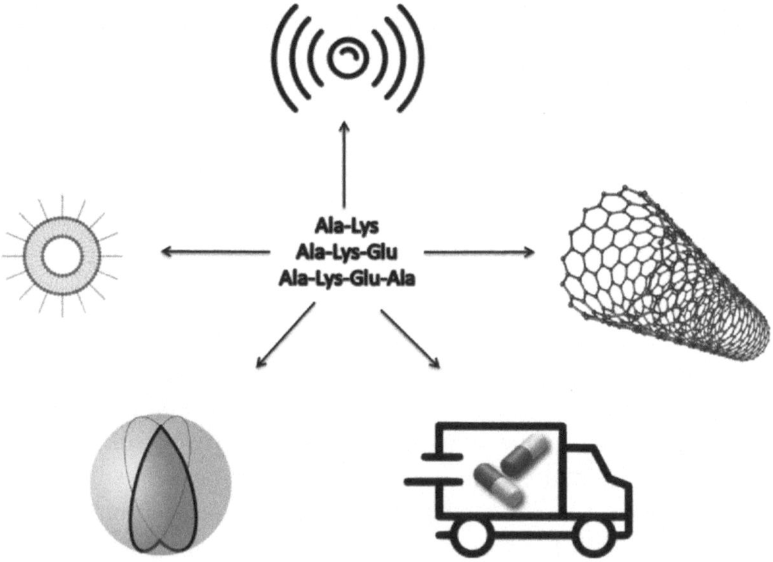

Figure 7.1 Possible self-assembly routes of peptide amphiphile systems. In secondary structures, amphiphilic peptides can be formed by contact and intra-contact, e.g., electrostatic interactions with hydrogen bonding, which can tend to be aggregated into larger spherical assemblies, fibrils, nanotubes, micelles, and drug delivery vesicles.

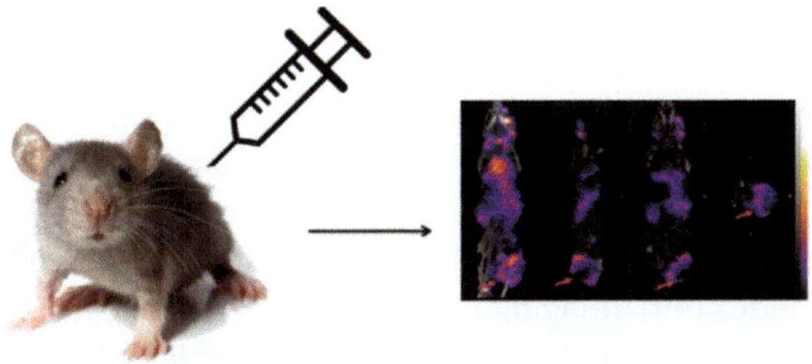

Figure 7.2 A representative model nanoprobe peptide targeting system for in vivo cancer imaging applications.

aggregate via non-covalent interaction, such as; hydrophobic and hydrogen bonding. Peptides self-assembly into many shapes of well-arranged nanostructures with a fundamental advantage in nanomedicine [51]. Peptide self-assembly routes are depicted in Fig. 7.1.

Ultrashort peptides (less than four amino acids) due to their economic efficiency, fast preparation, and they are ideal for large-scale production. Such ultrashort peptides shape a three-dimensional structure, like hydrogels, into various shapes, including the nanotube or the nanofiber. Certain ultrashort peptide hydrogels were used as bioimaging tools in bionanotechnology [19, 35]. The process of using peptides in molecular imaging is depicted in Fig. 7.2.

Visualization of the physiological, structural arrangement of the peptides on the cellular and subcellular level, a contrast agent holds great potential in developing them as therapeutics [12]. Usually, a targeted imaging system comprises a ligand, an imaging moiety, and a linker. These samples are closely linked and easily visible samples. Additionally, desirable medicinal properties are high reliability and consistency in clinical settings, weak immunogenicity, and toxicity. Thanks to the vital role of peptides in cellular functions and intercell interaction, it is exceptional as an imaging method [27, 28]. The FDA suggests peptides with residues of fewer than 100 amino acids that permit rapid blood clearance and non-target tissue clearance [32]. Peptides typically have strong affinities and selectivity and are stable even to nanomolar concentrations. Besides, somatostatin (SST), VIP vasoactive intestinal peptide, Arg Gly asp (RGD), and BBN/GRP have been developing for tumor receptor imaging increased somatostatin has also been created [29, 30].

Peptides usually are liable for chemical modification. Amino acids are subject to cyclization and PEGylation chemical modifications. The introduction of unnatural amino acid is utilized on peptides to enhance their metabolic stabilities and favorable pharmacokinetics [49]. Imaging markers for in vivo imaging systems are individually or partially conjugated with peptides. The subsequent composition shifts include an absolute chance that peptides may lose their binding ability and organic function.

Moreover, the creation of a chemical structure to minimize the contact between the active binding site and non-natural alteration, including an interpolate of the correct linker [11]. The processing of peptide-based imaging samples primarily involves three steps: (1) the detection and targeting of biological receptors; (2) the design and synthesis of peptide analogs for the optimization of biological activity and metabolic behavior; and (3) the chemical interaction between fluorescent group and peptide. Nonetheless, the peptide dependent study is ineffective as visualization and therapeutic efficacy testing as this relies on in-vivo selectivity when peptides may precisely attach to their target, stability, and toxicological effects in-vivo [47].

Targeting peptide molecules can be chosen primarily in three separate ways [34]; (2) chemical synthesis and structure-based computational engineering and design [11, 43]; and (3) screening of peptide libraries [43]. A phage display is a conventional method, widely used because of its benefits such as easy handling and multiple different peptides that can be screened appropriately [18], on the basis of a select peptide which connects with the desired target from a phage library [45]. Every phage clone shows one peptide, and up to 10^9 peptides can be displayed in the entire library. So unbound phages have been washed, and people with required binding behavior are retrieved and eventually restudied through competitive elution. There are several methods of affinity screening to improve the probability of receiving peptides with a strong affinity with the targeted molecules [18]. Mostly these recognition measures were subject to in vivo molecular imaging chemical and biological assessments.

7.2.2 Imaging Applications

In order to be used as in vivo signal sources, peptides and peptide analogs must be labeled with imaging moieties. Any of the widely-used photo moieties contain organic visual imaging eyes, PET and SPECT imagery radionuclides, and MR selecting chelators Gd^{3+} [30]. The working peptide is restricted to the lysine side chains group μ-amino, the arginine side chains group guanidinium, aspartic acid, glutamic acid group carboxy, cysteine thiol, and tyrosine phenol [3, 39]. Ordinarily, peptide modification uses carbodiimide chemistry using (EDC/NHS), maleimide conjugation to thiol groups, and diazonium modification of the phenol on tyrosine [20].

The peptides were selectively used as peptide visualization agents [31]. PET/SPECT images focused on radionuclides may be used to classify the imaging agents at the distribution of micromolar to picomolar. Radionuclides will be marked on peptides for selectivity purposes in order to facilitate peptide dependent tests for PET/SPECT imaging. Of the application of PET imaging techniques, 99 mTc, ^{123}I, and ^{111}I in were produced of peptide labeling of SPECT imaging and ^{18}F, ^{64}Cu, and ^{68}Ga isotopes [15]. The binding activity of a single peptide increases considerably via a scalable reaction.

NPs have unique physical and chemical properties themselves. QDs are more transparent than standard transparent colors and are built for visual imagery higher photostability than typical fluorescent coloration [50]. Peptides will subsequently be bound to nanomaterial in three ways: covalent conjugation, electrostatic adsorption, and specific noncovalent binding [42].

Several research papers suggest that the conjugation of different peptides to the surface of nanoparticles is a critical way to create multimodal imaging agents directly for targets. The formation of chemical connections will influence the target capability [14, 46]. Electrostatic adsorption is unfussy for the application of the similarity of oppositely charged nanomaterials to NP bioconjugation. As a guide to assembling negatively charged 11-mercaptoundecanoic acid-modified AUNPs, Wagner et al. also used positive charge peptide VW05 (H_2N-AbzLKELERKLKELERKELERKL-COOH). The pH of a solution can control the assembly/disassembly of hybrid materials, as NP-peptide interaction is dependent on electrostatic forces. Nevertheless, it is necessary to remember that the peptide can be readily separated from the complex through pH disturbance, and the usage of such in vivo picture probes is forbidden through ionic strength.

Strong noncovalent interaction between peptides and nanomaterials must be used to preserve peptides' bioactivity. For instance, stable kinds of complexes are produced between amino acids, especially histidine, donated by an electron to transition metals such as Ni(II) and Zn(II). Choi et al. used peptide that has a three-fold tandem repeat conjugated to ZnO binding motive (RPHRK) and a versatile linker (GGDA) [16]. In the end, the peptide conjugation into NPs relies on peptide selectivity (e.g. small molecular weight and relative proportional instability). Besides, there are specific requirements to be set down for optimum construction: (1) decrease the hydrodynamic size of target-NP conjugate products; (2) monitor the direction of the peptide show and peptide separation; (3) chemical diversity elimination during preparation [21].

7.2.3 Protein Cages

Proteins such as antibodies, enzymes, and intermediate proteins have long amino acid sequences and particular roles and diverse three-dimensional (3D) structures. It is considered as a smart building block in nanotechnology due to its unique characteristics. Such properties that protein cages possess; selectivity toward their target, self-assemble from repetitive building blocks, as shown in Fig. 7.3. They are also highly monodisperse (structure and composition), and biocompatible [17]. Non-covalent associations such as hydrogen links, electrostatic, hydrophobic, and van der Waals mediate self-assembly observed by protein cages. Despite the comparatively weak ties, nonetheless, they are together responsible for the richness and stability observed in many biological systems [37]. Furthermore, proteins are amphipathic macromolecules containing both nonpolar and polar amino acids. The solvent is located in the hydrophilic regions, and the hydrophobic regions are oriented towards the inner surfaces of the protein, creating a semi-closed ring. The 20 naturally occurring amino acids, used as protein

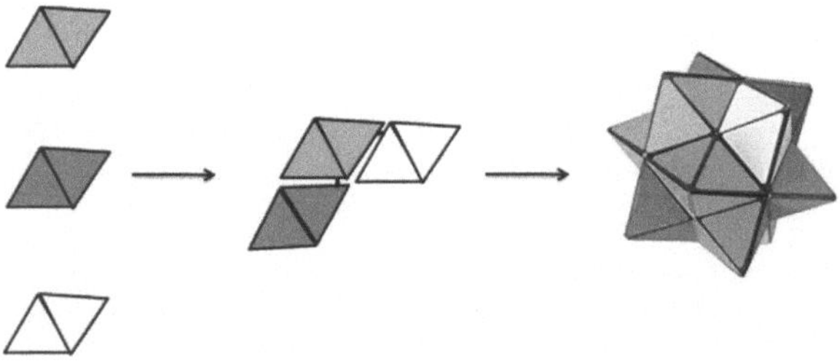

Figure 7.3 Schematic drawings of the protein assemblies from simple repetitive units to generate high rachial complex structures.

bases, have special chemical features that facilitate complicated interactions such as macromolecular identification and precise catalytic activity of enzymes. These characteristics make proteins especially appealing to biosensor technologies because they can detect and react in vivo, disease-related analytes. Besides, because of their well-defined geometries, physiological state assembly, and manipulation through protein engineering approaches, the use of protein to biomedical applications is especially relevant [54].

Protein-based nanomaterials have fewer biological barriers, such as opsonization, immune clearance, circulatory systems, and safety in clinical usage. Based on its unique properties and complex 3D structure, it has been used widely in medicine, biological engineering, etc. There are two major types of protein-based nanomaterials; nanoparticles and nanofibers, which have been used in drug delivery and cell recognition [48].

The accessibility of genomic data through bioinformatic tools made the screening of protein-based NPs much easier. However, the limited amount of proteins like non-viral PNPs such as ferritin, Hsp, vault proteins, and virus-like particulates have been reported [4–7]. The following table summarizes the role, structure, and development of protein-based nanoparticles (PNPs).

7.2.4 Virus-Like Particles (VLPs)

Virus-like particles (VLPs) are multi-protein complexes that emulate viral conformation and organization. VLPs consist of one or more structural units of viruses that are installed into a spherical or cylindrical structure. These can be designed to display highly

Table 7.1 Summarizes some of the viral-based nanoparticles that are currently under investigation for various pharmaceutical applications

	PNP	Organism	Biological function	Size, diameter	Geometry
Viruses and virus-like particles	CCMV	Cowpea chlorotic mottle virus capsid protein	Plant virus	28 nm	Icosahedral
	CPMV	Cowpea mosaic virus	Plant virus	28 nm	Pseudo icosahedral
	HBc	Hepatitis B virus capsid protein	Human virus	30–34 nm	Icosahedral
	MS2	Enterobacteriaceae	Bacteriophage	26 nm	Icosahedral
	P22	*Salmonella typhimurium*	Bacteriophage	60 nm	Icosahedral
	Qβ	*E. Coli*	Bacteriophage	28 nm	Icosahedral
Nonviral PNPs	Dps mini-ferritin	Archaea, bacteria	Oxidative and starvation responses	9 nm	Tetrahedral
	E2	*Bacillus stearothermophilus*	The core of the pyruvate dehydrogenase multienzyme complex	24 nm	Dodecahedral
	Encapsulin	Archaea, bacteria	Oxidative stress response	20–40 nm	Icosahedral
	Ferritin maxi-ferritin	Archaea, bacteria, eukarya	Iron storage	12 nm	Octahedral
	Hsp	Archaea, bacteria, eukarya	Chaperone	12 nm	Octahedral
	Lumazine synthase	Archaea, bacteria, eukarya	Mediates the biosynthesis of riboflavin	15.4 nm	Icosahedral
	Vault protein	Eukarya	Signaling and immune responses	40 nm	39-fold dihedral

Table 7.2 Classification of different virus-like particles based on the number of viral surface proteins and the existence of lipid envelopes or not

VLPs	Characteristic	Example
Nonenveloped VLPs	Unenveloped single-layered VLPs consisting of one protein	Hepatitis B core antigen VLPs
	Single-layered nonenveloped VLPs assembled by two proteins	SARS coronavirus VLPs
	Two-layered nonenveloped VLPs assembled by two proteins	Papillomavirus L1 and L2 VLPs
	Two-layered nonenveloped VLPs assembled by multiple proteins	FMDV-VLP
	Triple protein composed of multi-layered VLPs	Bluetongue virus
Enveloped VLPS	One protein composed of single-layered VLPs	Matrix protein 2 (M2e) VLPs influenza virus ectodomain
	One protein composed of single-layered VLPs	Hantaviruses VLPs
	One protein composed of single-layered VLPs	Hepatitis C VLPs
	Two proteins composed of multi-layered VLPs	SARS coronavirus VLPs

ordered heterologous epitopes to maximize the antigenic complexity for vaccine development [8]. Similar to viruses, different epitopes are shown on VLPs can also be prepared, stored, and introduced to strong immune responses by antigen-presenting cells [6]. The created particles are self-assembled from at minimum one viral protein produced by a stable system of host expression, including mammalian cells, insect cells, yeasts, bacterium, and non-host systems [44].

VLPs are self-assembled into complex, supramolecular systems with well-described surfaces within and outside. In protein cage attachment and disassembly, and on viral stability, the interface between subunits is crucial [1, 2]. Sphere protein frames originating from tightly structured and commonly utilized plants and bacteria in nanomedicine [4].

7.2.4.1 Cowpea chlorotic mottle virus (CCMV)

CCMV is arranged into a 28 nm outer diameter and an 18 nm inner diameter, icosahedral plant virus. CCMV has a triple genome and requires three morphologically corresponding to bipartite particles packaging the entire genome, RNA1 and RNA2 each is packaged separately, and RNA 3 is co-packaged with a fourth RNA [21].

7.2.4.2 Cowpea mosaic virus capsid protein (CPMV)

CPMV is a plant virus 28 nm with icosahedral. The genome comprises two distinct RNA sequences in a positive sense. The capsid consists of 60 copies of a tiny (S) protein sub-unit and 60 copies of a large (L) subunit that occupies two different regions. CPMV device constraints, however, involved constructing a CPMV cage out of separate blocks of subunits and rendering empty virus-like particles without the genome RNA infections impossible for reproduction [40, 41].

7.2.4.3 Enterobacteriaceae (MS2)

Bacteriophages MS2, 26 nm, icosahedral, and the genome is a positive-sense RNA coding for four proteins, the protein coating main, the protein maturing protein considered to be protein A that mediates bacterial pili phage connection, replication and the protein lyse [22].

7.2.4.4 Q β

Bacteriophage Qβ has an icosahedral diameter of 25 nm, and a single positive-sense RNA strand comprises of the genome. As MS2, it supports three separate dimer confirmation forms. Disulfides bind-CP to the hexamer, which results in an extremely stable framework The dissociation temperature falls from 85–100°C to ~40°C after the disulfide links have been extracted [13, 26].

7.2.4.5 Brome mosaic virus (BMV)

BMV is around 28 nm, positive-sense RNA icosahedral virus. BMV is somewhat close to CCMV, which shares a sequence identity of 70% amino acids. Much like CCMV, the genome comprises of three distinct RNA sequences in three distinct virions. The protein encoded by RNA1 and RNA2 is 1a and 2a, essential to replicate genomic and to transcribe a fourth RNA minus RNA3 line. RNA3 also encodes a protein essential for cell-to-cell movement, which directs the host infection. CCMV expressed that the structural transformation depends on the solution's pH and ionic power. The initial condition at higher PH and ionic power at naturally occurring pH, the conformational changes stabilized by Mg^{+2}. At higher salt and neutral pH, the capsid disassembles and the RNA genome precipitates.

Bacteriophage P22 is a 60 nm double-stranded DNA phage that infects *Salmonella typhimurium*, icosahedral. P22 developed into a robust platform based on self-assembles in vitro, including purification protein coating [23].

7.2.5 Non-viral Protein Cages

7.2.5.1 Enterobacteria phage (E2)

E2 is derived from the pyruvate-dehydrogenase-multienzyme complex's E2 central domain (dihydrolipoamide acetyltransferase). The protein consists of 24 subunits in *E. coli* or 60 subunits *Geobacillus stearothermophilus* that self-assembly into a hollow system of a cubic central core, or an icosahedron with 12 pores of five nm openings [33]. Pyruvate decarboxylase (E1), dihydrolipoamide acetyltransferase (E2), and dihydrolipoamide dehydrogenase are the enzymes that are unique to pyruvate dehydrogenase (E3). Because the E1 and E3 are not separated, the latter shapes the systemic nucleus of inherent durability in the presence of the E2, which enables long-term survival under harsh environments. The modulation of the exterior surface with functional ligands does not inhibit the assembly of the core domain E2, as is the case with other nanocages, including viral capsids. This function is the same as its natural condition in which the E2 domain is linked to two distinct folding protein domains on its top [33]. E2 protein nanocage can be genetically or chemically engineered in a similar way to other natural protein (viruses) to enable multiple functions such as drug loading and precise delivery. Ren et al. have shown that

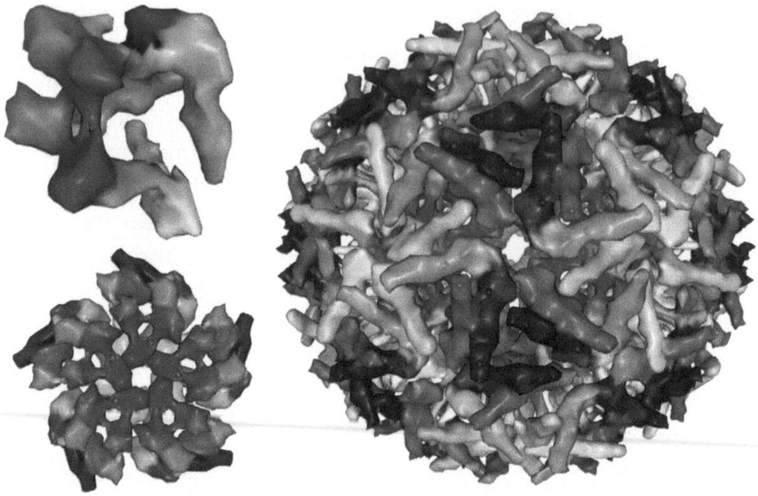

Figure 7.4 PyMOL illustrations of the monomeric unit of E2 protein to form pentameric assemblies with the fully assembled E2 protein structure. Images were generated from protein databank 1Z8Y.

the E2 protein can be balanced for loading drug molecules on both external and internal surfaces, thus exhibiting functional epitopes simultaneously on the protein surface [38].

7.2.5.2 Encapsulin

The first bacterial organelles to conduct these roles were carboxysomes. As defined in 1973, carboxysomes have a cross-sectional length of 100–150 nm, and a protein coat of 6–10 specific protein. They include RuBisCo and carbohydrase enzymes for autotrophic CO_2 fixation in bacteria. A specific carboxysomes sub-unit consists of 90 amino acids, with α/β-fold. In 1994, the smallest carboxysomes, a protein-based organelle of the hyperthermophilic bacterium *Thermotoga Maritima*, were identified as "linocins," measuring 20 to 24 nm in diameter. These forms in protein have been renamed encapsulins [25].

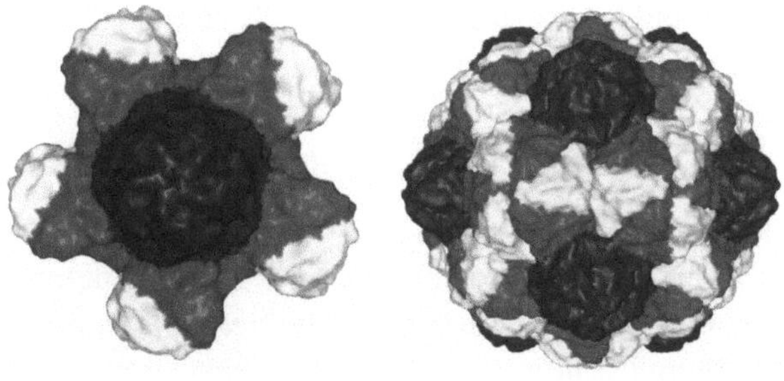

Figure 7.5 PyMOL protein assembly of encapsulins with the pentameric unit presentation. Images regenerate from protein data bank ID 4PT2.

7.2.5.3 Ferritin (maxi-ferritin)

Ferritin consists of the same or homologous subunits of 24 monomers, which shape hollow spheres of octahedral symmetry. The hydrous ferric oxide mineral core with fluctuating phosphate has a cavity defense of up to 4500 Fe atoms.

A four-helix (A, B, C, and De helixes) set with a small fifth helix (E helix) at the c-terminus composite maxi-ferritin. Every subunit communicates with six contiguous monomers in the octahedral cage system with three kinds of symmetry dependent interfaces [33].

7.2.5.4 Dps 4 (mini-ferritin)

It is mini-ferritin, oligomers, and form 12-meric nanocage by self-assembly, composed of identical monomers that assemble into a hollow, tetrahedrally symmetric. Moreover, it has a lower iron storage capacity. Dps isolated from *E. coli* in starvation conditions. Furthermore, it can protect the bacteria from the oxidative damage and form a quite stable complex with DNA without apparent sequence specificity. Dps structural analog to maxi-ferritins based on its crystal structure (hollow protein with 32 tetrahedral core group symmetry).

In addition, Dps's primary role is to detox iron and to weigh its diameter by up to 9 nm. It has an iron center up to iron ions—totaling 4.5 nm. It has a central cavity of roughly 4.5 nm. Dps folds into a quadruple bundle (A, B, C, and D helices), and the loop produces a small helix almost orthogonally to the axis and is found outside the protein cage installed.

7.2.5.5 Lumazine synthase

Lumazine synthase consists of 60 identical icosahedral (T = 1 geometrical arrangement) subunits with 15.4 nm outer and 9 nm inner diameters, each isolated from *Hyperthermophila Aquifex aeolicus*. Originally, lumazine synthase catalyzes the last stage of riboflavin biosynthesis in the cell. Outside the cell, the architecture of its hollow region is used as a blueprint for cargo protein encapsulation, including green fluorescent proteins. As a modeling nanoscale for T1, it conjugates the capacity of the contrast agent MR in-vivo with the magnet field power of 7 T, as been evaluated. Lumazine synthase is formed by 60 equal subunits of icosahedral capsid (T = 1 state) structures and diameters of 15.4 nm and 9 nm in each of the exterior and internal designs [52].

7.2.5.6 Vault protein

Naturally occurring nanoparticles build from a protein shell that is made up of multiple copies of major vault protein (MVP), involved in nucleo-cytoplasmic transport. More than 70% of the vault mass, consisting of two extra protectins, is produced from the 100 kDa MVP, the polyvalent vault (ADP-ribose) (vPARP) polymerase, and the telomerase-associated protein (TEP1), besides several short untranslated RNAs (vRNA). Moreover; Vaults have rolled in many functions in the regulation of cellular processes such as; transportation, signal transmissions, and immune response. Vaults are promising as a delivery system due to its stability, targeted, biocompatibility, nonimmunogenic, monodispersity, and large internal volume, which are able to encapsulate hundreds of thousands of molecules or complete proteins [55].

7.2.6 Natural Polymers

These molecules occur in nature and can be extracted, such as; polysaccharides, proteins, and polyesters derived from both plant and animal. Become a significant attraction in biomedical applications due to its properties, including; biocompatible, mechanical properties, and biodegradable [9].

7.3 Conclusions, Outlook and Future Scopes

Various protein-based nanomaterials have been developed in biological environments to support essential processes. Such robust molecular systems are typically self-assembled and motivate several artificial proteins nanocages. Such nanomaterials have recently been recognized to build synthetic nanosystems. Protein-based NPs may also be reconstructed with innumerable reports describing the binding to their inner and outer surfaces of a wide variety of functional small molecules and inorganic/organic molecules. Different techniques have been developed for genetic modification and/or bioconjugation to build protein-based nanosystems for drug delivery, polymeric nanostructures, vaccines and biocatalysis applications. It is a question of self-assembly, disassembly, and reassembly of protein cages, along with understanding factors influencing their surface pores. Thus, it is necessary for controlled load/release and movement of cargo molecules within and out of the protein-based cavity to understand underlying PNP self-assembling structures and the pores opening and closing processes. This understanding would facilitate the integration of protein-based systems through drug delivery applications, which improve its therapeutic output and eliminate adverse and the release of a medication on precise cells. Bionanomaterials in healthcare have been more and more involved to address a set of problems connected with pharmacological intervention substances in recent advancements of nanotechnology. For diagnostic and therapeutic uses, engineered nanomaterials with distinct physical and chemical characteristics were examined. By conjunction with NPs, therapeutic agents suppress systemic toxicities, maximize bioavailability, extend the half-life of circulation, decrease immunogenicity and improve drug delivery. In guided imaging and treatment, multifunctional nanoparticles have proven very successful. While many drug delivery mechanisms based on nanoparticles have been produced and/or undergoing clinical trials and/or medicinal examination, a few of such drug carriers are used in the medical sector. This is because of certain health conditions such as high load capacities of drugs and a short circulatory half-life. As a consequence of the discovery of related targets and the development of new materials, smart multifunctional NPs are the most successful candidates as drug carriers in nanomedicine. These molecules would facilitate the advancement of individual drug delivery technologies by increasing patients' quality of life.

References

1. Aljabali, A. A., Barclay, J. E., Butt, J. N., Lomonossoff, G. P., and Evans, D. J. (2010). Redox-active ferrocene-modified Cowpea mosaic virus nanoparticles. *Dalton Transactions*, **39**(32), 7569–7574, doi: 10.1039/c0dt00495b.
2. Aljabali, A. A., Barclay, J. E., Lomonossoff, G. P., and Evans, D. J. (2010). Virus templated metallic nanoparticles. *Nanoscale*, **2**(12), 2596–2600, doi: 10.1039/c0nr00525h.
3. Aljabali, A. A., Barclay, J. E., Steinmetz, N. F., Lomonossoff, G. P., and Evans, D. J. (2012). Controlled immobilisation of active enzymes on the Cowpea mosaic virus capsid. *Nanoscale*, **4**(18), 5640–5645, doi: 10.1039/c2nr31485a.
4. Aljabali, A. A., and Evans, D. J. (2014). Polyelectrolyte-modified cowpea mosaic virus for the synthesis of gold nanoparticles. *Methods in Molecular Biology*, **1108**, 97–103, doi: 10.1007/978-1-62703-751-8_7.
5. Aljabali, A. A., and Evans, D. J. (2014). Templated mineralization by charge-modified cowpea mosaic virus. *Methods in Molecular Biology*, **1108**, 89–95, doi: 10.1007/978-1-62703-751-8_6.
6. Aljabali, A. A., Sainsbury, F., Lomonossoff, G. P., and Evans, D. J. (2010). Cowpea mosaic virus unmodified empty viruslike particles loaded with metal and metal oxide. *Small*, **6**(7), 818–821, doi: 10.1002/smll.200902135.
7. Aljabali, A. A., Shah, S. N., Evans-Gowing, R., Lomonossoff, G. P., and Evans, D. J. (2011). Chemically-coupled-peptide-promoted virus nanoparticle templated mineralization. *Integrative Biology*, **3**(2), 119–125, doi: 10.1039/c0ib00056f.
8. Aljabali, A. A., Shukla, S., Lomonossoff, G. P., Steinmetz, N. F., and Evans, D. J. (2013). CPMV-DOX delivers. *Molecular Pharmaceutics*, **10**(1), 3–10, doi: 10.1021/mp3002057.

9. Aljabali, A. A. A., Bakshi, H. A., Hakkim, F. L., Haggag, Y. A., Al-Batanyeh, K. M., Al Zoubi, M. S., et al. (2020). Albumin nano-encapsulation of piceatannol enhances its anticancer potential in colon cancer via downregulation of nuclear p65 and HIF-1α. *Cancers (Basel)*, **12**(1), 113, doi: 10.3390/cancers12010113.
10. Aljabali, A. A. A., and Evans, D. J. (2018). Internal deposition of cobalt metal and iron oxide within CPMV eVLPs. *Methods in Molecular Biology*, **1776**, 189–201, doi: 10.1007/978-1-4939-7808-3_12.
11. Andersson, L., Blomberg, L., Flegel, M., Lepsa, L., Nilsson, B., and Verlander, M. (2000). Large-scale synthesis of peptides. *Peptide Science*, **55**(3), 227–250.
12. Arcangeli, C., Circelli, P., Donini, M., Aljabali, A. A., Benvenuto, E., Lomonossoff, G. P., and Marusic, C. (2014). Structure-based design and experimental engineering of a plant virus nanoparticle for the presentation of immunogenic epitopes and as a drug carrier. *Journal of Biomolecular Structure and Dynamics*, **32**(4), 630–647, doi: 10.1080/07391102.2013.785920.
13. Bundy, B. C., and Swartz, J. R. (2011). Efficient disulfide bond formation in virus-like particles. *Journal of Biotechnology*, **154**(4), 230–239.
14. Cai, W., Shin, D.-W., Chen, K., Gheysens, O., Cao, Q., Wang, S. X., et al. (2006). Peptide-labeled near-infrared quantum dots for imaging tumor vasculature in living subjects. *Nano Letters*, **6**(4), 669–676.
15. Chatalic, K. L., Kwekkeboom, D. J., and de Jong, M. (2015). Radiopeptides for imaging and therapy: a radiant future. *Journal of Nuclear Medicine*, **56**(12), 1809–1812.
16. Cho, N.-H., Cheong, T.-C., Min, J. H., Wu, J. H., Lee, S. J., Kim, D., et al. (2011). A multifunctional core–shell nanoparticle for dendritic cell-based cancer immunotherapy. *Nature Nanotechnology*, **6**(10), 675.
17. DeFrates, K., Markiewicz, T., Gallo, P., Rack, A., Weyhmiller, A., Jarmusik, B., and Hu, X. (2018). Protein polymer-based nanoparticles: fabrication and medical applications. *International Journal of Molecular Sciences*, **19**(6), 1717.
18. Deutscher, S. L. (2010). Phage display in molecular imaging and diagnosis of cancer. *Chemical Reviews*, **110**(5), 3196–3211.
19. Di Fabrizio, E., Schlücker, S., Wenger, J., Regmi, R., Rigneault, H., Calafiore, G., et al. (2016). Roadmap on biosensing and photonics with advanced nano-optical methods. *Journal of Optics*, **18**(6), 063003.
20. Erathodiyil, N., and Ying, J. Y. (2011). Functionalization of inorganic nanoparticles for bioimaging applications. *Accounts of Chemical Research*, **44**(10), 925–935.

21. Field, L. D., Delehanty, J. B., Chen, Y., and Medintz, I. L. (2015). Peptides for specifically targeting nanoparticles to cellular organelles: quo vadis? *Accounts of Chemical Research*, **48**(5), 1380–1390.
22. Fu, Y., and Li, J. (2016). A novel delivery platform based on Bacteriophage MS2 virus-like particles. *Virus Research*, **211**, 9–16.
23. Fuller, M. T., and King, J. (1982). Assembly in vitro of bacteriophage P22 procapsids from purified coat and scaffolding subunits. *Journal of Molecular Biology*, **156**(3), 633–665.
24. Gao, X., and Matsui, H. (2005). Peptide-based nanotubes and their applications in bionanotechnology. *Advanced Materials*, **17**(17), 2037–2050.
25. Giessen, T. W., and Silver, P. A. (2017). Widespread distribution of encapsulin nanocompartments reveals functional diversity. *Nature Microbiology*, **2**(6), 1–11.
26. Golmohammadi, R., Fridborg, K., Bundule, M., Valegård, K., and Liljas, L. (1996). The crystal structure of bacteriophage Qβ at 3.5 Å resolution. *Structure*, **4**(5), 543–554.
27. Hoelder, S., Clarke, P. A., and Workman, P. (2012). Discovery of small molecule cancer drugs: successes, challenges and opportunities. *Molecular Oncology*, **6**(2), 155–176.
28. Hong, H., Goel, S., Zhang, Y., and Cai, W. (2011). Molecular imaging with nucleic acid aptamers. *Current Medicinal Chemistry*, **18**(27), 4195–4205.
29. Igarashi, H., Fujimori, N., Ito, T., Nakamura, T., Oono, T., Nakamura, K., et al. (2011). Vasoactive intestinal peptide (VIP) and VIP receptors-elucidation of structure and function for therapeutic applications. *International Journal of Clinical Medicine*, **2**(4), 500–508, doi: 10.4236/ijcm.2011.24084.
30. Lee, S., Xie, J., and Chen, X. (2010). Peptide-based probes for targeted molecular imaging. *Biochemistry*, **49**(7), 1364–1376.
31. Maecke, H. (2005). Radiolabeled peptides in nuclear oncology: influence of peptide structure and labeling strategy on pharmacology. In Bogdanov, A. A., and Licha, K. (eds.) *Molecular Imaging*. Ernst Schering Research Foundation Workshop, vol. 49. Springer, Berlin, Heidelberg, pp. 43–72.
32. Massoud, T. F., and Gambhir, S. S. (2003). Molecular imaging in living subjects: seeing fundamental biological processes in a new light. *Genes & Development*, **17**(5), 545–580.
33. Milne, J. L., Wu, X., Borgnia, M. J., Lengyel, J. S., Brooks, B. R., Shi, D., et al. (2006). Molecular structure of a 9-MDa icosahedral pyruvate

dehydrogenase subcomplex containing the E2 and E3 enzymes using cryoelectron microscopy. *Journal of Biological Chemistry*, **281**(7), 4364–4370.

34. Nagpal, R., Behare, P., Rana, R., Kumar, A., Kumar, M., Arora, S., et al. (2011). Bioactive peptides derived from milk proteins and their health beneficial potentials: an update. *Food & Function*, **2**(1), 18–27.
35. Ni, M., and Zhuo, S. (2019). Applications of self-assembling ultrashort peptides in bionanotechnology. *RSC Advances*, **9**(2), 844–852.
36. Niemeyer, C. M. (2001). Nanoparticles, proteins, and nucleic acids: biotechnology meets materials science. *Angewandte Chemie International Edition*, **40**(22), 4128–4158.
37. Petrov, A., and Audette, G. F. (2012). Peptide and protein-based nanotubes for nanobiotechnology. *Wiley Interdisciplinary Reviews: Nanomedicine and Nanobiotechnology*, **4**(5), 575–585.
38. Ren, D., Dalmau, M., Randall, A., Shindel, M. M., Baldi, P., and Wang, S. W. (2012). Biomimetic design of protein nanomaterials for hydrophobic molecular transport. *Advanced Functional Materials*, **22**(15), 3170–3180.
39. Roberts, M., Bentley, M., and Harris, J. (2002). Chemistry for peptide and protein PEGylation. *Advanced Drug Delivery Reviews*, **54**(4), 459–476.
40. Sainsbury, F., Saunders, K., Aljabali, A. A., Evans, D. J., and Lomonossoff, G. P. (2011). Peptide-controlled access to the interior surface of empty virus nanoparticles. *ChemBioChem*, **12**(16), 2435–2440, doi: 10.1002/cbic.201100482.
41. Sainsbury, F., Saxena, P., Aljabali, A. A., Saunders, K., Evans, D. J., and Lomonossoff, G. P. (2014). Genetic engineering and characterization of Cowpea mosaic virus empty virus-like particles. *Methods in Molecular Biology*, **1108**, 139–153, doi: 10.1007/978-1-62703-751-8_11.
42. Sapsford, K. E., Algar, W. R., Berti, L., Gemmill, K. B., Casey, B. J., Oh, E., et al. (2013). Functionalizing nanoparticles with biological molecules: developing chemistries that facilitate nanotechnology. *Chemical Reviews*, 113(3), 1904–2074.
43. Sasubilli, R. (2006). Solid-phase synthesis of peptides and peptide mimetics using urethane and backbone amide linker strategies. ProQuest Dissertations, UMI Number: 1438058, University of Missouri, Kansas City, Missouri.
44. Saunders, K., Sainsbury, F., and Lomonossoff, G. P. (2009). Efficient generation of cowpea mosaicvirus empty virus-like particles by the proteolytic processing of precursors in insect cells and plants. *Virology*, **393**(2), 329–337.

45. Smith, G. P. (1985). Filamentous fusion phage: novel expression vectors that display cloned antigens on the virion surface. *Science*, **228**(4705), 1315–1317.
46. Sosabowski, J. K., and Mather, S. J. (2006). Conjugation of DOTA-like chelating agents to peptides and radiolabeling with trivalent metallic isotopes. *Nature Protocols*, **1**(2), 972.
47. Sun, X., Li, Y., Liu, T., Li, Z., Zhang, X., and Chen, X. (2017). Peptide-based imaging agents for cancer detection. *Advanced Drug Delivery Reviews*, **110**, 38–51.
48. Tsai, P.-S., Yang, Y.-M., and Lee, Y.-L. (2006). Fabrication of hydrophobic surfaces by coupling of Langmuir—Blodgett deposition and a self-assembled monolayer. *Langmuir*, **22**(13), 5660–5665.
49. Uhlig, T., Kyprianou, T., Martinelli, F. G., Oppici, C. A., Heiligers, D., Hills, D., et al. (2014). The emergence of peptides in the pharmaceutical business: from exploration to exploitation. *EuPA Open Proteomics*, **4**, 58–69.
50. Wang, B., He, X., Zhang, Z., Zhao, Y., and Feng, W. (2013). Metabolism of nanomaterials in vivo: blood circulation and organ clearance. *Accounts of Chemical Research*, **46**(3), 761–769.
51. Wang, J., Liu, K., Xing, R., and Yan, X. (2016). Peptide self-assembly: thermodynamics and kinetics. *Chemical Society Reviews*, **45**(20), 5589–5604.
52. Wei, Y., Kumar, P., Wahome, N., Mantis, N. J., and Middaugh, C. R. (2018). Biomedical applications of lumazine synthase. *Journal of Pharmaceutical Sciences*, **107**(9), 2283–2296.
53. Wen, A. M., Lee, K. L., and Steinmetz, N. F. (2020). Plant virus-based nanotechnologies. In Norris, P., and Friedersdorf, L. (eds.) *Women in Nanotechnology*. Women in Engineering and Science. Springer, Cham, pp. 57–69.
54. Yoo, J.-W., and Mitragotri, S. (2010). Polymer particles that switch shape in response to a stimulus. *Proceedings of the National Academy of Sciences*, **107**(25), 11205–11210.
55. Yu, K., Yau, Y. H., Sinha, A., Tan, T., Kickhoefer, V. A., Rome, L. H., et al. (2017). Modulation of the vault protein-protein interaction for tuning of molecular release. *Scientific Reports*, **7**(1), 1–10.

Chapter 8

Green Nanomaterials: Synthesis, Properties and Spectroscopic Applications

Muammer Din Arif,[a] Md Asadur Rahman,[b] Md Mahmudul Haque Milu,[c] Abu Bakar Siddik,[d] and Md Enamul Hoque[b]

[a]*Department of Industrial and Production Engineering, Military Institute of Science and Technology (MIST), Mirpur Cantonment, Dhaka 1216, Bangladesh*
[b]*Department of Biomedical Engineering, Military Institute of Science and Technology (MIST), Mirpur Cantonment, Dhaka 1216, Bangladesh*
[c]*Department of Biomedical Engineering, Jashore University of Science and Technology (JUST), Jashore 7408, Bangladesh*
[d]*Department of Biomedical Engineering, Khulna University of Engineering & Technology (KUET), Khulna 9203, Bangladesh*
enamul1973@gmail.com, enamul1973@bme.mist.ac.bd

The negative consequences of run-away industrialization, accumulation of wastes and the need for environmental remediation have sparked the growth in green nanomaterials research. Such nanomaterials are the offspring of multi-disciplinary collaboration involving green chemistry, nanotechnology and materials engineering. Often biomedical and agricultural sciences are enlisted to aid in the preparation. As such, the scope of green nanomaterials and their applications is tremendous. This chapter engages in an

Nanomaterials for Spectroscopic Applications
Edited by Kaushik Pal

ISBN 978-981-4877-69-5 (Hardcover), 978-1-003-16033-5 (eBook)
www.jennystanford.com

exhaustive review of state-of-the-art research into green nanomaterials synthesis, their custom-tailored structure-based properties and the associated applications in optoelectronics, semiconductors, biomedical, therapy, pharmaceutical and water treatment, etc. In the synthesis part, a discussion is included highlighting the mimicry of nature or natural resources, like plants and micro-organisms, for the preparation of environmentally benign nanomaterials, nanostructures, nanocomposites, etc. Then, the properties section gives a detailed elaboration of the primary forces and natural phenomena which govern the behavior and properties of nanoparticles. Indeed, their properties are very different, often exotic, compared to their macroscopic materials. Next, the applications section explains how the unique properties of green nanomaterials are being employed in diverse areas for the benefit of mankind and the elimination of environmental wastes. Finally, the chapter is concluded with a look into the barriers in the up-take of these nanomaterials for full-scale commercialization focusing on environmental and ethical concerns. Additionally, to serve as a guideline, recommendations and future scope of research work involving green nanomaterials are included.

8.1 Introduction

Rapid industrialization and the associated development of engineered materials, with specific properties, have brought about drastic changes to modern civilization. Along with their benefits; however, they have brought attendant problems, i.e. pollution and ethical concerns. The uncontrolled accumulation of polythene and other plastics, greenhouse gas emissions, global warming, lead poisoning, etc. are some of its negative manifestations. Therefore, the 20th and the 21st centuries have witnessed the ever-growing interest and drive towards greener alternatives. It has now become a question of the future survival of the human species.

8.1.1 Green Chemistry and Green Nanotechnology

As a direct consequence of the drive towards a greener future, two fields of science and technology have advanced at a breathtaking pace: green chemistry and green nanotechnology. The fundamental tenet of green chemistry is sustainability. It is an area of modern chemistry that focuses on the design of products and processes that minimize or eliminate the utilization and/or generation of environmentally detrimental substances. It strives to decrease the consumption of non-renewable resources and to develop technologies for preventing or remedying pollution.

The second applied discipline is green nanotechnology. Like green chemistry, it attempts to address pollution and over-consumption of non-renewable resources. It emphasizes the use of nanotechnology to improve sustainability and to develop products that are eco- and health-friendly. As a result, green nanotechnology has championed the replacement of existing products with novel nanoproducts that have low environmental loading and can often remedy causes of pollution, such as effluent treatment and heavy metal removal from contaminated soils. Nanotechnology research increased in the last decade and its current applications are in nanoscale components for devices, advanced sensors and novel materials (Marchiol, 2012).

The merger of the above two disciplines of green chemistry and green nanotechnology with materials science and engineering has led to the emergence of new products: green nanomaterials. As their name suggests, green nanomaterials are either prepared via green processes from green, often renewable sources, with the application, if possible, of renewable energy. Frequently, nanomaterials being used for green applications in industrial, commercial, medical, or household domains are also termed green nanomaterials. The overarching goals of green nanomaterials—namely, more resource-efficiency and inherently safer designs of materials, products, etc.—have been pursued in a wide range of contexts.

The scope of properties and applications of green nanomaterials, is thus, tremendous. Therefore, this book chapter will focus on specific areas of research and development of the green production of nanomaterials, some of their more preferred structure-based properties and the associated applications in disparate disciplines. With this aim, the chapter is divided into three main sections. The first section discusses the recent innovations in green synthesis, with the special focus of the biosynthesis technique. The second section elaborates on the structure-based useful properties of nanomaterials, where the fundamental forces of nature and physical theories have very different outcomes compared to the macroscopic world. The third section highlights some of the recent multi-disciplinary applications of green nanomaterials, especially in optoelectronics and spectroscopy. The chapter is summed up with a look into future barriers and prospects of green nanomaterials with a special focus on environmental and ethical issues.

8.2 Synthesis of Green Nanomaterials

Metallic nanoparticles (MNPs) and other nanostructures have long been produced via chemical or physical means. The chemical synthesis of nanoparticles (NPs) entails three major ingredients: a source, a reducing agent, and a stabilizing or capping agent. While physical methods employ techniques such as planetary ball milling, plasma deposition, pulverization, lithography, etc. (Salunkhe et al., 2014). Most of these physiochemical processes are expensive, often involve the use of hazardous chemicals (as reducing and dispersing agents, solvents, etc.), may pose biological or health risks, high-end equipment and usually require the expenditure of considerable energy. Thus, to meet the growing demands of nanomaterials in a sustainable and eco-friendly way, green synthesis of such materials has gained popularity. These alternative environmentally benign and sustainable methods produce what are known as green nanomaterials (Narayanan and Sakthivel, 2011).

There are many existent and under-development green synthesis methods. They are preferred due to their low environmental loading, zero-emission, reduced energy usage, employment of natural reagents, utilization of non-toxic dispersing substances, harmless reducing media, choice of eco-friendly solvents and precursors, use of natural processes and biocompatibility. Often these green methods mimic nature or enlist living organisms, such as micro-organisms and plants, for green NPs production. Biosynthetic processes result in nanostructures with better-defined dimensions and morphologies compared to many physicochemical synthesis processes. In such green approaches, bacteria, yeasts, fungi, algae, viruses and even biomolecules replace chemical or physical production techniques for industrial-scale synthesis of nanomaterials while playing dual roles as reducing and dispersing agents. Hence, the remaining sub-sections discuss the state-of-the-art in the production of green NPs and nanostructures by using some of these biotic techniques and agents. The final sub-section discusses some of the ethical and environmental concerns arising from the existent or intended production and use of green nanomaterials.

8.2.1 Biogenic Synthesis of Nanoparticles

Current research endeavors have tried to harness the biogenic synthesis pathways of micro-organisms, namely: fungus, yeast, algae, bacteria, virus, etc., for the green synthesis of nanomaterials. For instance, Sharma et al. (2019) compared the bioreduction routes of different algae which have been exploited to produce green NPs from agro wastes. Such efforts recycle waste materials, economize synthesis processes and minimize the requirement of using hazardous chemicals. Another research group, Pathak et al. (2019), reviewed the existing biotechnical methods for synthesizing green nanomaterials from a special type of photosynthetic micro-organism known as cyanobacteria. This species can produce several biomolecules that facilitate the synthesis and stabilization of NPs. On the other hand, Narayanan and Han (2017) discussed viral platforms for NPs production. They elaborated on icosahedral plant viruses, which are sphere-shaped, as an effective viral nanofactory. Also, they highlighted virus nanoparticles (VNPs) and virus-like particles (VLPs), sourced from plants, which can be employed in nano(bio)technological applications. These natural or synthetic platforms are highly versatile in material synthesis and molecular entrapment. Specifically, plant virus capsids have biotemplates that can be enlisted to produce novel green nanostructures. For example, sphere-shaped plant viruses can produce capsid proteins that self-assemble to form icosahedral 3D nanostructures with intricate nanoarchitecture, demonstrating mono-dispersity and symmetry. Furthermore, viral DNA and protein engineering of icosahedral viruses can be harnessed. These viruses have different sizes, interiors, exteriors and interfaces between their coat protein (CP) subunits which could be taken advantage of to synthesize functional green nanomaterials and nanostructures via biomineralization, encapsulation, infusion and controlled self-assembly.

8.2.2 Plant-Based Synthesis of Nanoparticles

The various biotic processes for the synthesis of NPs discussed so far involve micro-organisms (bacteria, fungi, yeasts, algae, viruses, etc.). Although these biogenic and bioreduction processes hold promise, there are some associated risk factors. First, the employment of micro-organisms entails biohazards related to pathogenicity. Second, large cultures and bioreactors are required for sustaining and growing these microbes. Finally, there is the chance of accidental mutation, especially in VNPs, which might alter their function or cause them to multiply at an unpredictable rate. Thus, synthesis of NPs using plants and plant-based extracts (phytogenic fabrications) is a greener way and is preferred by many in the scientific community (Pal et al., 2019; Hamzah, 2020; Danhalilu et al., 2019).

The advantages of using plants or plant-derived substances for the biosynthesis of NPs are:

(1) No requirement for toxic and expensive reagents. Bio-reduction of metal ions via biomolecules from botanical sources (e.g. enzymes, proteins, amino acids, vitamins, polysaccharides, etc.) is eco-friendly, yet chemically complex.
(2) Easy and economic availability of plant-based extracts makes biosynthesis of NPs from plants viable for industrialization.
(3) Diverse species of plants are available for the biosynthesis of NPs.
(4) In general, NPs synthesized from botanical sources are more stable and their rate of synthesis is faster than those from micro-fauna.
(5) Plant-based biogenic processes do not need high pressures, temperatures or clean-rooms; negating the need for sophisticated laboratory facilities as is common for micro-organism based synthesis of NPs.

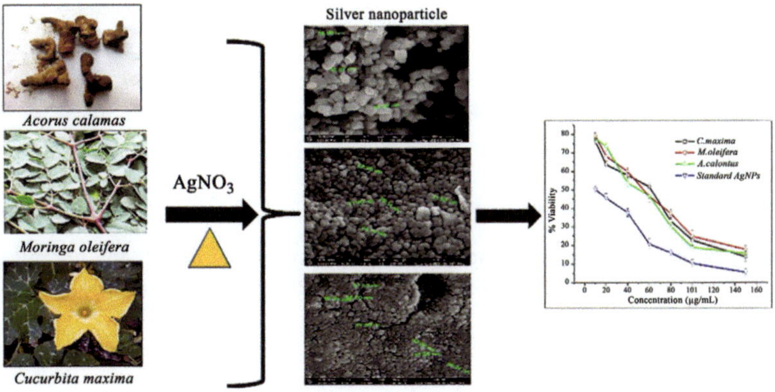

Figure 8.1 The outcome of using different plant species for the biogenic synthesis of AgNPs, reprinted from Nayak et al., 2015, copyright (2015), with permission from Elsevier.

The efficacy of plants and plant extracts in reducing metal ions from solutions has been discussed since the turn of the 20th century, but in the past few decades, plants have been exploited for MNP synthesis. Biomolecules in botanical extracts are very efficient in the bioreduction of metallic cations to NPs in a single-step process without the demand for high energy. Silver (Ag) (Nayak et al., 2015; Fig. 8.1) and gold (Au) NPs have been under special focus in phytogenic fabrication due to the NPs' biomedical, bioimaging, optoelectronic, semiconductor, antibacterial, etc. uses. Others have applied plant-based bioreduction processes for environmental remediation, effluent treatment and toxic heavy metal recovery.

The extant research literature abounds with reports of research investigating the plant-based synthesis of green nanomaterials, nanostructures and nanocomposites. For example, the synthesis of AgNPs by Emeka et al. (2014), AuNPs by Karuppiah et al. (2015), iron (Fe) NPs by Lin et al. (2015), palladium (Pd) NPs by Nasrollahzadeh et al. (2018), metal oxides (i.e. zinc oxide, ZnO) NPs by Sangeetha et al. (2011), titanium dioxide (TiO_2) NPs by Nasrollahzadeh and Sajadi (2015), iron III oxide (Fe_3O_4) NPs by Venkateswarlu et al. (2014), nanocomposites by Venkateswarlu and Yoon (2015), and quantum dots (QDs) by Sachdev and Gopinath (2015).

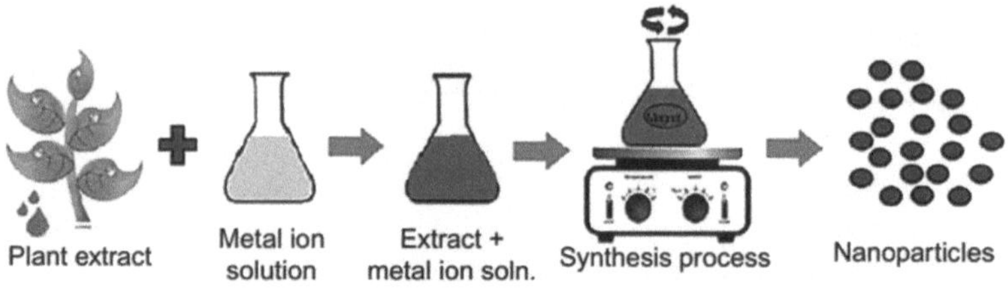

Figure 8.2 Typical approach to the green synthesis of NPs via botanical extracts, reprinted from Shamaila et al., 2016, copyright (2016), with permission from Elsevier.

The above list of works demonstrates that botanical extracts from disparate species can be utilized in NP production. The popular botanical extracts include coffee, tea leaves, vitamins, biodegradable polymers, amino acids, banana juice, wines, saccharides, etc. Figure 8.2 is a schematic representation of the most common approach to synthesizing green NPs from plant extracts. It can be seen from the diagram that these processes do not entail the need for high energy sources or toxic chemicals. Figure 8.3 is a schematic representation showing how plant-based biomolecules aid in the formation of NPs by serving as capping agents.

Indeed, the cornucopia of plant species means that there are still great opportunities for developing greener biosynthesis methods which can be tailored for full-scale industrialization and commer-

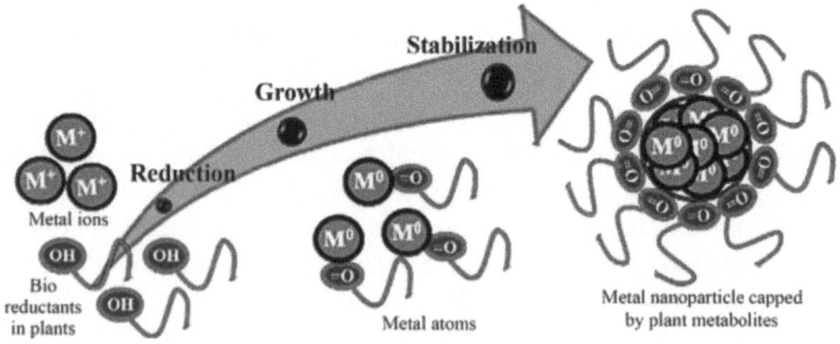

Figure 8.3 Illustration of the mechanism for green synthesis of metal NPs by plant metabolites, reprinted from Shamaila et al., 2016, copyright (2016), with permission from Elsevier.

cialization. Along these lines, Ketabchi et al. (2016) investigated the optimization of parameters for the sonosynthesis of cellulose nanoparticles (CNPs) from the plant fiber kenaf. The researchers used a very popular and effective statistical design of experiment tool known as central composite design (CCD) to optimize the process parameters, making CNP production more streamlined and efficient. They identified the three main phases (i.e. 3 factors) in CNP synthesis: NaOH dosage, the quantity of $NaClO_2$ and sonication time. The response variables for their CCD model were: CNP size, quality and temperature for degradation. The statistical approach found quadratic polynomial and 2 factor interaction (2FI) models for the responses with more than 95% confidence level. Subsequent morphological, thermal analysis, and Fourier transform infrared (FTIR) spectroscopy revealed the progressive removal of non-cellulosic constituents from the product mixture. Besides, TEM analysis was used to measure the fiber's diameter variation from $\sim$170 μm to $\sim$100 nm. From their modeling and critical analysis, the researchers suggested that the optimal parameters were: 0.2 g of NaOH/4 g of fiber (1st stage), 5 ml of $NaClO_2$/4 g of fiber (2nd stage) and 20 min of sonication (3rd stage).

The above discussion on the green synthesis of NPs indicates that there are numerous plant sources. Figure 8.4 demonstrates

Figure 8.4 Common botanical sources, including parts of plants and plant extracts, which are commonly used for green NP synthesis, reprinted from Nasrollahzadeh et al., 2019, copyright (2019), with permission from Elsevier.

some of the common plants and their extracts as sources for green NP production. The overall payback from investments in such research is 2-fold; the acquisition of new knowledge and the commercial benefits of enlisting the prepared NPs in diverse applications: delivery platforms for drugs, therapeutics, pharmaceuticals, cosmetics, semiconductors, optoelectronics, biosensors, treatment of cancerous tumors, food packaging, solar panels, nanocomposites, environmental remediation, effluent treatment, water purification, catalysis, spectroscopic analysis, antibacterial, antifungal and antiviral agents, etc. (Nasrollahzadeh et al., 2019). An elaborate discussion on some of these current uses of green nanomaterials and structures is included in section 8.4 of this book chapter.

8.2.3 Plant-Based Synthesis of Nanostructures

The recent years have witnessed many innovative works on the use of plants and their extracts for green nanoparticles synthesis (Nasrollahzadeh et al., 2019). Nevertheless, there is another group of nanomaterials, called nanostructures, whose intricate and special architecture based physical and chemical properties call for more in-depth review. The preparation of nanostructures via plant-based biogenic routes leads to high reduction rates of precursors, and well-defined structures and morphologies, etc. of the nanomaterials. Thus, research studies have identified many plants which can be used in green nanomaterials and nanostructure synthesis: *Euphorbia peplus L.* (Sajjadi et al., 2017), *Centella asiatica L.* (Nasrollahzadeh et al., 2018), *Calliandra haematocephala* (Raja et al., 2017), *Eucalyptus* (Mo et al., 2015), etc. Along these lines, a very recent research work by Subbiah et al. (2020) demonstrated that ZnO, zinc-manganese oxide ($Zn_{0.98}Mn_{0.02}O$) and zinc-manganese-magnesium oxide ($Zn_{0.96}Mn_{0.02}Mg_{0.02}O$) nanostructures can be successfully produced from *Ocimum tenuiflorum's* leaf extracts. Subsequent XRD analysis revealed that Mg co-doping with $Zn_{0.98}Mn_{0.02}O$ significantly reduced the crystallite size from 46 to 34?nm. This was attributed to the difference in ionic sizes of the constituent elements. The researchers further verified the existence of Mn and Mg in the Zn-Mn-Mg-O network via infrared (IR) spectral analysis. Their characterization tests led them to claim that the synthesized hybrid nanostructure possessed enhanced absorbance and tuning of energy gap due to the intra-lattice ionic substitution by Mn/Mg cations. Such properties of nanomaterials are in high demand in optoelectronics and photocatalytic applications.

Yet another research group, Siddiqui et al. (2020), used eugenol (4-allyl-2-methoxyphenol) extracted from *O. sanctum* (Tulsi) leaves as a reducing agent in the biosynthesis of copper di-oxide (CuO) nanoflowers (NFs). Moreover, their characterization tests revealed that the green nanomaterial had crystalline single-phase growth with a monoclinic structure. Figure 8.5 is a graphical summary of the process used and the nanostructure of NFs synthesized. According to the figure, the possible growth mechanism is schematically illustrated by Fig. 8.5a. The hydroxide (OH^-) anions coordinate with copper (II) (Cu^{2+}) cations and determine the rate of the reaction for alkaline solutions. This results in nucleation sites, and thus, the growth of CuO micellar structures (Fig. 8.5b). The resultant complex ionic structures form symmetric 3D crystalline nanostructures among themselves via van der Waals forces and hydrogen bonding. When observed by transmission electron microscopy (TEM) these 3D nanomorphologies show up as nanoflowers (NFs) (Figs. 8.5c and e).

The actual electron exchange phenomena which give rise to the CuO NFs were also explained by the researchers. The Tulsi extract, eugenol, has a chemical structure with replaceable hydrogen atoms and a neighboring donor in the oxygen of the o-methoxy group. Consequently, two eugenol molecules generally share electrons with one copper (Cu) atom to form Cu^{2+} $(\text{eugenol})^{2-}$ complex. This electron transfer phenomenon is dictated by the active reduction of Cu^{2+} ions via acid-base reactions and results in the formation of nanoparticles as summarized by Eqs. 8.1 and 8.2 below:

$$Cu^{2+}(\text{eugenol})^{2-} + H_2O \rightarrow Cu(OH)_2 + 2(H - \text{eugenol}) \quad (8.1)$$

$$Cu(OH)_2 + 2(H - \text{eugenol})(\Delta) \rightarrow CuO + 2(H - \text{eugenol}) \quad (8.2)$$

The researchers further demonstrated that the NFs could degrade methylene blue, a common and toxic industrial dye, with 90% efficiency. Also, the NFs displayed a strong barrier against *E. coli* (27 ± 2 mm) at 100 $\mu g\ mL^{-1}$ concentration. These results, the

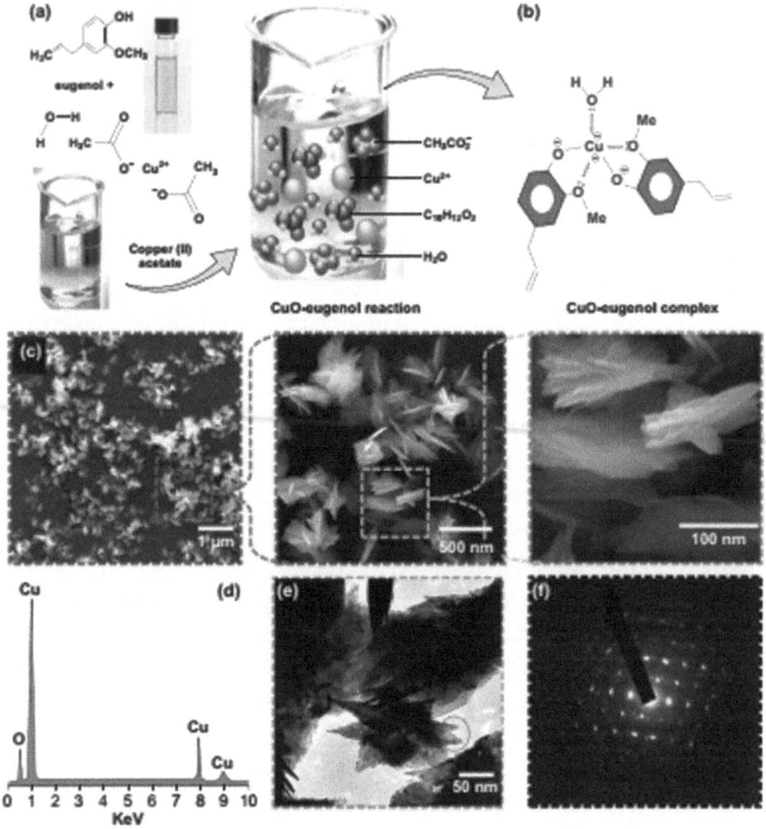

Figure 8.5 (a, b) Illustration of the growth of nanocrystals due to copper–eugenol complex formation. (c) FESEM of eugenol-assisted CuO NFs formation for various magnifications. (d) EDX spectral analysis of the NFs. (e) TEM image of the NFs, and (f) SAED pattern of eugenol-assisted CuO NFs, reprinted from Siddiqui et al., 2020, copyright (2020), with permission from Springer Nature.

researchers suggested, indicated that the green NFs have many useful environmental and medicinal applications.

Xu et al. (2019) experimentally investigated the use of plant fibers, e.g. cellulose, as nanoscaffolds and stabilizing agents for the preparation of nanocomposites and nanostructures. They looked into the rapid biosynthesis of stable AgNPs using cellulose nanocrystals (CNCs), i.e. dialdehyde cellulose nanocrystal (DACNC),

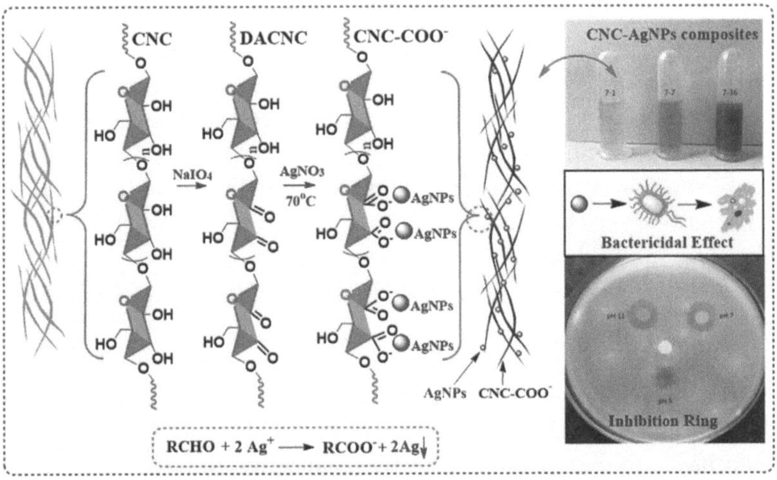

Figure 8.6 Schematics of the chemical reaction which results in CNC-AgNPs from CNC and $AgNO_3$ and picture of its antibacterial potency, reprinted from Xu et al. (2019), copyright (2019), with permission from Springer Nature.

which served the dual role of reducing and stabilizing agent. The periodate CNCs and bioreduced Ag ions resulted in cellulose nanocrystal–AgNPs (CNC–AgNPs) composite with a significant percentage of AgNPs in the scaffold of CNC. Subsequently, UV–Vis spectroscopy, Fourier infrared spectroscopy (FT-IR), X-ray diffraction (XRD), atomic absorption spectroscopy (AAS), thermogravimetric analysis (TGA) and TEM were enlisted to characterize the prepared nanocomposite. TEM analysis showed that CNC–AgNPs had spherical AgNPs with a maximum dimension of 30 nm deposited on the CNCs' surfaces. Moreover, the as-prepared nanocomposite demonstrated effective antibacterial potency against common disease-causing strains: *Escherichia coli* and *Staphylococcus aureus*. This led the researchers to claim that their nanocomposite could be integrated into materials used for making food packagings. Figure 8.6 illustrates the synthesis process for CNC–AgNPs and also a photograph demonstrating the antibacterial efficacy of the prepared nanocomposite.

Wahab et al. (2018) critically analyzed current research on the synthesis and medicinal prospects of AgNP-containing biopolymer nanocomposites. They highlighted that many research works validated the antibacterial activity of these green nanomaterials, which were largely associated with the NP's size, shape, surface area and medium. However, the authors commented that the broad-acting and potent antibacterial activity of this hybrid metal-biopolymer nanocomposite are based on the multifaceted mechanism of action that is not yet completely understood; neither is the NPs toxicity level for medical usage.

Another use of botanical extracts for nanomaterial synthesis is in the preparation of useful catalysts. Such preparations can minimize contaminants, costs and environmental impacts. Thus, recently there have been many research initiatives focused on the production of natural support-based metallic nanostructure synthesis to be used as effective nanocatalysts via biomimetic processes. These nanocatalysts have high surface area to volume ratios and can detoxify or reduce hazardous dyes and compounds in the environment or the human body.

Gautam et al. (2020) have published an exhaustive and in-depth review concerning the biogenic syntheses of nanocatalysts and their utility in neutralizing harmful synthetic organic pollutants. Of special note is the substitution of volatile and hazardous reagents by cellular extracts from plants in the phytogenic fabrication of nanocatalysts. Plant extracts commonly employed include vitamins, alkaloids, flavonoids, ketones, tannins, terpenoids, aldehydes, polysaccharides, phenolics and amino acids (Nath and Banerjee, 2013). These can be sourced from a plant's stem, seed, tissues, flowers, leaves, etc. The bioreduction abilities vary based on the particular species of plant, and the diversity and content of the extracted biomolecules. The fundamental process for biosynthesis plant-based nanocatalysts is simple and robust: the metal salt solution is stirred with the chosen plant extract at desired pH, either with or without mechanical agitation, and the resulting nanocatalysts are stabilized by mild heating.

In their review, Gautam et al. (2020) highlighted two plant species: *Brassica juncea* and *Medicago sativa*, metallophytes which can accumulate Au when exposed to $KAuCl_4$ solution. The stored AuNPs can be found in the plants' vascular bundle, cortical tissue and often in the epidermis. The authors further stated that phytogenic synthesis of nanocatalysts is more reliable than by any other biogenic route as these nanomaterials are characterized by high surface area, ease of tunability, recyclability, etc. Finally, the authors listed some of the environmental organic pollutants which can be reduced by nanocatalysts: polychlorinated biphenyls (PCBs), synthetic azo dyes, pesticides, phenolics, polycyclic aromatic hydrocarbons (PAHs), surfactants, etc. Owing to their complex chemical structures, these recalcitrant pollutants are highly soluble and persistent, causing detriment to ecosystems.

8.2.4 Synthesis of Green Nanoceramics and Nanocomposites

Green nanoceramics and nanocomposites have recently attained special interest of the scientific community. The synergy of properties of two or more substances, along with their eco-friendly and biocompatibility aspects, make these group of nanomaterials especially suitable for niche applications, such as in biomedical, drug delivery, tissue engineering, antimicrobial activity, bone and hip implants, etc.

Golieskardi et al. (2019) detailed the synthesis of a special bioceramic by doping 3 mol% Yttria-stabilized zirconia (3Y-TZP) with aluminium oxide (Al_2O_3) and cerium di-oxide (CeO_2) NPs. The effect of the dopants on the 3Y-TZP's microstructural stability was investigated for a temperature range from 1250°C to 1550°C. The results indicated that the bioceramic's mechanical strength depended on the sintering temperature and dopant amount. Furthermore, the optimum sintering temperature was 1450°C at which point all 3Y-TZP samples attained greater than 98% of their theoretically predicted density of 6.1 g/cm^3. Subsequent tests also revealed that the bioceramic of 3Y-TZP possessed Vickers hardness of 10.9 GPa and fracture toughness (K_{IC}) of 10 MPa.m$^{1/2}$ for optimum dopant concentration. Finally, the biocompatibility of 3Y-TZP bioceramic was looked into via cell culture study employing osteoblast cells. The results demonstrate a high percentage of cell attachment and proliferation, confirming the biocompatibility of the prepared bioceramic.

Another study on bioceramics was conducted by Ragurajan et al. (2018). The researchers investigated the effect of doping 3 mol% Yttria stabilized tetragonal zirconia polycrystalline (3Y-TZP) ceramic samples with Al_2O_3 and (manganese di-oxide) MnO_2 NPs at a temperature range of 1250°C to 1550°C and 2 hours holding time. They found that Al_2O_3 and MnO_2 NPs inclusions improved the mechanical properties and aging resistance of the bioceramic. The optimum dopant level was 0.6 wt% Al_2O_3 and 0.4 wt% MnO_2. Also, microscopic analysis showed that the tetragonal grain size of the ceramic was not affected by dopant addition. The synthesis process could ensure a maximum Vickers hardness of 11.6 GPa, fracture toughness (K_{IC}) of 9.8 MPa $m^{1/2}$, the flexural strength of 900 MPa and Young's modulus of 210 GPa for the doped bioceramic. Furthermore, retardation of low-temperature degradation of 3Y-TZP ceramic by doping was investigated by exposure to Ringer's solution at 37°C. It was found that the degradation rate and weight loss of (0.6 wt% Al_2O_3 and 0.4 wt% MnO_2) doped ceramic samples were 0.66% after 8 weeks, making it significantly less than undoped specimens of the ceramic. Finally, the authors suggested that Al_2O_3 and MnO_2 NPs infused 3Y-TZP bioceramic was a promising material for hip implants and other biomedical applications on account of its enhanced mechanical properties and better aging resistance.

Padmanabhan et al. (2019) performed a comparative study to understand the effect of lithium (Li) substitution in hydroxyapatite (HAP) NPs, preparing them using a sonochemical synthesis process. The hybrid-NP, i.e. lithium substituted hydroxyapatite (Li-HAP) was compared against pure HAP for their antimicrobial and hemolytic activities. X-ray diffraction (XRD), SEM, TEM and FTIR spectroscopy were used to characterize both types of NPs. The XRD analysis showed the existence of the HAP phase in both pure HAP and Li-HAP specimens. Besides, it was found that the crystallite size was smaller in Li-HAP. Moreover, FTIR analysis identified the presence of many functional groups (e.g. hydroxyl, phosphate, etc.). The authors further claimed that the general rod and plate structure of HAP was transformed, upon doping, into agglomerated needle-shaped crystals. Subsequent antimicrobial, hemolytic and MTT investigations employing human pathogens, human blood and human bone cells, respectively, evidenced that the biocompatibility and cytotoxicity of Li-HAP samples were better than pure HAP.

Biocompatible ceramic was also synthesized and investigated by Michael et al. (2016a). They studied hydroxyapatite (HA) [$Ca_{10}(PO_4)_6(OH)_2$], a bioactive ceramic with superior osteoconductive properties and produced nanohydroxyapatite (NHA) by employing a precipitation method assisted by ultrasonication. The characteristics of the nanoceramic aid in its easy integration into bone tissue without activating an immune reaction. Therefore, the researchers claimed that this bioceramic can serve as an excellent load-bearing bone implant.

In a similar study, Michael et al. (2016b), synthesized NHA from calcium nitrate tetrahydrate and diammonium hydrogen phosphate using ultrasonication assisted precipitation technique. Next, they investigated three independent process parameters: temperature (T) (70, 80 and 90°C), ultrasonication time (t) (20, 25 and 30 min), and amplitude (A) (60, 65 and 70%) to optimize the production process. Response surface methodology (RSM) using 3 factors and 5 levels of CCD were employed for the optimization. The responses of the model were analyzed using field-emission SEM and Brunauer–Emmett–Teller (BET) equipment. The results showed that the thermal stability and porosity of the as-prepared NHA were improved after calcination which could be due to the removal of impurities from the NHA powder. The authors concluded that the optimized results could benefit both researchers and practitioners interested in synthesizing and using this biocompatible nanocomposite.

Hoque et al. (2018) fabricated biomimetic bone tissue engineering scaffold using polylactic acid (PLA) and calcium carbonate ($CaCO_3$) composites, doped with AgNPs. Next, they examined the effect of injection parameters on the porosity rate of both single and hybrid nanocomposite scaffolds. It was found that the inclusion of 1 wt% of $CaCO_3$ and 0.5 wt% AgNPs did not hamper the flowability of the viscous polymer. SEM demonstrated the improved surface morphology and pore structure of the prepared samples. Subsequent compression tests revealed the coupled effect of $CaCO_3$ NPs on the scaffold's mechanical properties. The compressive modulus of PLA/$CaCO_3$/AgNP scaffold reached values of 0.38 GPa and 0.28 GPa, at porosities of 40% and 50%, respectively. The researchers concluded that the resulting mechanical stiffness of the NP infused hybrid scaffolds was greater than single PLA scaffolds for the same porosity level.

An exhaustive review of recent literature on bone implants from nanomaterials, i.e. nanofibers and nanocomposites, was conducted by Michael et al. (2016c). The authors propounded that bones, natural nanocomposites of collagenous and calcium phosphates mainly in the form of HA nanocrystallites, experience continuous loading and unloading cycles. This often causes fatigue-related brittle-fractures and breakdowns. Therefore, to address such issues, nanocomposites, especially polymers reinforced with nanoceramics, could be enlisted as load-bearing structures for bone implants. Furthermore, the authors discussed the state-of-the-art nanocomposite preparation techniques being investigated worldwide. They also focused on the effect of different nanofillers on the physicomechanical properties of nanocomposites in relation to that of natural human bone. Finally, the review updated the findings related to nanocomposites that meet the physicomechanical properties (strength and elasticity) as well as biocompatibility requirements for load-bearing bone implants.

Nuge et al. (2017) demonstrated an efficient method for preparing electrospun gelatin doped with MNPs derived from precursor metal salts. They reasoned that the doping of metallic ions with proven antibacterial properties into gelatin from electrospinning (Ge-espun) could help address inflammatory reaction and infection. For the synthesis of the hybrid nanoparticle incorporated biopolymer, the researchers used Silver Nitrate ($AgNO_3$), hydrated iron nitrate ($Fe(NO_3)_3.9H_2O$), hydrated copper nitrate ($Cu(NO_3)_2.3H_2O$), hydrated zinc nitrate ($Zn(NO_3)_2.6H_2O$) and hydrated nickel nitrate ($Ni(NO_3)_2.6H_2O$) as precursors. These metal salts were reduced to MNPs by acetic acid. The subsequent agglomeration of the resulting NPs was prevented by the matrix of gelatin polymer. Electrospinning of both neat Ge-espun and metal-NPs/Ge-espun was then carried out and resulted in smooth fibers having diameters of about 280 nm and ~40 to 150 nm, respectively. Further, the antibacterial activity of the prepared MNPs/Ge-espun fibers was tested for different metal loading by measuring the inhibition of colony-forming units. Results achieved demonstrated that a broad spectrum of antibacterial activity was achieved by Ag/Ge-espun, followed by Fe/Ge-espun and Zn/Ge-espun. On the other hand, Cu/Ge-espun and Fe/Ge-espun demonstrated activity against gram-positive bacteria which the authors recommended should be explored further against antibiotic-resistant bacteria.

8.2.5 Ethical and Environmental Concerns Related to Green Nanomaterials

The above sub-sections elaborated on some of the prevalent techniques being employed in the green synthesis of nanomaterials, nanocomposites, nanocatalysts and nanostructures. Special focus was given to the use of biogenic processes; especially phytogenic fabrication. Such biosynthesis offers many advantages and promises a greener world. However, there are some challenges and concerns, both ethical and environmental, regarding the production of green nanomaterials which need to be addressed before commercialization. One challenge is determining the correct balance between price, scalability and applicability for the production and use of green NPs. Micro-organism-based synthesis routes may be too expensive in some cases and there is always the possibility of unintentional environmental contamination. This is especially true for VNPs and VLPs. Another issue is the genetic manipulation of plants and the exploitation of plant extracts for developing nanomaterials. In this case, ethical and responsible use of botanical resources needs to be established to prevent run-away commercialization. Furthermore, experiments involving ecosystems and animals to determine the cytotoxicity of a prepared NP involve ethical considerations. Moreover, clinical trials to determine the toxicities of certain NPs for therapeutic use, e.g. Ag and Au NPs, can invoke moral and legal codes. Also, the application of green nanomaterials in the domestic sector, i.e. in food packages, containers, reagents, etc., needs to be scrutinized by proper scientific and legal entities before marketing. Thus, it can be surmised that there are many environmental, health care and safety, ethical or societal implications, and uncertainties related to the synthesis and subsequent utilization of green nanomaterials which need to be addressed via scientific research since, till date, there is a serious dearth of such studies in the academia.

8.3 Engineered Properties of Green Nanomaterials

Green nanomaterials are widely used in nanotechnology to increase environmental sustainability, eliminating the negative alliance of nanomaterials. Since nanomaterials contain different shapes than their bulk materials, the former has some unique engineered structural, optoelectrical and physicochemical properties compared to the latter. For this reason, green nanomaterials have found niche applications in disparate disciplines. Some of these structure-derived useful properties are elaborated on in the following sub-sections.

8.3.1 Structure-Based Properties of Functional Semiconductor Green Nanomaterials

The macroscopic structure-based properties of a substance (such as boiling and melting points, reactivity, conductivity, catalysis power, etc.) are quantified through a wide study on bulk sample materials. However, this is not true for the nanoscale of that material (Ejima et al., 2017). For instance, the boiling point of water in bulk quantity presents an interesting case. One mole of water contains 6.023×10^{23} molecules. When the molar boiling point of water is measured, it is true for any quantity of water molecule (Lower, 2017), because it is an average value. Yet, this concept is not working for some materials when the nanoscale regime is reached. Again, the visible color of gold is golden. Nonetheless, this is only true for macro and micro scale of gold. In the case of nanoscale gold material, the color turns into ruby red. These changes occur because the physical properties of materials are size-dependent.

In the case of the nanometer scale, the properties of materials become size-dependent. There are four ways in which nanoscale materials are different from their bulk materials (Eckelman et al., 2008).

8.3.1.1 Dominance of the electromagnetic force

There exist four basic forces in nature which are electromagnetic force, gravitational force, strong and weak nuclear force. The gravitational force is a function of mass and distance between two objects. According to the law of gravitation, the force is directly proportional to the multiplication of the masses and inversely proportional to the square of the distance between the mass centers. However, nanoscale particles possess very negligible masses; this gravitational force cannot create value in the case of nanoscale materials.

On the other hand, the electromagnetic force depends on the charge and distance of two particles, and this force is not affected by mass. It acts based on the charge and magnetic properties of nanoscale materials. When the charge or magnetization is increased between two particles, this electromagnetic force also increases. Besides, this force decreases when the distance between the particles increases. Nanoscale materials have negligible mass so there are no effects of gravitational force on nanoscale materials rather nanoscale materials depend on the electromagnetic force. There are two other forces—strong and weak nuclear forces. These forces act on the particles which are at a short distance and composed of a nucleus and therefore, this force also becomes negligible in nanomaterials, as they are considerably bigger than atoms. Therefore, gravitational and other nuclear forces turn to be negligible meanwhile, in nanoscale materials electromagnetic forces become dominant.

Influence of the four basic forces on the different scales of materials (Nano Sense, 2005) is given in the following Table 8.1.

8.3.1.2 Quantum effects

Quantum mechanics is a new trend to describe the motion and energy of nanomaterials instead of classical mechanics. Our everyday objects are greater in size and possess a much smaller velocity than the velocity of light. Classical mechanics can play an excellent role in describing the natural behaviors and phenomena of these bulk or macro materials. But in the case of nanoscale materials,

Table 8.1 Influences of the four fundamental forces of nature and their dependence on size, copyright from Nano Sense, 2005

Scale	Gravitational force	Electromagnetic force	Strong nuclear force	Weak nuclear force
Cosmic scale $\geq 10^7$m	✓	✓		
Macroscale $10^{-2} - 10^6$m	✓	✓		
Microscale $10^{-3} - 10^{-7}$m	✓	✓		
Nanoscale $10^{-8} - 10^{-9}$m		✓		
Sub-atomic scale $\leq 10^{-10}$			✓	✓

quantum mechanics is used instead of classical mechanics; as their diminutive size means that quantum effects become prominent. The following phenomena can be easily described with quantum mechanics in nanomaterials.

- **Quantization of energy**: Electrons can only exist in an atom at discrete energy levels. If the light spectrum emitted by an energetic atom is observed, it is observed that it is a composition of different single lines. These individual lines are of different colors, i.e. characteristic spectral lines. According to the conservation of energy, when an electron falls from a higher to a lower energy state, it emits a photon whose resulting energy is equal to the exact energy difference of the two levels. If electrons were not restricted to the discrete energy levels, the spectrum from an excited atom would be a continuous spread of colors. Classical mechanics predicted that electrons can only exist in discrete energy level but quantum mechanics describes it well and named it quantization of energy (Abd Ellah and Abouelmagd, 2017). This discrete preference in absorption and emission of light manifests itself as plasmonic resonance in NPs.
- **Wave-particle, the duality of the matter**: In the 17th century, two rival theories about the properties of light were proposed

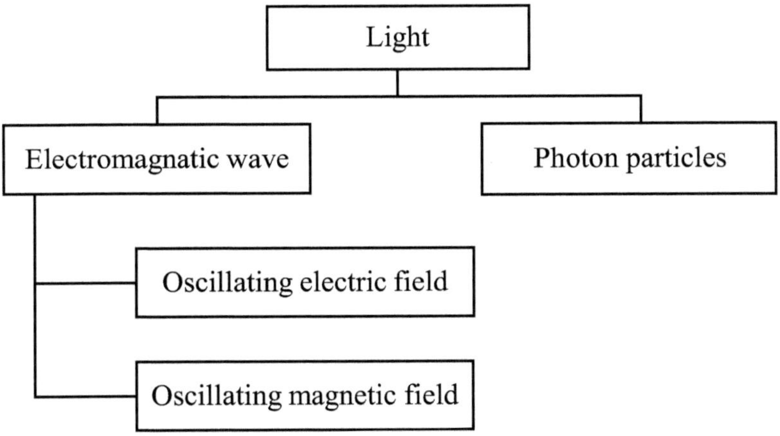

Figure 8.7 The dual character of light.

(as shown in Fig. 8.7). The Dutch astronomer Huygens published a theory of light in 1690. He believed that light is a longitudinal wave and this wave propagates along a medium called 'aether.' In 1704, Sir Issac Newton proposed another theory for light. Newton proposed that light was shot out from a source in small particles, that is, photons. So, the dual character of light can be represented as in Fig. 8.7.

The wave property can be described mathematically as (Prasad, 2004):

Wave number, $\bar{\upsilon} = \dfrac{\upsilon}{c} = \dfrac{1}{\lambda}$

Here, λ = wavelength, it is the distance between two successive peaks of the wave.

υ = frequency, the number of full waves passing through a point per second.

$c = \lambda \times \upsilon$ = speed of light and

$\bar{\upsilon}$ = it is the number of complete wavelengths per cm.

On the contrary, according to the particle property of light, the energy of photon E can be written as $h\upsilon$.

$$n = \frac{E}{h\upsilon} = \text{number of photons}$$

$$p = \frac{h}{\lambda} = \frac{h\upsilon}{c} = \text{momentum of each photon and}$$

$$h = 6.63 \times 10^{-34}\ \text{Js} = \text{Planck's constant}$$

In the case of reflection, refraction, polarization, and interference, the wave nature of light is sufficient to explain the observed behavior. On the other hand, there arise some conditions where the wave model fails and the particle model of light is used. One of the phenomena is known as the photoelectric effect. According to the photoelectric effect, which was explained by Albert Einstein, when light falls on the surface of a metal, some electrons of this metal are knocked out. Wave property of light cannot explain this phenomenon but the particle property can describe it well. It is inflexible to envisage for humans the concept of 'matter-wave' but in the case of nanoparticles, it is conceivable to perceive wave-like behavior and their interactions with incident electromagnetic waves can only be explained via the corpuscular theory of light.

- **Quantum tunneling**: This phenomenon occurs at nanoscale where a particle often violates the law of classical mechanics and penetrates the particle's potential barrier which is higher than its kinetic energy. In quantum mechanics, therefore, a nanoparticle is often described by its probability wave. For a barrier, according to classical physics, it is passable if the body has enough kinetic energy to go over the barrier. But if the particle does not have enough energy, the probability of identifying it is null on the other side of the barrier. However, in quantum mechanics, the probability of finding a particle whose energy is less than the barrier's is finite (Trixler, 2013; Zhang, 2015; Singh, 2016). This can be explained by assuming a virtual tunnel through the barrier (Note: the thickness of the barrier must be comparable to the wavelength of the particle). The phenomenon regarding quantum tunneling is illustrated in Fig. 8.8. So, it is only observed in nanoscale materials. This idea is the basis for the principle of imaging nanostructured surfaces called the 'scanning tunneling microscopy.' This instrument can also be utilized as a nanofabrication tool.

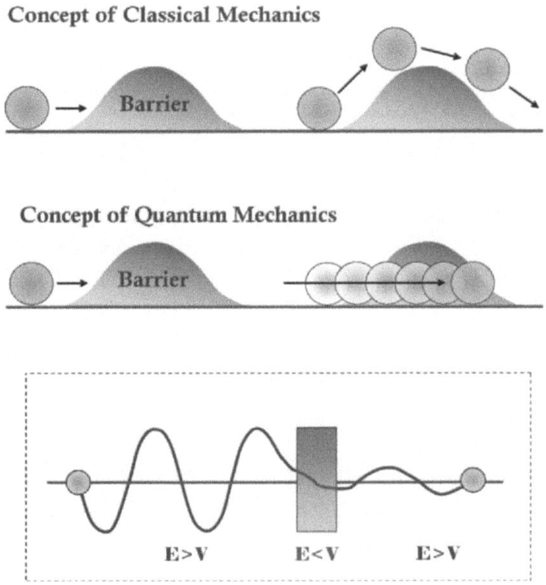

Figure 8.8 Illustration of the phenomenon of quantum tunneling.

8.3.1.3 Surface to volume ratio

In the case of nanomaterials, the surface area to volume ratio is an incremental phenomenon as shown in Fig. 8.9. Hence, a greater portion of the material is exposed to the surrounding materials for a potential reaction as its size is reduced. Surface performs different functions: it preserves things in and out, allows material or

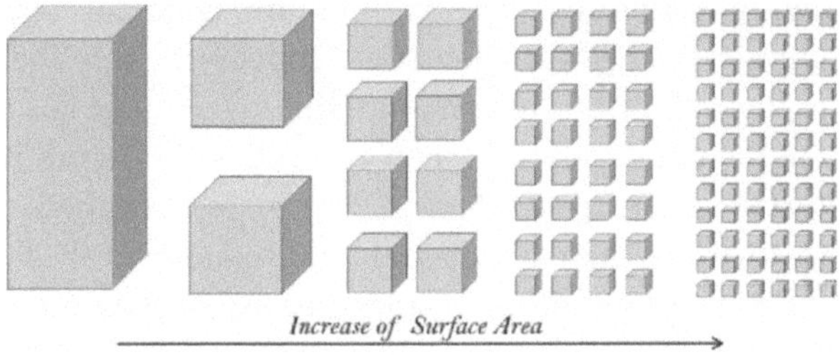

Figure 8.9 Incremental surface to volume ratio phenomenon in nanomaterials.

Table 8.2 Size, volume and the collective areas of different structures

Size of cube's side	Volume	Collective surface area
1 m	1	6
0.1 m	1000	60 m^2
0.01 m	10^6	600 m^2
0.001 m	10^9	6000 m^2
1 nm	10^{27}	6000 km^2

energy flow across the interface, can terminate or initiate a chemical reaction, etc.

Therefore, the volume remains the same but the potential surface area is increased. So, the surface area to volume ratio is increased for these smaller blocks (Kattan and Books, 2011) as can be seen from Eq. 8.3.

$$\text{Surface area to volume ratio} = \frac{\text{Surface area}}{\text{Volume ratio}} \tag{8.3}$$

Now as seen from Table 8.2, if a cube of 1 m^3 volume is progressively cut into smaller and smaller pieces then the collective surface area to volume ratio is increased.

8.3.1.4 Random molecular movement

In nanomaterials, molecular movement occurs frequently for their small size nature. So, more reactions can be possible. Tiny particles keep moving randomly in solutions or colloidal suspensions due to kinetic energy transfer from collisions with the matrix's molecules/atoms. At the macroscale, this movement is not noticeable; but at the nanoscale, it is seen everywhere. This motion makes nanomaterials and supramolecular structures to move around in space, rotate about their bonds and vibrate along with their bonds (Concord Consortium, 2005; Liyanage et al., 2016). In macroscale, this bulky movement is not so influential but the properties of NPs differ randomly due to this random movement. This is known as the Brownian motion of nanoparticles.

8.3.2 Optoelectrical and Semiconductor Properties of Green Nanomaterials

The optoelectrical properties of green NPs are interdependent. Some noble metal NPs have properties that are size-dependent and demonstrate a strong ultra-violet visible extinction band which is not visible in case of bulky materials. This excitation band is termed as localized surface plasma resonance (LSPR). The peak wavelength regarding the LSPR spectrum depends upon the size, shape, and interparticle spacing of the NPs. Also, LSPR is dependent on its dielectric properties and local environment including the substrate, solvents and adsorbents (Eustis and El-Sayed, 2006). For the rusty color of blemished glass or door windows, gold colloidal NPs are responsible, while AgNPs are typically yellow (Ahmed and Ikram, 2016). The *d* orbitals electrons of the surface in these NPs are freely transportable through the nanomaterial. The mean free path of Au and Ag is greater than the size of the aforesaid NPs. As a result, no scattering occurs in the bulk materials of Au and Ag but the NPs of these materials set into standing resonance conditions that are responsible for LSPR.

Several studies showed that AgNPs absorb electromagnetic radiation through LSPR. The wavelength of absorbance is NP's size-dependent (Smith and Nie, 2010; Khan et al., 2019). Table 8.3 shows the relation between the wavelength and size of NPs synthesized by different reagents.

Table 8.3 The relations between the wavelength and size of NPs synthesized by different reagents, copyright from Khan et al., 2019

Nanoparticle	Size	Synthesized by reagent	SPR finding wavelength
Silver (Ag)	Spherical	Glucose reduction	400 nm
Silver (Ag)	Spherical	NaOH reduction	420 nm
Silver (Ag)	7 nm Spherical	Gallic acid using chemical reduction	410 nm
Silver (Ag)	29 nm Spherical	Gallic acid using chemical reduction	425 nm
Silver (Ag)	89 nm Spherical	Gallic acid using chemical reduction	490 nm

Based on SP, the colors of nanomaterials differ from the bulk materials. If the NPs' dimension becomes comparable to the wavelength of light (in this condition, the frequency of light becomes close to SP), the light energy is absorbed (Zhang, 2018) and the relation can be demonstrated as SP frequency = 0.58 × plasma frequency of bulk material. Consequently, in the case of bulk material, the plasma frequency reaches in the ultra-violet region, the SP frequency of NP approaches in the visible region and the colors are seen.

NPs are also widely used for their unique and potential electronic properties in the electronics industry. AgNPs have some unique features such as conductance, resistance, resistivity, etc. so that they can be used in electronic devices rather than using bulky materials. The electrical conductance of AgNPs depends on its size which is usually selected from 4 to 12 nm and is grown in-ceramics. DC electrical resistance of AgNPs is measured in the temperature range of 80–300 K. The surface resistivity of AgNPs also increases linearly with the temperature ranging from 120–300 K. Using AgNPs as conductive fillers in electronically conductive adhesives (ECA) decreases the resistivity (Ma et al., 2016). These and similar findings indicate the structure-based electrical properties of AgNPs which have thus been exploited in electronic devices and semiconductors.

8.3.3 Mechanical and Thermal Properties of Green Nanomaterials

As stated above, the mechanical, thermal and other physicochemical properties of NPs are different from their bulk materials. NPs are normally used for potential applications in photonics, energy conversion, antifogging and optoelectrical characteristics.

Mechanical properties: When the size of the particle approaches nanoscale then the de Broglie wavelength of the charge carrier, the boundary conditions are destroyed and the atomic density is changed. Only for having this character, NPs can be used as drug delivery carriers by adsorbing on matrix materials (An et al., 2019). NPs have different mechanical properties compared to micro materials and bulk materials so that it can be used for surface modification for many devices, tribology and nanomanufacturing of devices. For investigating the precise mechanical nature of NPs, several mechanical properties like hardness, elasticity, stress, strain, friction, adhesion, etc. were studied by An et al. (2019).

Mechanical forces are generated from typical forces around the NPs. One of them is van der Waals attraction force. It is one of the weak attraction forces that play a vital role in the mechanical properties of NPs. This force includes three parts (Guo et al., 2013): one is the orientation force resulting from permanent dipole moment of polar molecules, the second one is the induced dipole moment and the third one is London dispersion forces which exist in polar and non-polar molecules.

At the nanoscale surface and interface, mechanical force is dominant. NPs have a greater surface to volume ratio compared to their micro or bulky substances. Relative to the microstructural metals and alloys, the nanostructural metals have higher grain boundary volumes. Thus, grain boundary plays an important role in the mechanical properties of NPs. The high density of incoherent interfaces or other lattice defects is seen for the changes in the grain size of materials.

Thermal properties: The melting point is one of the thermal properties of materials. According to the definition of melting temperature for macroscopic materials, the melting temperature of materials is size-independent for example, an ice cube and a glacier both melt at the same temperature. However, the case is quite different in nanoscale. Low melting temperature is a remarkable property of metal NPs because of their thermodynamic effect size. When a nanocrystal's size is decreased, surface energy is increased and the melting point is decreased. So, the melting point of NPs is different from their bulky forms. There are some ways to study the thermal properties of metal NPs. Differential scanning calorimeter along with thermogravimeter is commonly used to examine the thermal properties of AgNPs (Ma et al., 2016). Since the surface to volume ratio is small in a bulk material, the surface effects associated with the melting point are generally ignored. But it is not a negligible matter in the case of nanoscale materials. Several studies have already proven it. Recent studies investigated and found that the melting point of AgNPs (size $= 4 \sim 50$ nm) decreases proportionally with the size of NPs (Ankireddy et al., 2013).

8.3.4 Environmental Issues Pertaining to Green Nanomaterials' Properties

Studies related to the properties of green nanomaterials also expand on their eco-friendly, health care and safety, ethical or societal issues. A comprehensive review of related extant literature demonstrates that there is still a severe scarcity of such research studies. Before full industrialization, any uncertainties arising from the unique properties of green nanomaterials need to be evaluated by designed and controlled experiments in aquatic, aerial and terrestrial ecosystems. For medical applications, experimental trials on animals, especially mammals and primates, and clinical trials need to be undertaken. These experiments may be hampered by specific legal and ethical codes existent in a particular country. As such the pace of nanotechnology and nanosynthesis is much faster than the legislation and enforcement of regulations. These legal matters also need to be addressed before enjoying the fruits of nanoscience and nanotechnology. Therefore, collaboration among scientists, practitioners and legal entities is of paramount importance.

8.4 Applications of Green Nanomaterials

The above discussions on the synthesis and structure-based unique properties of green nanomaterials have given the reader considerable indications of the scope of their applications. The research literature is rich with such studies and endeavors, most of which are multi-disciplinary. It is outside the scope of this book chapter to engage in any exhaustive discussion of the manifold uses of green nanomaterials. Thus, the following sub-sections focus on the current applications of green nanomaterials in certain specific fields, i.e. functional materials, semiconductors, optoelectronics and spectral analysis. Where possible, future scopes of green nanomaterials are elaborated on with extrapolation from present research findings.

8.4.1 Applications as Functional Materials

Thrombin (TB) can be classified as a significant protein in molecular biology. Its activity and concentration are vital signs of pathological and physiological illness. So, a simple but relatively fast scheme for the detection of TB, with both high sensitivity and high selectivity, is definitely of huge importance in clinical applications.

For this purpose, a novel anodic electrochemiluminescent (ECL) aptasensor for finding TB has been made using carbon dot capped gold nanoflower (CD/AuNF) nanohybrid synthesis (You et al., 2018). The prepared CD/AuNF nanohybrids have gold nanoflowers in the center and a coating of carbon dots on the exterior. The hybrids are of uniform size ranging from 50 to 60 nm. Self-assembly along with biomolecules of AuNFs and the exceptional ECL activities of the CDs make CD/AuNF nanohybrids capable of biosensing. The specificity of the aptasensor was measured by reacting it with several other potential interferences like adenosine triphosphate (ATP), human IgG (IgG), bovine serum albumin (BSA), L-glutamate (L-Glu), L-cysteine (L-Cys) and lysosome (Lys). The sensor showed excellent selectivity for TB with a wide limit from 0.5 nM to 40 nM, and a low detection range ($S/N = 3$) of 0.08 nM.

Mostly CDs are required to be hybridized with additional nanomaterials for biosensing application, for detection of biomolecules, and/or signal amplification (Dong et al., 2015). Nanogold materials exhibit brilliant functionalization and biocompatibility.

As nuclear technology grows, vast numbers of radionuclides eventually go to the natural world. Therefore, for health and safety concerns, radionuclides need to be separated from drainage. Due to their high specific surface regions, broad amounts of binding sites, numerous functional groups, controllable pore sizes and ease of simple surface adjustments, nanomaterials are considered possible candidates for the efficient and systematic removal of radionuclides from aqueous solutions.

Developmental defects, birth abnormalities, obesity and many forms of cancer may be directly triggered if radionuclides from radioactive sources are inhaled. Uranium (U(VI)) release may result in severe biological harm such as toxic hepatitis, degradation of the skin, histopathological system harm 90, loss of kidney function and cancerous tumor (Song et al., 2017). Strontium (Sr(II)) has an important role in hepatocellular carcinoma diagnosis and the care of sixty other tumors of the liver (Banerjee et al., 2016; Bevara et al., 2018). Cobalt (Co(II)) can lead to several medical conditions, including low blood pressure, inflammation of the lungs, paralysis and diarrhea. Hence, one of the main environmental remediation priorities today is the detection and subsequent elimination of these radionuclides from polluted water. However, effective and selective methods for eliminating radionuclides from wastewater remain very much to be established (Wen et al., 2017).

Superior sorption efficiency, high selectivity and ecological sustainability are attractive features of modern nanomaterials including graphene oxides (GO), covalent organic frameworks (COF), carbon nitride (C_3N_4), metal-organic frameworks (MOFs), nanoscale zerovalent iron (NZVI), Mxenes, carbon nanofiber (CNF) and carbon nanotubes (CNTs). GO, a significant graphene derivative has a practical base plane with large enough oxygen-containing groups and edges as epoxy, hieroglyph and carboxy. GO nanosheets have a strong adsorption potential in preconcentrations for radionuclides in broad quantities of aqueous liquids, given the oxygen-containing functional groups and wide surface areas (Wang et al., 2016; Lingamdinne et al., 2019; Koduru et al., 2019; Lingamdinne et al., 2017).

MOFs offer the ability to be customized for different applications by adjusting their pores and geometry. This function may have major impacts on the production of structural isomer molecules adsorbent-based separation processes, which typically rely on the subtle match of adsorbates and micropore adsorbents (Li et al., 2018; Gao et al., 2019). Thanks to their attractive transparent structural properties and functionality, COFs have achieved considerable attention. The architectures and large surface areas are flexible, and are ideal for radionuclide pre-concentration (Salonen et al., 2017).

The treatment of a variable number of radionuclides by manageable particle size, strong reactivity and ample reactive surface space was successfully achieved by NZVI (Zou et al., 2016; Chen et al., 2018). Because it was calculated that C_3N_4 is super hard and diamond equivalent, C_3N_4 has stimulated significant interest (Shen et al., 2015). The structural units (-NH_2/-NH-/=N-) in g-C_3N_4 enable radionuclide deletion at the sample sites. Such groups even demonstrate outstanding radionuclide sorption by complexation or redox reactions (Hu et al., 2015).

Utilizing adequate etching agents from MAX phases at room temp, the MXens are typically made with specific grading of A element layers. MXenes are regarded as an outstanding tool for the solidification of radionuclides because of their strong physical and chemical durability, their hydrophilic properties and their environmentally-friendly nature (Wang et al., 2019). CNFs and CNTs are carbon-based nanomaterials, which have been thoroughly investigated in the removal of extremely soluble, heavy sorption and specific radionuclides after surface alteration (Xiao et al., 2013; Wang et al., 2005). The technique for synthesizing new nanomaterials allows various forms of older nanomaterials to be synthesized in broad volume at a low cost, which would allow nanomaterials to be cleaned up in the light of ecological radionuclide emissions.

8.4.2 Applications in Semiconductor Industry

Metal-organic chemical vapor, chemical vapor transport and pulsed laser deposition, are the most popular synthesis of the ZnO nanowires (You et al., 2018). Such processes can yield high-end, single crystalline, multi-micron length nanowires. These methods, however, involve high temperatures of around 450–900°C and also have more constraints on the option of substrates, the uniformity of the samples and poor product return. Simultaneously, the solution methods impact the ability for scaling up and smooth operations to manufacture high-density nanowires because of low growth temperatures below 350°C (Zhao et al., 2015). Thus, green nanosynthesis is a preferred option; as discussed previously.

ZnO nanowire FETs were studied for the efficiency of their devices and image detection (Jain et al., 2016; Al-Sabahi et al., 2016). There have recently been many reports on the function of surface conditions and on passivation for the ZnO nanowire FET's transportation properties, especially on the impact of nanowire size and surface roughness linked to surface trapping. The harshness of the interface plays a major role in transistors for electronic transport.

The nanostructures of ZnO are often produced and rendered as photodiodes. The research and industrial uses of nanoscale photodiodes are becoming rapidly abundant (Marie et al., 2015; Kim et al., 2014). An innovative analysis of ZnO revealed that ZnO heterojunction photodiodes composed of nanowire core and shell frameworks p-Si and n-ZnO nanostructures. In several photodiode applications the conformal cover made of n-type ZnO layer, which encircles p-type silicon (Si) nanowire, has been used. In contrast to many other planar thin-film photodiodes, such photodiodes emit improved ultraviolet (UV) and visible spectra.

Light-harvesting technologies are now becoming increasingly relevant because the need for renewable energy sources is growing gradually. In the last few years, photovoltaics have been studied extensively and proved to be a key forthcoming supply of energy. The light transfer performance and long-term reliability of semiconductive equipment trigger some of the problems in this area (Polman and Atwater, 2012). The long-term loss of reliability related to semiconductor excessive heating and environmental factors, including moisture and UV radiation, are known to produce chip output degradation. The solar-thermophotovoltaic strategy will address a variety of issues linked to straight exposure to semiconductor materials. A solar-thermophotovoltaic system can be used to bring a medium layer up to high temperatures by utilizing a great absorber equipped for large uptake of solar radiation. The emitter may be powered using natural, atomic or waste heat sources in a thermophotovoltaic instrument (Bauer, 2013).

Higher temperature operations will pose new problems for material degradation in these products. Use of the metals with higher melting temperatures was suggested for the absorber component (Wu et al., 2012). One-dimensional photonic crystal frameworks have been established for emitters. Moreover, experimental tests tend to be quite problematic because many of the necessary physical properties are missing in data.

The studies on the subject are mainly abstract debates after many years. A complete solar-thermophotovoltaic system with a performance rate of 3.2% has lately been installed (Lenert et al., 2015).

The spectral heterogeneity of the absorber and fairly low operating temperatures were the key limiting factors in this groundbreaking research owing to material constraints. Refractory materials can overcome all issues and expand scientific limits dramatically. It has proved that large absorption can be accomplished through a metamaterial design consisting of a modeled rectangular ring in visible and near-infrared regions that results in impedance matching and reduced reflection (Li et al., 2014).

Moreover, titanium nitride (TiN) can be designed to improve the impedance matching due to its flexible optical properties, utilizing some ideal absorber configurations for effective energy extraction from light. Lately, a plan was tabled to suggest prototypes for specific emitters that require products with large melting points, including TiN (Molesky et al., 2012). TiN is a very attractive plasmonic substance, a unique optical property of nanomaterials, with a melting point near 2900°C. However, if illuminated for 5 s at 550 nm with the pulsed lasers, both absorbers have a strong absorbance. TiN nanostructuring ability has been further checked and measured to Au. After research, scanning electron microscopy (SEM) proves that TiN nanostructures could survive high-intensity pulses in which Au nanostructures fail (Li et al., 2014). For nonlinear plasmonics, resilience under high-intensity laser illumination is essentially a desirable feature (Barnes et al., 2003).

Tip-based plasmonics is indeed a large field where metallic NPs were considered to be valuable. In addition to uses in science like near-field scanning optical microscopy (NSOM) or certain local field enhanced signal measurements, other industrial implementations like HAMR are often deemed viable. The accomplishment of greater data storage concentrations includes the lowering of bit sizes of magnetic storage technologies, which poses additional problems, including data losses in very tiny particles due to thermal stability issues. Products with low anisotropy can be used as a remedy, however low field strengths are needed.

Plasmonic NP-based tips have the benefit of concentrating electromagnetic energy at sub-wavelength in a small amount and with strong field intensity, and heating efficiency. Nonetheless, because of difficult operational environments, long-term device reliability persists as a concern. Alternate architectures will reduce the thermal load on NFT but long-term reliability is still a problem (Xiong et al., 2014). The nanostructured near field transducer should demonstrate chemical consistency and corrosion-resistance for long service life and durability (Lindquist et al., 2013; Kryder et al., 2008) and HAMR technologies will eventually be pursued for alternate plasmon materials (Zhou et al., 2014).

TiN is renowned for being high temperature immune to degradation and long-lasting. The optical properties of the substance have only recently been seen to alter considerably with increasing temperatures of up to 375°C (Sundari et al., 2013). It demonstrates that in the near-infrared band where diode laser technology is highly resourceful, TiN's optical efficiency is also comparable to the equivalent AuNPs (Guler et al., 2015). Thus, the development is one step closer to actual-life devices with near-field transducer models based on the TiN nanostructures.

8.4.3 Applications in Optoelectronics

Sensors based on surface plasmon resonance (SPR) phenomenon have been inspired by the scientifically linked study in recent years. SPR sensors often deliver several benefits relative to current traditional diagnostic instruments, including fast planning, no marking requirement, high precision, low-cost responsiveness and real-time sensing ability. The advancement of optical sensors has intrigued over the last two decades because of their capacity to conduct remote sensing and numerous detections with one system and electromagnetic interference.

To learn the biological and physiological roles, and also to improve medical diagnostic techniques it is important to identify clinically significant molecules, such as enzymes, hormones, nucleic acids, cells and medications. These molecules perform various biological functions, such as the storage and transmission of genetic material, biological activity control, molecular transport and catalytic reaction. Therefore, the diagnosis of diseases may often be seen as the detection of biomarkers (Labib et al., 2016).

Thanks to identification based on the difference in the refractive index, unique binding is measured because it happens without time-consuming altered washing and this feature decreases time-to-result relative to other technologies. Therefore, the identification is not dependent on specific molecules labeling. Due to the fact that a target label may change its binding affinity and kinetic characteristics and therefore raise reaction efficiency and method costs. In comparison, other medicinal analytes may be specifically identified by SPR. This may be used to boost efficiency, selectivity and/or the resolution of refractive indexes. Reliable methods to identify them in sophisticated blends of clinically appropriate proteins with highly significant characteristics to diagnose the degree of associated diseases are required. Protein identification contributes to molecular therapy and tests a medical response as well. For clinical analyses, the SPR sensors are hence very appropriate (Laocharoensuk, 2016; Ouellet et al., 2012).

PR sensors provide an innovative medium for accelerated research which can be used for cancer detection, prognosis and drug surveillance. They have shown evidence for complex functional research, molecular signature profiling, tumor cell enrichment and purification (Kirsch et al., 2013). By current developments in practical nanomaterials and by providing tunable plasmon surface characteristics intended for particular purposes, the art of synthesizing SPR-based sensory substances has been greatly enhanced. The movement towards using label-free and reusable controlled nanomaterials has vastly enhanced the sensing capability of specific cancer biomass, and multiplexed analytical platforms (Ranjan et al., 2017).

Eight microfluidic channels have been built for 32 sites of cancer biomarker testing, for example, a prostate-specific antigen and alpha-fetoprotein up to 500 pg/ml has been developed (Aćimović et al., 2014). The UV polymerization process for the identification of prostate-specific antigen was also employed using the prostate-specific antigen-imprinted SPR sensor methodology for micro-contact imprinting (Ertürk et al., 2016). They identified a prostate-specific antigen in the range 0.1–50 ng/ml and considered the limit of detection (LOD) to be 91 pg/ml for normal prostate-specific antigen mixture.

The SPR sensor has indeed been built to provide a point-of-care sensor of full prostate antigen identification in human serum samples (Uludag and Tothill, 2012). Sandwich research on an antimicrobial-modified SPR sensor to measure the different amounts of total prostate-specific antigen within individual serums was also accomplished. The integrated SPR immunoassay sensor was effectively used as a clinical device for patient serum samples.

In vivo diagnosis and treatment, immunoglobulins (Ig) have acquired considerable popularity amongst health authorities across the globe. It has been suggested that IgG was found on serum utilizing polymer NPs attached to the SPR sensor (Türkoğlu et al., 2012). Kinetic tests were also carried out with specific IgG (0.05–2.0 mg/ml) samples. The reflectivity value was improved by 0.3 to 2.5 in comparison with IgG, which grew through the IgG tests from 0.05 to 2.0 mg/ml.

The identification of cardiac biomarkers plays a major role in the treatment and threat of homogeneity in the chest and severe myocardial pain patients. The triggers for diagnosing acute myocardial infarction are myoglobin, creatine kinase-myocardial band and cardiac troponins. An SPR sensor along with the advantages of synthetic molecular cloning receptors has been built (Osman et al., 2013). In the concentration range between 0.1 μg/ml and 10 μg/ml, a myoglobin printed polymeric film was developed from an SPR sensor interacting with protein solutions. Once the myoglobin test solution hit the SPR sensor, they recorded a fast reaction.

The amyloid β-protein oligomerization is known to be the earliest activity of Alzheimer's disease. There is indeed a risk that sequence-specific antibodies may be unnecessarily utilized by current testing approaches to counter toxic fibril and monomeric amyloid β-protein. Attention is paid to producing further specifically identified antibodies to neurotoxic amyloid β-protein oligomers, for example, conformation-specific antibodies (Murakami et al., 2016). Researchers noticed that in Alzheimer's patient's cerebrospinal fluid, the ratio of harmful to overall amyloid β-protein was greater than in enzyme-linked immunosorbent assay determined control molecules.

A self-assembled single layer of protein chimera was identified as a new strategy to show an array of micro-electronic antibodies for the identification of influenza nucleoprotein (Brun et al., 2015). The technical and functional characteristics of the biological interface by SPR defined their process. They also demonstrated that the non-covalent configuration could allow the disconnection of binding antibodies irreparable by targeting an antibody feature by chemical cross-linking in situ.

Researchers have produced a molecularly imprinted water-compatible polymer foil comprising sufficient monomers, cross-linker, nanoparticles of polystyrene and the urinary testosterone prototype (Zhang et al., 2014). The molecular encoded polymer film and non-printed film have been used for the calculation of indeterminate unspecific adsorption in artificial urine and human urine.

Dengue is the main disease triggered by dengue virus that is placed in the host and known internationally as one of the most significant vector-borne diseases by the female *Aedes aegypti* mosquito. As a fast SPR-based approach, anti-dengue identification of the virus in serum samples was proposed (Jahanshahi et al., 2014). They reported that they could rapidly and conveniently use their method, defined as 10 min identification of the immunoglobulin M-based dengue diagnostic test. As ligands on SPR receptors, they used various dengue virus serotypes. They found 83–93 percent sensitivity and required to indicate SPR angle difference for increasing dengue serotype in specimens to be 100 percent specificity to 1 ml.

Researchers also developed a plasmon-coupling assay for methotrexate detection (Yockell-Lelièvre et al., 2016). The gold nanoparticles were immobilized in the sensor mounted in a compact four-channel localized SPR reader and free methotrexate/folic acid rivalry. Their sensor system enabled methotrexate to be calculated by plasmonic spectroscopy as per their total internal reflection. The tremendous variations in the plasmon-coupling method will render color adjustments to the human eye for methotrexate noticeable.

SPR will soon be an effective, robust and innovative framework for regular clinical research while also expanding broad perspectives for more advances in personalized medicine and point-of-care.

8.4.4 Applications in Spectral Analysis

Surface-enhanced Raman scattering (SERS) is largely used to identify detectable amounts of biological substances in food safety, biomedical and safety applications using noble metal-based NPs like Au and Ag to radically improve Raman scattering from molecules positioned near the metal surface. The usual objective by-products (surfactants, anions) pollute classic NPs produced with colloidal chemistry, which could have adverse consequences on several of the investigated living specimens (cells, bacteria) and therefore limit bio identification accuracy.

New ultrapure laser synthesized nanomaterials focused on locomotive SERS as a traveling bacteria recognition sample like AuNPs and AuSi nanostructures were researched (Kögler et al., 2018). It demonstrates that Au-based nanomaterials can effectively boost Raman signals in Rhodamine 6G models while the improvement factor depends on the contents of Au in the NP composition. It has been demonstrated that the identification of two types of bacteria (*Listeria innocua* and *Escherichia coli*) was more accurate through the observed improvements and cleanliness of laser-synthesized nanomaterials. The findings obtained support less worrying biological systems experiments, which are focused on the strong biocompatibility of ultrapure laser-synthesized materials.

Advanced imaging and labeling technologies are important for advancement in the realization of cell biology and disease treatment. Researchers have reacted to that requirement by investigating new multi-layered nanocomposites (MNCs) utilizing plasmonic NPs as super contrast photoacoustics (PA), adsorbing dyes and near-infrared fluorescent agents in small, non-absorbing shells (Novoselova et al., 2019). The design proof has been conducted with a mixture of biodegradable polymers, including gold nanorods (GNRs) and indocyanine green (ICG). In MNCs with the gold-ICG interface, they showed synergistic PA impacts which cannot be accomplished with ICG and GNR.

The researchers also studied non-linear PA and spaser, to detect ultrasharp PA and emission peaks. The usage of the synergetic interaction of plasmonic modes with a close absorbing environment enables low-toxicity multimodal MNCs with distinct thermal, plasma, and acoustic properties in the area to make such in vivo advancement in PA cytometry and near-infrared spasers.

Utilizing fullerene nanomaterials and bovine serum as the signal probe, an extremely sensitive and sandwiched electrochemiluminescence (ECL) aptasensor has been suggested for the detection of insulin (Wang et al., 2018). In this sense, the nano-C_{60}/BSA@luminol nanocomposites Ferrocenecarboxylic acid@aptamer II is initially synthesized and added to the dual amplification of the luminol-hydrogen peroxide (luminol-H_2O_2). As a matter of fact, the high luminol loading and additional strength of ECL can be accomplished by nano-C_{60}/BSA with a huge surface area and benefiting biocompatibility. Due to the existence of large luminol, the acquired nano-C_{60}/BSA@luminol nanohybrid was used as ECL signal samples and performed a significant role in identifying aptamer I. Besides, it is significant that Fc has a higher catalytic impact of H_2O_2, which has created both a better ECL and more agitated state luminol. In conjunction, a broad linear margin of 0.0001 to 1000 ng/ml and a low detection limit of 0.04 pg/ml (S/N = 3) and correlation coefficient (R^2) of 0.993 was provided by the aspirated ECL aptasensor. The newly proposed ECL aptasensor should provide new insight on active insulin detection and indicate promising opportunities for bioanalysis applications.

Atherosclerosis is an aggregation of lipids and fibrotic content as a result of the aging and unhealthy living style that thicken and toughen the artery's walls. Both arterial plaques are also liable for problems that may contribute to life threats such as strokes and myocardial infarction. Effective detection of the condition and recognition of abnormal atherosclerotic plaques tend to be difficult. The formation of atherosclerotic plaques is an inflammatory mechanism that in both early and late phases of the condition induces the aggregation of macrophages in the artery walls. A group of researchers proposed a new SERS technique for detecting a staged development of atherosclerosis (Dugandžić et al., 2018). It was focused on the usage, by way of SERS, of the labeled golden NPs. The reported findings shape the foundation of a possible animal model in vivo research. The synthesis of branched gold NPs in the existence of PVP K-55 and trisodium citrate has been documented as a result of reductions in tetrachloroauric acid with hydrazine hydrate.

Methods for evaluating biomolecules are focused on fast and responsive immunoreactions (Wu et al., 2016). In this research, a lysozyme detection system for leukemia, meningitis and other kidney problems via the SPR sensor was developed for possible diagnosis without the need for a procedure that saves time and money. The sensor was developed with the hydrophobic NPs PMAPA by changing the gold coating, which was defined by ATR–FTIR, AFM and WCA. The sensor has been commonly utilized in the real-time labeling process and additional practical monomer for the identification of lysozyme in aqueous lysozyme and egg white mixture. The double-character SPR sensor is capable of creating a hydrophobic matrix and a working supplier category 1. The updated SPR sensors were then effectively used to identify lysozymes selectively in the SPR device in the stationary process (Saylan et al., 2017). The lysozyme output of SPR sensors was 4.2 times higher, and myoglobin 3.2 times greater. The two SPR sensor responses indicate evidently that the hydrophobic mechanism of interaction gives the hydrophobic lysozyme noteworthy characteristics of the SPR sensor. The lysozyme is therefore identified and measured precisely in all-natural sources, egg white and aqueous solutions, by PMAPA's SPR detector.

8.4.5 Environmental Concerns and Remedies Pertaining to the Application of Green Nanomaterials

Conventional adsorbents are generally constrained in their effectiveness by their lack of selectivity, their surfaces or active sites and the lack of adsorption. With their incredibly high-specific region and related areas of sorption, a small intra-particle diffusion space, tunable pore size and surface chemistry, nanoadsorbents offer significant enhancements in this sense. There can be primarily three methods: carbon-based, metal-based and polymeric nanoadsorbents (Xue et al., 2016).

The quality of organic removal was good with carbon-based nanoadsorbents. Owing to their adsorbing ability for rapid metal ions, oxidized carbon nanotubes may be extremely effective for the easy removal of heavy metals. Effective and efficient, low-cost adsorbents for heavy metals such as lead, arsenic, mercury, cadmium, copper, nickel and chromium are nanoadsorbents derived from a metal or its oxide such as titanium dioxide, iron oxide and alumina.

The latest studies have concentrated on dendrimers and engineered adsorbents efficient in extracting both organic and hazardous metals with respect to polymer nanoadsorbents. Their inner shells could be hydrophobic for organic compound sorption, whereas the outer parts can be modified for heavy metal adsorption (e.g. hydroxyl or amine-terminated). However, the external layer alteration of anionic polysaccharide oxidized carbon nanotubes is an effective means of growing the adsorbance capacity of carbon nanotubes for eliminating cationic compounds from the environment (Xiong et al., 2008; Duman et al., 2016).

Owing to dwindling freshwater supplies, desalination could potentially become a significant source of fresh water in the immediate future. Conventional desalination techniques are now used, such as reverse osmosis, but these are expensive due to the high energy requirements. A variety of energy-efficient alternatives of the most promiscuously protein-polymer biomimetic membranes, thin sheet nanocomposite membranes, aligned carbon nanotubes, etc. may play an extremely important role in developing nanotechnology based solutions to water desalination.

A large number of membrane nanotechnology studies concentrate on establishing synergism by attaching nanomaterials to the surface of the polymer or inorganic membranes; these materials are usually metal oxides (Al_2O_3, TiO_2, zeolites). This technique aims to lower fouling by increasing membrane hydrophilicity and water permeability.

Other processes like deodorizing, defouling, disinfection and self-cleaning are integrated by these membranes. Systems may be implemented on the market in the future, but global level-up manufacturing, feasible desalination efficiency, and long-term stabilization are among the most vital difficulties to be perceived before their effective commercialization. Recent developments have included the complicated mathematical estimation studies of the reliability of nanofiltration membranes under different regulatory conditions (Agboola et al., 2014).

Hydride materials such as sodium aluminum hydride ($NaAlH_4$), ammonia borane (NH_3BH_3) and lithium borohydride ($LiBH_4$) have been identified. Graphene, carbon aerogels and nanoporous silica scaffolds, among others, have been investigated as porous materials due to their large surface area and pore depth. These will allow a major contribution to solid-state hydride-based storage of hydrogen reinforced content by supplying a functional support network along with a hydride enclosure and a percolated heat transfer network (Khalid et al., 2016; Yadav et al., 2015).

Oxygen is a strong oxidant, and while it is the strongest electron acceptor in breathing, certain levels can be lethal to bacteria; especially of the anaerobic variety. H_2O_2 consists of the breathing spread which absorbs O_2 in the manufacture of free radicals in several systems. The formation of hydroxyl radicals is almost definitely the reason for the oxidation of deoxyribonucleic acid, protein and membrane lipids by hydrogen peroxide. Reactive oxygen species can therefore impair bacteria to lose their membranes slowly, with their oxidative stress impacting their roles and ability for communication (Dasgupta and Ramalingam, 2016). Nano-Ag is the antibiotic nanomaterial most commonly used. Its high antimicrobial properties, sufficiently broad antimicrobial range, fairly low toxicity and simple usage render water disinfection and microbial control an excellent option. Primarily metallic NPs such as silver, copper, gold, bismuth, and also certain alloys (e.g. Au–Ag), have been the focus of antimicrobial research. Various oxide NPs, such as TiO_2, CuO, ZnO, Fe_3O_4, and also iron oxide composite materials have demonstrated antimicrobial action, primarily against bacteria. Many of these metals and metal oxides are currently painted on different materials for different purposes such as food packaging, sterilized containers, etc. (Ranjan and Ramalingam, 2016).

As for macronutrient nanofertilizers (N, P, K, Mg, S, etc.), Chhipa's review is noted where a comprehensive set of this latest method of nanofertilization, developed for successfully targeted distribution, is rendered evident by nanofertilizers (TiO_2, SiO_2, carbon nanotubes) (Chhipa, 2016). Recent work has centered on creating nanocontainers from starch to provide the plant tissues with nutrients, biostimulants and crop defense molecules. This method has the direct benefit of being biocompatible, biodegradable and non-toxic to plants, animals and climate.

Micro-organisms have been used, but techniques of synthesis are sluggish and constrained such that nanomaterials have no unique sizes and shapes (Park et al., 2015). Fungal bioproduction of NPs provides a range of functional benefits relative to other micro-organisms, such as yeast and algae. Fungi are becoming more commonly recognized in the world, with higher metal concentrations than the bacterial ones being accepted and ample extracellular redox proteins being secreted, thus decreasing the solubility of metal ions to their insoluble state. Thus, their mycelia could be utilized as micro-plants for the extraction of heavy metals from polluted soil.

The majority of studies to date are devoted to extracting silver nanoparticles, maybe as they are simple to manufacture and are essential in the disinfection sciences. There is still some concern in the green synthesis of the AuNPs, although in far less than Ag. Also, AuNPs have significant value in useful applications, albeit to a far lesser extent than Ag. 'Greener' techniques have also been used with other base metals (Fe, Pt, Pd, Cu), their derivatives (alloys or core-shells), oxides, and salts; with the aim of utilizing them in various environment remedial applications.

Nevertheless, many green NPs still exists today with significant limitations such as those which have only been taken out in research labs on a tiny scale, whereas pilot plant or industrial processing of nanomaterials utilizing natural resources has not been researched fully. Therefore, it is time to take steps to scale the planning processes (Shukla and Iravani, 2017). On the other side, there is often a significant drawback to the loss of influence regarding the final properties of nanomaterials. Efforts are then required to optimize reaction conditions, to isolate and purify the nanoparticles formed and their separation by size and shape (Shukla and Iravani, 2017). Both these scenarios need to be considered and researched further before exploitation of their products, i.e. green nanomaterials.

8.5 Conclusions

This chapter endeavored to paint a balanced picture of green nanomaterials whilst discussing about its synthesis, specific properties and some current applications. However, it is easy to be misled into believing that the barriers to the full-scale use of green nanomaterials are inconsequential. On the contrary, the existent hurdles are very real and often appear to be insurmountable.

The reader has already been engaged with a brief, yet exploratory, discourse on the prevailing environmental, legal and ethical concerns about the production and employment of green nanomaterials. Most nations of the world have realized these issues and are struggling belatedly to interpret and implement new rules and regulations regarding nanomaterials. Often, these hastily enacted laws are reactionary and may cause companies and research institutions to be more risk-averse; contemplating addressing the side-effects of nanotechnology instead of proactively developing novel green nanomaterials. Thus, efforts and resources are being channeled away from the research and development areas of nanomaterials. This is a major cause for concern which must be resolved by the concerted efforts of academics, researchers, practitioners, legal professionals and government representatives.

The scarcity of in-depth research summarizing the state-of-the-art findings in green nanotechnology and the potential toxicity aspects of NPs is another impediment to the growth of green nanomaterials. NPs by definition are lightweight and not visible to the naked eye, thus posing problems with their containment and detection. The long-term health effects of MNPs inhalation and the impact of NPs on the environment need to be further scrutinized. Only then can effective regulations be implemented via informed decisions.

Irrespective of the existing impedances, the growth in green nanomaterials technology has thrived. Sustainable synthesis of NPs, especially by phytogenic and myogenic (fungi based) routes, has ensured a variety of green nanomaterials custom-tailored for different applications. From the present research trend, it seems that plant-based synthesis of nanomaterials and botanical extracts driven nanostructure production are the champions. In the near to mid-term future, this green production method will surely dominate. The prospects for the long run is still indeterminate due to the exponential rate of innovations in this field. However, it is the view of the authors that enhanced forms of biogenic processes will be enlisted for nanomaterials and hybrid-nanostructure generation in the distant future.

As for the property based applications of green nanomaterials, the potentials are limitless. Green nanomaterials have already become indispensable in numerous multi-disciplinary applications: agriculture, biotechnology, biomedical, bioimaging, drug delivery, optoelectronics, semiconductors, pharmaceuticals, cosmetics, household devices, lab-on-chip and even in quantum computing. An extrapolation based on such trajectory shows that green nanomaterials' uses will become ubiquitous. In the near to mid-term future, it will be in most 'smart' equipment used in day-to-day life. In the long-term, it may be found in our bodies, repairing organ damage or coursing through our vascular system to neutralize pathogens. Overall, green nanomaterials are set to change the very landscape of the 21st century.

References

Abd Ellah, N. H., and Abouelmagd, S. A. (2017). Surface functionalization of polymeric nanoparticles for tumor drug delivery: approaches and challenges. *Expert Opinion on Drug Delivery*, **14**(2), 201–214, doi: 10.1080/17425247.2016.1213238.

Aćimović, S. S., Ortega, M. A., Sanz, V., Berthelot, J., Garcia-Cordero, J. L., Renger, J., et al. (2014). LSPR chip for parallel, rapid, and sensitive detection of cancer markers in serum. *Nano Letters*, **14**(5), 2636–2641, doi: 10.1021/nl500574n.

Agboola, O., Maree, J., Kolesnikov, A., Mbaya, R., and Sadiku, R. (2014). Theoretical performance of nanofiltration membranes for wastewater treatment. *Environmental Chemistry Letters*, **13**(1), 37–47, doi: 10.1007/s10311-014-0486-y.

Ahmed, S., and Ikram, S. (2016). Biosynthesis of gold nanoparticles: a green approach. *Journal of Photochemistry and Photobiology B: Biology*, **161**, 141–153.

Al-Sabahi, J., Bora, T., Al-Abri, M., and Dutta, J. (2016). Controlled defects of zinc oxide nanorods for efficient visible light photocatalytic degradation of phenol. *Materials*, **9**(4), 238, doi: 10.3390/ma9040238.

An, L., Zhang, D., Zhang, L., and Feng, G. (2019). Effect of nanoparticle size on the mechanical properties of nanoparticle assemblies. *Nanoscale*, **11**, 9563–9573.

Ankireddy, K., Iskander, M., Vunnam, S., Anagnostou, D. E., Kellar, J., and Cross, W. (2013). Thermal analysis of silver nanoparticles for flexible printed antenna fabrication. *Journal of Applied Physics*, **114**(12), 124303.

Banerjee, D., Kim, D., Schweiger, M. J., Kruger, A. A., and Thallapally, P. K. (2016). Removal of TcO4− ions from solution: materials and future outlook. *Chemical Society Reviews*, **45**(10), 2724–2739, doi: 10.1039/c5cs00330j.

Barnes, W. L., Dereux, A., and Ebbesen, T. W. (2003). Surface plasmon subwavelength optics. *Nature*, **424**(6950), 824–830, doi: 10.1038/nature01937.

Bauer, T. (2013). *Thermophotovoltaics Basic Principles and Critical Aspects of System Design*. Springer Berlin.

Bevara, S., Giri, P., Achary, S. N., Bhallerao, G., Mishra, R. K., Kumar, A., et al. (2018). Synthetic Na/K-birnessite for efficient management of Sr(II) from nuclear waste. *Journal of Environmental Chemical Engineering*, **6**(6), 7200–7213, doi: 10.1016/j.jece.2018.11.021.

Brun, A. P. L., Soliakov, A., Shah, D. S. H., Holt, S. A., Mcgill, A., and Lakey, J. H. (2015). Engineered self-assembling monolayers for label free detection of influenza nucleoprotein. *Biomedical Microdevices*, **17**(3), doi: 10.1007/s10544-015-9951-z.

Castro, L., Blázquez, M. L., Muñoz, J. A., González, F., García-Balboa, C., and Ballester, A. (2011). Biosynthesis of gold nanowires using sugar beet pulp. *Process Biochemistry*, **46**(5), 1076–1082.

Chen, Z., Wei, D., Li, Q., Wang, X., Yu, S., Liu, L., et al. (2018). Macroscopic and microscopic investigation of Cr(VI) immobilization by nanoscaled zero-valent iron supported zeolite MCM-41 via batch, visual, XPS and EXAFS techniques. *Journal of Cleaner Production*, **181**, 745–752, doi: 10.1016/j.jclepro.2018.01.231.

Chhipa, H. (2016). Nanofertilizers and nanopesticides for agriculture. *Environmental Chemistry Letters*, **15**(1), 15–22, doi: 10.1007/s10311-016-0600-4.

Concord Consortium (2005). Brownian motion [Web based computer simulation]. Retrieved from http://mw.concord.org/modeler1.3/mirror/thermodynamics/brown.html.

Danhalilu, R. L., Kankara, A. I., and Batagarawa, S. M. (2019). Green synthesis of metallic nanoparticles using leaf extract of *Calotropis* species and their applications: a review. *Current Journal of Applied Science and Technology*, **32**, 1–10.

Dasgupta, N., and Ramalingam, C. (2016). Silver nanoparticle antimicrobial activity explained by membrane rupture and reactive oxygen generation. *Environmental Chemistry Letters*, **14**(4), 477–485, doi: 10.1007/s10311-016-0583-1.

Dong, Y., Wu, H., Shang, P., Zeng, X., and Chi, Y. (2015). Immobilizing water-soluble graphene quantum dots with gold nanoparticles for a low potential electrochemiluminescence immunosensor. *Nanoscale*, **7**(39), 16366–16371, doi: 10.1039/c5nr04328j.

Dugandžić, V., Drikermann, D., Ryabchykov, O., Undisz, A., Vilotijević, I., Lorkowski, S., et al. (2018). Surface enhanced Raman spectroscopy-detection of the uptake of mannose-modified nanoparticles by macrophages in vitro: a model for detection of vulnerable atherosclerotic plaques. *Journal of Biophotonics*, **11**(12), doi: 10.1002/jbio.201800013.

Duman, O., Tunç, S., Bozoğlan, B. K., and Polat, T. G. (2016). Removal of triphenylmethane and reactive azo dyes from aqueous solution by magnetic carbon nanotube-κ-carrageenan-Fe3O4 nanocomposite. *Journal of Alloys and Compounds*, **687**, 370–383, doi: 10.1016/j.jallcom.2016.06.160.

Eckelman, M. J., Zimmerman, J. B., and Anastas, P. T. (2008). Toward green nano: E-factor analysis of several nanomaterial syntheses. *Journal of Industrial Ecology*, **12**(3), 316–328, doi: 10.1111/j.1530-9290.2008.00043.x.

Ejima, H., Richardson, J. J., and Caruso, F. (2017). Metal-phenolic networks as a versatile platform to engineer nanomaterials and biointerfaces. *Nano Today*, **12**, 136–148, doi:10.1016/j.nantod.2016.12.012.

Emeka, E. E., Ojiefoh, O. C., Aleruchi, C., Hassan, L. A., Christiana, O. M., Rebecca, M., Dare, E. O., and Temitope, A. E. (2014). Evaluation of antibacterial activities of silver nanoparticles green-synthesized using pineapple leaf (Ananas comosus). *Micron*, **57**, 1–5.

Ertürk, G., Özen, H., Tümer, M. A., Mattiasson, B., and Denizli, A. (2016). Microcontact imprinting based surface plasmon resonance (SPR) biosensor for real-time and ultrasensitive detection of prostate specific antigen (PSA) from clinical samples. *Sensors and Actuators B: Chemical*, **224**, 823–832, doi: 10.1016/j.snb.2015.10.093.

Eustis, S., and El-Sayed, M. A. (2006). Why gold nanoparticles are more precious than pretty gold: noble metal surface plasmon resonance and its enhancement of the radiative and nonradiative properties of nanocrystals of different shapes. *Chemical Society Reviews*, **35**(3), 209–217.

Gao, Q., Xu, J., and Bu, X.-H. (2019). Recent advances about metal–organic frameworks in the removal of pollutants from wastewater. *Coordination Chemistry Reviews*, **378**, 17–31, doi: 10.1016/j.ccr.2018.03.015.

Gautam, P. K., Shivalkar, S., and Samanta, S. K. (2020). Environmentally benign synthesis of nanocatalysts: recent advancements and applications. In *Handbook of Nanomaterials and Nanocomposites for Energy and Environmental Applications*. Springer Nature Switzerland, pp. 1–19.

Golieskardi, M., Satgunam, M., Ragurajan, D., Hoque, M. E., Ng, A. M. H., and Shanmugenantha, L. (2019). Advanced 3Y-TZP bioceramic doped with Al_2O_3 and CeO_2 potentially for biomedical implant applications. *Materials Technology*, **34**(8), 480–489.

Guler, U., Kildishev, A. V., Boltasseva, A., and Shalaev, V. M. (2015). Plasmonics on the slope of enlightenment: the role of transition metal nitrides. *Faraday Discussions*, **178**, 71–86, doi: 10.1039/c4fd00208c.

Guo, D., Xie, G., and Luo, J. (2013). Mechanical properties of nanoparticles: basics and applications. *Journal of Physics D: Applied Physics*, **47**(1), 013001.

Hamzah, N. A. (2020). Biosynthesis of silver nanoparticles using extract of Ziziphus leaves. *Plant Archives*, **20**(1), 2243–2248.

Hoque, M. E., Daei, G., Mahmoud, J., and Khalid, M. (2018). Next generation biomimetic bone tissue engineering matrix from poly (L-lactic acid) Pla/calcium carbonate composites doped with silver nanoparticles. *Current Analytical Chemistry*, **14**(3), 268–277.

Hu, R., Wang, X., Dai, S., Shao, D., Hayat, T., and Alsaedi, A. (2015). Application of graphitic carbon nitride for the removal of Pb(II) and aniline from aqueous solutions. *Chemical Engineering Journal*, **260**, 469–477, doi: 10.1016/j.cej.2014.09.013.

Jahanshahi, P., Zalnezhad, E., Sekaran, S. D., and Adikan, F. R. M. (2014). Rapid immunoglobulin M-based dengue diagnostic test using surface plasmon resonance biosensor. *Scientific Reports*, **4**(1), doi: 10.1038/srep03851.

Jain, R., Dhanjai, and Sinha, A. (2016). Graphene-zinc oxide nanorods nanocomposite based sensor for voltammetric quantification of tizanidine in solubilized system. *Applied Surface Science*, **369**, 151–158, doi: 10.1016/j.apsusc.2016.02.077.

Karuppiah, C., Palanisamy, S., Chen, S. M., Emmanuel, R., Muthupandi, K., and Prakash, P. (2015). Green synthesis of gold nanoparticles and its application for the trace level determination of painter's colic. *RSC Advances*, **5**(21), 16284–16291.

Kattan, P. I. (2011). *Ratio of Surface Area to Volume in Nanotechnology and Nanoscience* (Basic Nanomechanics Series). Petra Books.

Ketabchi, M. R., Khalid, M., Ratnam, C. T., Manickam, S., Walvekar, R., and Hoque, M. E. (2016). Sonosynthesis of cellulose nanoparticles (CNP) from kenaf fiber: effects of processing parameters. *Fibers and Polymers*, **17**(9), 1352–1358.

Khalid, F., Dincer, I., and Rosen, M. A. (2016). Analysis and assessment of an integrated hydrogen energy system. *International Journal of Hydrogen Energy*, **41**(19), 7960–7967, doi: 10.1016/j.ijhydene.2015.12.221.

Khan, I., Saeed, K., and Khan, I. (2019). Nanoparticles: properties, applications and toxicities. *Arabian Journal of Chemistry*, **12**(7), 908–931.

Kim, K., Utashiro, K., Abe, Y., and Kawamura, M. (2014). Structural properties of zinc oxide nanorods grown on Al-doped zinc oxide seed layer and their applications in dye-sensitized solar cells. *Materials*, **7**(4), 2522–2533, doi: 10.3390/ma7042522.

Kirsch, J., Siltanen, C., Zhou, Q., Revzin, A., and Simonian, A. (2013). Biosensor technology: recent advances in threat agent detection and medicine. *Chemical Society Reviews*, **42**(22), 8733, doi: 10.1039/c3cs60141b.

Koduru, J. R., Karri, R. R., and Mubarak, N. M. (2019). Smart materials, magnetic graphene oxide-based nanocomposites for sustainable water purification. *Sustainable Polymer Composites and Nanocomposites*, 759–781, doi: 10.1007/978-3-030-05399-4_26.

Kögler, M., Ryabchikov, Y. V., Uusitalo, S., Popov, A., Popov, A., Tselikov, G., et al. (2018). Bare laser-synthesized Au-based nanoparticles as nondisturbing surface-enhanced Raman scattering probes for bacteria identification. *Journal of Biophotonics*, **11**(7), doi: 10.1002/jbio.201700225.

Kryder, M. H., Gage, E. C., McDaniel, T. W., Challener, W. A., Rottmayer, R. E., Ju, G., et al. (2008). Heat assisted magnetic recording. *Proceedings of the IEEE*, **96**(11), 1810–1835, doi: 10.1109/JPROC.2008.2004315.

Labib, M., Sargent, E. H., and Kelley, S. O. (2016). Electrochemical methods for the analysis of clinically relevant biomolecules. *Chemical Reviews*, **116**(16), 9001–9090, doi: 10.1021/acs.chemrev.6b00220.

Laocharoensuk, R. (2016). Development of electrochemical immunosensors towards point-of-care cancer diagnostics: clinically relevant studies. *Electroanalysis*, **28**(8), 1716–1729, doi: 10.1002/elan.201600248.

Lenert, A., Bierman, D. M., Nam, Y., Chan, W. R., Celanović, I., Soljačić, M., and Wang, E. N. (2015). Addendum: a nanophotonic solar thermophotovoltaic device. *Nature Nanotechnology*, **10**(6), 563–563, doi: 10.1038/nnano.2015.117.

Li, J., Liu, Y., Ai, Y., Alsaedi, A., Hayat, T., and Wang, X. (2018). Combined experimental and theoretical investigation on selective removal of mercury ions by metal organic frameworks modified with thiol groups. *Chemical Engineering Journal*, **354**, 790–801, doi: 10.1016/j.cej.2018.08.041.

Li, W., Guler, U., Kinsey, N., Naik, G. V., Boltasseva, A., Guan, J., et al. (2014). Refractory plasmonics with titanium nitride: broadband metamaterial absorber. *Advanced Materials*, **26**(47), 7959–7965, doi: 10.1002/adma.201401874.

Lin, J., Weng, X., Jin, X., Megharaj, M., Naidu, R., and Chen, Z. (2015). Reactivity of iron-based nanoparticles by green synthesis under various atmospheres and their removal mechanism of methylene blue. *RSC Advances*, **5**(87), 70874–70882.

Lindquist, N. C., Jose, J., Cherukulappurath, S., Chen, X., Johnson, T. W., and Oh, S.-H. (2013). Tip-based plasmonics: squeezing light with metallic nanoprobes. *Laser & Photonics Reviews*, **7**(4), 453–477, doi: 10.1002/lpor.201209044.

Lingamdinne, L. P., Choi, Y.-L., Kim, I.-S., Yang, J.-K., Koduru, J. R., and Chang, Y.-Y. (2017). Preparation and characterization of porous reduced graphene oxide based inverse spinel nickel ferrite nanocomposite for adsorption removal of radionuclides. *Journal of Hazardous Materials*, **326**, 145–156, doi: 10.1016/j.jhazmat.2016.12.035.

Lingamdinne, L. P., Koduru, J. R., and Karri, R. R. (2019). A comprehensive review of applications of magnetic graphene oxide based nanocomposites for sustainable water purification. *Journal of Environmental Management*, **231**, 622–634, doi: 10.1016/j.jenvman.2018.10.063.

Liyanage, D. D., Thamali, R. J., Kumbalatara, A. A. K., Weliwita, J. A., and Witharana, S. (2016). An analysis of nanoparticle settling times in liquids. *Journal of Nanomaterials*, **2016**, 7061838.

Lower, S. (2017). A gentle introduction to water and its structure. Retrieved from http://www.chem1.com/acad/sci/aboutwater.html.

Ma, H., Zeng, J., Harrington, S., Ma, L., Ma, M., Guo, X., and Ma, Y. (2016). Hydrothermal fabrication of silver nanowires-silver nanoparticles-graphene nanosheets composites in enhancing electrical conductive performance of electrically conductive adhesives. *Nanomaterials*, **6**(6), 119.

Marchiol, L. (2012). Synthesis of metal nanoparticles in living plants. *Italian Journal of Agronomy*, **7**, e37.

Marie, M., Mandal, S., and Manasreh, O. (2015). An electrochemical glucose sensor based on zinc oxide nanorods. *Sensors*, **15**(8), 18714–18723, doi: 10.3390/s150818714.

Michael, F. M., Khalid, M., Ratnam, C. T., Chee, C. Y., Rashmi, W., and Hoque, M. E. (2016a). Sono-synthesis of nanohydroxyapatite: effects of process parameters. *Ceramics International*, **42**(5), 6263–6272.

Michael, F. M., Khalid, M., Ratnam, C. T., Rashmi, W., Hoque, M. E., and Ketabchi, M. R. (2016b). Nanohydroxyapatite synthesis using optimized process parameters for load-bearing implant. *Bulletin of Materials Science*, **39**(1), 133–145.

Michael, F. M., Khalid, M., Walvekar, R., Ratnam, C. T., Ramarad, S., Siddiqui, H., and Hoque, M. E. (2016c). Effect of nanofillers on the physico-mechanical properties of load bearing bone implants. *Materials Science and Engineering: C*, **67**, 792–806.

Mo, Y. Y., Tang, Y. K., Wang, S. Y., Lin, J. M., Zhang, H. B., and Luo, D. Y. (2015). Green synthesis of silver nanoparticles using eucalyptus leaf extract. *Materials Letters*, **144**, 165–167.

Molesky, S., Dewalt, C. J., and Jacob, Z. (2012). High temperature epsilon-near-zero and epsilon-near-pole metamaterial emitters for thermophotovoltaics. *Optics Express*, **21**(S1), doi: 10.1364/oe.21.000a96.

Murakami, K., Tokuda, M., Suzuki, T., Irie, Y., Hanaki, M., Izuo, N., et al. (2016). Monoclonal antibody with conformational specificity for a toxic conformer of amyloid β42 and its application toward the Alzheimer's disease diagnosis. *Scientific Reports*, **6**(1), doi: 10.1038/srep29038.

Nano Sense (2005). Unique properties at the nanoscale [Student materials]. Retrieved from https://nanosense.sri.com/activities/sizematters/properties/SM_Lesson3Student.pdf.

Narayanan, K. B., and Han, S. S. (2017). Icosahedral plant viral nanoparticles-bioinspired synthesis of nanomaterials/nanostructures. *Advances in Colloid and Interface Science*, **248**, 1–19.

Narayanan, K. B., and Sakthivel, N. (2011). Green synthesis of biogenic metal nanoparticles by terrestrial and aquatic phototrophic and heterotrophic eukaryotes and biocompatible agents. *Advances in Colloid and Interface Science*, **169**(2), 59–79.

Nasrollahzadeh, M., and Sajadi, S. M. (2015). Synthesis and characterization of titanium dioxide nanoparticles using Euphorbia heteradena Jaub root extract and evaluation of their stability. *Ceramics International*, **41**(10), 14435–14439.

Nasrollahzadeh, M., Atarod, M., Sajjadi, M., Sajadi, S. M., and Issaabadi, Z. (2019). Plant-mediated green synthesis of nanostructures: mechanisms, characterization, and applications. In *Interface Science and Technology*, Vol. 28, Elsevier, pp. 199–322.

Nasrollahzadeh, M., Issaabadi, Z., and Sajadi, S. M. (2018). Green synthesis of Pd/Fe_3O_4 nanocomposite using *Hibiscus tiliaceus L.* extract and its application for reductive catalysis of Cr (VI) and nitro compounds. *Separation and Purification Technology*, **197**, 253–260.

Nasrollahzadeh, M., Sajjadi, M., and Sajadi, S. M. (2018). Biosynthesis of copper nanoparticles supported on manganese dioxide nanoparticles using *Centella asiatica L.* leaf extract for the efficient catalytic reduction of organic dyes and nitroarenes. *Chinese Journal of Catalysis*, **39**(1), 109–117.

Nath, D., and Banerjee, P. (2013). Green nanotechnology–a new hope for medical biology. *Environmental Toxicology and Pharmacology*, **36**(3), 997–1014.

Nayak, D., Pradhan, S., Ashe, S., Rauta, P. R., and Nayak, B. (2015). Biologically synthesised silver nanoparticles from three diverse family of plant extracts and their anticancer activity against epidermoid A431 carcinoma. *Journal of Colloid and Interface Science*, **457**, 329–338.

Novoselova, M. V., Bratashov, D. N., Sarimollaoglu, M., Nedosekin, D. A., Harrington, W., Watts, A., et al. (2019). Photoacoustic and fluorescent effects in multilayer plasmon-dye interfaces. *Journal of Biophotonics*, **12**(4), doi: 10.1002/jbio.201800265.

Nuge, T., Tshai, K. Y., Lim, S. S., Nordin, N., and Hoque, M. E. (2017). Preparation and characterization of CU-, FE-, AG-, ZN-and NI-doped gelatin nanofibers for possible applications in antibacterial nanomedicine. *Journal of Engineering Science and Technology*, **12**(3), 68–81.

Osman, B., Uzun, L., Beşirli, N., and Denizli, A. (2013). Microcontact imprinted surface plasmon resonance sensor for myoglobin detection.

Materials Science and Engineering: C, **33**(7), 3609–3614, doi:10.1016/j.msec.2013.04.041.

Ouellet, E., Lund, L., and Lagally, E. T. (2012). Multiplexed surface plasmon resonance imaging for protein biomarker analysis. *Microfluidic Diagnostics Methods in Molecular Biology*, 473–490, doi: 10.1007/978-1-62703-134-9_30.

Padmanabhan, V. P., TSN, S. N., Sagadevan, S., Hoque, M. E., and Kulandaivelu, R. (2019). Advanced lithium substituted hydroxyapatite nanoparticles for antimicrobial and hemolytic studies. *New Journal of Chemistry*, **43**(47), 18484–18494.

Pal, G., Rai, P., and Pandey, A. (2019). Green synthesis of nanoparticles: a greener approach for a cleaner future. In *Green Synthesis, Characterization and Applications of Nanoparticles*. Elsevier, pp. 1–26.

Park, T. J., Lee, K. G., and Lee, S. Y. (2015). Advances in microbial biosynthesis of metal nanoparticles. *Applied Microbiology and Biotechnology*, **100**(2), 521–534, doi: 10.1007/s00253-015-6904-7.

Pathak, J., Ahmed, H., Singh, D. K., Pandey, A., Singh, S. P., and Sinha, R. P. (2019). Recent developments in green synthesis of metal nanoparticles utilizing cyanobacterial cell factories. In *Nanomaterials in Plants, Algae and Microorganisms*. Academic Press, pp. 237–265.

Polman, A., and Atwater, H. A. (2012). Photonic design principles for ultrahigh-efficiency photovoltaics. *Nature Materials*, **11**(3), 174–177, doi: 10.1038/nmat3263.

Prasad, P. N. (2004). *Introduction to Biophotonics*. John Wiley & Sons.

Ragurajan, D., Golieskardi, M., Satgunam, M., Hoque, M. E., Ng, A. M. H., Ghazali, M. J., and Ariffin, A. K. (2018). Advanced 3Y-TZP bioceramic doped with Al_2O_3 and MnO_2 particles potentially for biomedical applications: study on mechanical and degradation properties. *Journal of Materials Research and Technology*, **7**(4), 432–442.

Raja, S., Ramesh, V., and Thivaharan, V. (2017). Green biosynthesis of silver nanoparticles using Calliandra haematocephala leaf extract, their antibacterial activity and hydrogen peroxide sensing capability. *Arabian Journal of Chemistry*, **10**(2), 253–261.

Ranjan, R., Esimbekova, E. N., and Kratasyuk, V. A. (2017). Rapid biosensing tools for cancer biomarkers. *Biosensors and Bioelectronics*, **87**, 918–930, doi: 10.1016/j.bios.2016.09.061.

Ranjan, S., and Ramalingam, C. (2016). Titanium dioxide nanoparticles induce bacterial membrane rupture by reactive oxygen species

magnetic nanoparticles using Syzygium cumini seed extract. *Physica B: Condensed Matter*, **449**, 67–71.

Wahab, M. A., Islam, N., Hoque, M. E., and Young, D. J. (2018). Recent advances in silver nanoparticle containing biopolymer nanocomposites for infectious disease control: a mini review. *Current Analytical Chemistry*, **14**(3), 198–202.

Wang, X., Chen, C., Hu, W., Ding, A., Xu, D., and Zhou, X. (2005). Sorption of 243Am(III) to multiwall carbon nanotubes. *Environmental Science & Technology*, **39**(8), 2856–2860, doi: 10.1021/es048287d.

Wang, X., Chen, L., Wang, L., Fan, Q., Pan, D., Li, J., et al. (2019). Synthesis of novel nanomaterials and their application in efficient removal of radionuclides. *Science China Chemistry*, **62**(8), 933–967, doi: 10.1007/s11426-019-9492-4.

Wang, X., Fan, Q., Yu, S., Chen, Z., Ai, Y., Sun, Y., et al. (2016). Retracted: High sorption of U(VI) on graphene oxides studied by batch experimental and theoretical calculations. *Chemical Engineering Journal*, **287**, 448–455, doi: 10.1016/j.cej.2015.11.066.

Wang, Y., Sha, H., Ke, H., Xiong, X., and Jia, N. (2018). A sandwich-type electrochemiluminescence aptasensor for insulin detection based on the nano-C60/BSA@luminol nanocomposite and ferrocene derivative. *Electrochimica Acta*, **290**, 90–97, doi: 10.1016/j.electacta.2018.08.080.

Wen, T., Wang, J., Yu, S., Chen, Z., Hayat, T., and Wang, X. (2017). Magnetic porous carbonaceous material produced from tea waste for efficient removal of As(V), Cr(VI), humic acid, and dyes. *ACS Sustainable Chemistry & Engineering*, **5**(5), 4371–4380, doi: 10.1021/acssuschemeng.7b00418.

Wu, C., Iii, B. N., John, J., Milder, A., Zollars, B., Savoy, S., and Shvets, G. (2012). Metamaterial-based integrated plasmonic absorber/emitter for solar thermo-photovoltaic systems. *Journal of Optics*, **14**(2), 024005, doi: 10.1088/2040-8978/14/2/024005.

Wu, J., Hou, Y., Wang, P., Wang, Z., Li, Y., Wang, S., and Yang, M. (2016). Detection of lysozyme with aptasensor based on fluorescence resonance energy transfer from carbon dots to graphene oxide. *Luminescence*, **31**(6), 1207–1212, doi: 10.1002/bio.3092.

Xiao, N., Zhou, Y., Ling, Z., and Qiu, J. (2013). Synthesis of a carbon nanofiber/carbon foam composite from coal liquefaction residue for the separation of oil and water. *Carbon*, **59**, 530–536, doi: 10.1016/j.carbon.2013.03.051.

Xiong, J., Borg, Ø., Blekkan, E. A., and Holmen, A. (2008). Hydrogen chemisorption on rhenium-promoted γ-alumina supported cobalt catalysts. *Catalysis Communications*, **9**(14), 2327–2330, doi: 10.1016/j.catcom.2008.05.017.

Xiong, L., Liu, J., Bai, L., Li, X., Lin, C., and Lin, J.-F. (2014). Strength and structural phase transitions of gadolinium at high pressure from radial X-ray diffraction. *Journal of Applied Physics*, **116**(24), 243503, doi: 10.1063/1.4904747.

Xu, Q., Jin, L., Wang, Y., Chen, H., and Qin, M. (2019). Synthesis of silver nanoparticles using dialdehyde cellulose nanocrystal as a multi-functional agent and application to antibacterial paper. *Cellulose*, **26**(2), 1309–1321.

Xue, X.-Y., Cheng, R., Shi, L., Ma, Z., and Zheng, X. (2016). Nanomaterials for monitoring and remediation of water pollution. *Environmental Chemistry Letters*, **15**, 23–27, doi: 10.1007/s10311-016-0595-x.

Yadav, S., Tam, J., and Singh, C. V. (2015). A first principles study of hydrogen storage on lithium decorated two dimensional carbon allotropes. *International Journal of Hydrogen Energy*, **40**(18), 6128–6136, doi: 10.1016/j.ijhydene.2015.03.038.

Yockell-Lelièvre, H., Bukar, N., Toulouse, J. L., Pelletier, J. N., and Masson, J.-F. (2016). Naked-eye nanobiosensor for therapeutic drug monitoring of methotrexate. *Analyst*, **141**(2), 697–703, doi: 10.1039/c5an00996k.

You, X., Lin, W., Wu, H., Dong, Y., and Chi, Y. (2018). Carbon dot capped gold nanoflowers for electrochemiluminescent aptasensor of thrombin. *Carbon*, **127**, 653–657, doi: 10.1016/j.carbon.2017.11.024.

Zhang, B. (2018). *Physical Fundamentals of Nanomaterials*. William Andrew.

Zhang, P. (2015). Scaling for quantum tunneling current in nano- and subnano-scale plasmonic junctions. *Scientific Reports*, **5**, 9826.

Zhang, Q., Jing, L., Zhang, J., Ren, Y., Wang, Y., Wang, Y., et al. (2014). Surface plasmon resonance sensor for femtomolar detection of testosterone with water-compatible macroporous molecularly imprinted film. *Analytical Biochemistry*, **463**, 7–14, doi: 10.1016/j.ab.2014.06.014.

Zhao, X., Zhou, R., Hua, Q., Dong, L., Yu, R., and Pan, C. (2015). Recent progress in Ohmic/Schottky-contacted ZnO nanowire sensors. *Journal of Nanomaterials*, **2015**, 1–20, doi: 10.1155/2015/854094.

Zhou, N., Xu, X., Hammack, A. T., Stipe, B. C., Gao, K., Scholz, W., and Gage, E. C. (2014). Plasmonic near-field transducer for heat-assisted magnetic recording. *Nanophotonics*, **3**(3), doi: 10.1515/nanoph-2014-0001.

Zou, Y., Wang, X., Khan, A., Wang, P., Liu, Y., Alsaedi, A., et al. (2016). Environmental remediation and application of nanoscale zero-valent iron and its composites for the removal of heavy metal ions: a review. *Environmental Science & Technology*, **50**(14), 7290–7304, doi: 10.1021/acs.est.6b01897.

Chapter 9

Functional Green Nanomaterials Designing Approach to Spectroscopic Evaluation

Ahmed Emad,[a,b] Irene S. Fahim,[a,b] and Tawfik Ismail[a,c]

[a]*School of Engineering and Applied Sciences, Nile University, Sheikh Zayed, Giza, Egypt*
[b]*Engineering Systems Research Center, Nile University, Sheikh Zayed, Giza, Egypt*
[c]*Wireless Intelligent Networks Center, Nile University, Sheikh Zayed, Giza, Egypt*
isamy@nu.edu.eg

The challenges of producing clean energy while preserving our environment have been an objective that is being actively pursued around the world. Developing a way to produce clean energy from the sun is one of the most promising approaches. However, while there is a plethora of activities in the solar cell area there seems to be much less activity in related areas that will make it possible to reduce the overall cost of delivering energy and improve system efficiency through using green materials. The incorporation of a natural nanopolymeric layer to replace the conventional polymeric top layer will be introduced in this study. This nanolayer will enhance the electrical and absorption properties of the solar cell.

Nanomaterials for Spectroscopic Applications
Edited by Kaushik Pal

ISBN 978-981-4877-69-5 (Hardcover), 978-1-003-16033-5 (eBook)
www.jennystanford.com

9.1 Introduction

Nowadays, countries face a significant obstacle in supplying large amounts of energy from non-renewable sources due to the enormous increase in energy demand. China produces 30% carbon dioxide emissions, followed by the United States with 15% based on 2010 statistics [1]. The fact that controlling carbon emissions is a very tough issue. The contribution of carbon emissions by fossil fuels has increased by about 90% from 1990 to 2010. Carbon emissions are affected by several means as an industry, transportation, and heat production. Fossil fuels contribute to 21% of the total greenhouse gas emissions besides burning coal, transportation, and industry [2]. Consequently, the scope of energy harvesting has been shifted in the last decades from non-renewable energy to renewable energy. The large amounts of pollutants produced from different industries pushed recent studies to look for smart solutions. Utilizing solar energy is one of the most promising approaches for one of the cleanest energy sources. Since exploitation of light without transforming it into different forms is considered a passive technology, recent research tries to utilize solar energy and convert it into electricity.

A study held in 2007 shows the effectiveness of solar cells to suffice the world's consumption of electricity. Four percent of the surface area of the land would be enough to supply the world's usage of electricity. For example, less than 1% of European land would serve all the continent and supply the energy required for all the countries [3]. The use of solar cells has improved from 1.4 GW to 40 GW within 10 years, only from 2000 to 2010. This increase addresses the interest in countries to head to solar energy. Germany, the United States, and China have led the evolution of these numbers throughout the last years. China had a significant contribution to the photovoltaics industry, with 32.3% of total solar cells capacity worldwide. As per 2018 statistics, the United States and Germany share 11.5% and 8.3%, respectively [4]. In Europe, solar energy contributes to 4% of the electricity of the total continent consumption. Solar energy resources contribute to over 7% of the electricity generated in Germany, Italy, and Greece. In the worst-case scenario, it is predicted that at least 10% of the electricity needs in Europe will be provided via solar power by 2030. On the other hand, the African countries are making progress in the exploitation of solar energy, but still, better results can be obtained. Theoretically photovoltaics (PVs) in Africa can produce up to 660,000 TWh [5, 6].

Solar cells and photovoltaics industry are the current promising alternatives for fossil fuels and the rest of non-renewable energy. The nowadays point of research to supply electricity is the enhancement of two factors in the solar cells: the efficiency and cost. Single junction-based PVs have been the dominant and the commonly used structure for many years since the required fabrication process was not heavily cost in addition to the large resultant area for the cells. However, Shockley–Queisser started studying several techniques and structures in order to enhance the efficiency of the solar cell besides tracing down the heat flow and dynamics. Based on his reviews, technologies would never proceed over 31%. The design architecture of multi-junction connected to each other has evolved an outcome that surpasses the limit traced by Shockley–Queisser during his studies on single-junction based devices. The building materials for multi-junctions' structures could be semiconductors materials. The formulation of numerous single p-n junctions together results in multiple p-n junctions. It exhibits stability, better performance including efficiency. Each Incident light source with a specific wavelength would excite carriers in each junction with a net electric current for the whole multijunction cell. The primary aim of invoking the concept of numerous p-n junctions is occurring enhancement in the magnitude of absorbance in the cell. The region of the limited operating wavelength for solar cells needs to be as wide as possible. Since considering high absorbance for the cell active materials and a wide range for the operating wavelength of the incident source, this would reflect in high energy and power conversion efficiency. After different experimental results for multi-junction based solar cells, performance has been raised above 43% [8, 9]. The industrial tandem solar cells are commercially used with efficiency reached up to 30% under ordinary exposure to sunlight. The effect of focused sunlight sources pushes efficiency by up to 40%

[10]. Note that these results are achieved considering the high cost for manufacturing and many complex modifications.

Modifications are investigated in order to surpass Shockley–Queisser limit. These modifications either in fabrication or structure would maintain high efficiency [11]. The effectiveness of wide-bandgap materials in solar cells will reflect in enhancement for efficiency [12]. Wideband gap-based semiconductors owe an energy bandgap of more than 2.2 eV [12]. Gallium nitride (GaN) and silicon carbide (SiC) are examples of wide bandgap semiconductors of the third generation. The promising characteristics for wide bandgap materials are their highly shifted breakdown point of the electric field. WBG semiconductors are fabricated under high temperature. This is considered an advantage since these materials can resist for a long time under high temperature exposed to the cell [13, 14]. The strategy for WBG usage is decreasing the leakage current in the cell. Since the difference between the lower energy level in the conduction band and higher level in the valence band is higher than 2 eV, this would make the intrinsic carriers' concentration lower than usual. This is the principle for the positive effect caused by WBG materials.

In this chapter, the improvement of the electrical and optical properties of solar cells will depend on the functionalization of a thin natural polymeric layer with adding additives in the form of nanoparticles (i.e. the formation of the nanopolymeric layer) within the active layers in the current solar cell. This can be achieved by incorporating inorganic nanofillers dispersed at the nanoscale within natural polymeric matrices, which produce membranes of polymer nanocomposite (PNC) with enhanced electrical properties. The nanocomposite materials combine the benefits of the polymer matrix ranging from flexibility, polymer processability, and inorganic filler thermal stability [8].

9.2 Solar Cells

Solar cells or PVs industry is dramatically growing and developing in terms of manufacturing technologies, efficiencies, design, materials, and cost. The research points for solar cells are concerned with cost-

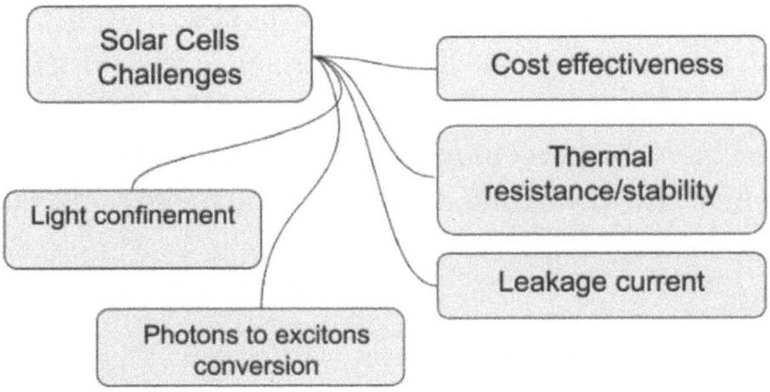

Figure 9.1 Solar cells challenges.

effectiveness, light trapping in the cell, absorption, minimization for leakage currents and thermal stability (Fig. 9.1). Different solar cell types are introduced as three generations. The first one is characterized by monocrystalline silicon wafers, followed secondly with amorphous silicon. Thirdly, the low-cost section of solar cells; organic-based photovoltaics (OPVs) [15]. Solar cells' common materials are mostly light materials. This lightweight compared with the net output power is considered to be an advantage for solar cells. This selectivity to control high power with low weight structures. Additionally, these light structures could be established commercially at any remote location required. This causes a reduction in many financial sectors in production. That's why it is recommended to use photovoltaic systems in many areas. As discussed before, PVs are built from two sections: P-type and N-type. These layers account for acceptors and donors for charge carriers. The intersection region between those two layers is called the depletion region and this phenomenon is responsible for the junction. The junction is the channel where electrons flow through P and N regions [16].

The main dominant semiconductor is the crystalline silicon. It possesses the interest of all commercial sectors with a percentage that surpasses 90% of the worldwide solar cells market. Since there are different structures for solar cells, the semiconductor single-junction contributes 25% of these industries. As discussed before, according to Shockley–Queisser analysis, the single junction could not be improved over a certain limit [4]. Modifications in a structure such as tandem structure which is composed of multiple junctions can push overcoming Shockley–Queisser limit. Practically, after very careful and accurate experiments, multi-junction solar cells have overcome the percentage of 40% under specific conditions of focusing light. However, in theory it could achieve more than 50% to contribute to crucial applications [17].

9.3 Types of Solar Cells

Solar cell technology is dominated by silicon solar cells. About 80% of the functioning cells are based on semiconducting materials. The other type of cells is either organic based cells or modified structures for semiconducting based cells. In this section, a brief description of the different types of solar cell materials is provided.

9.3.1 Inorganic Solar Cells

Solar cells are classified either organic or inorganic based on the nature of the active material (absorbance layer) of the p-n junction. Inorganic materials used in solar cells have high electron mobility, electrical conductivity, and thermal stability. Their characteristics give advantages over organic solar cells considering optical output power and reliability as shown in Fig. 9.2. A wide range of materials can be used for inorganic materials-based solar cells. Materials like indium gallium nitride (InGaN), gallium indium phosphide (GaInP), and gallium arsenide (GaAs) are commonly used as choices for active materials. III-nitride systems enable high-performance solar cells due to their WBG gaps. Several modifications are introduced for inorganic solar cells to enhance the power conversion efficiency. WBG gaps structures and photonic crystals are two main techniques used to allow maintaining higher light trapping performance.

Since WBG materials reduce the leakage current, they are designed for enhancing the behavior of the solar cells. Devices based on substrates like SiC show better performance [18]. The fabrication and preparation processes for such materials operating under very high temperatures can reach up to 1200°C. That's why the structure

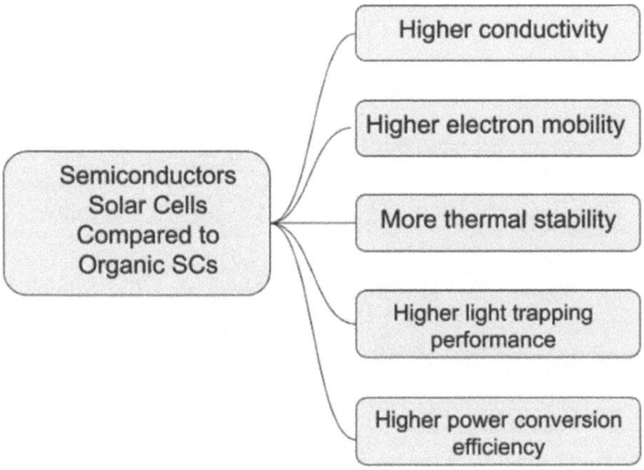

Figure 9.2 Comparison between semiconductors to organic solar cells.

developed using such materials can be stabilized firmly under high temperature compared with semiconductors like crystalline silicon [19].

9.3.2 Organic Solar Cells

The concept of green chemistry is applied in fabricating organic solar cells. It is the design and processing of chemical products that reduce or eliminates the use or production of harmful substances. Such a process produces tiny waste, generates environmentally byproducts, and uses energy efficiently. Green chemistry relates to the life cycle of the solar cell, including its design, manufacture, and usage [20]. Consequently, organic solar cells have been considered as an alternative for its low cost compared with inorganic ones. However, organic materials owe lower electrical conductivity, and hence carriers' mobility is low. Recent research is focusing on increasing power conversion efficiency for organic solar cells. Therefore, there are different structures and types for the establishment of organic solar cells. Each type exhibits a modification either in the structure or usage of different materials for enhancement of efficiency. Figure 9.3 summarizes the modification introduced to organic solar cells to enhance its efficiency.

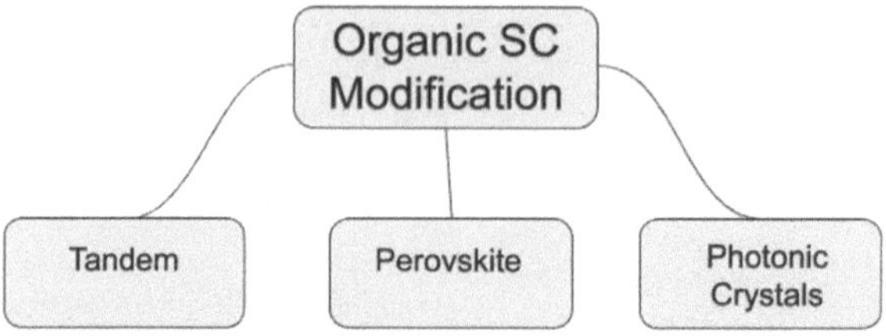

Figure 9.3 Modifications to organic solar cells.

9.3.3 Tandem Cells

The simultaneous increase in the number of cells will lead to higher conversion efficiency. The technique of implementing this approach is tandem solar cells. Tandem solar cells are a formulation for a multi-junction structure with a combination of two cells, at least. The two subcells are connected, with one at the top and the other at the bottom. A layer is positioned between the two cells to decrease the recombination of electrons and holes again [21]. The net current of the cell is the minimum short-circuit current between the two cells. The tandem cell's open-circuit voltage is the summation of the two connected cells together of an open circuit voltage as summarized in Fig. 9.4 [22]. This structure has produced more carriers through active material with such analysis.

9.3.4 Perovskite Solar Cells

Perovskite solar cells are a branch of solar cells based on materials with a specific orientation of crystal structure. Materials have the

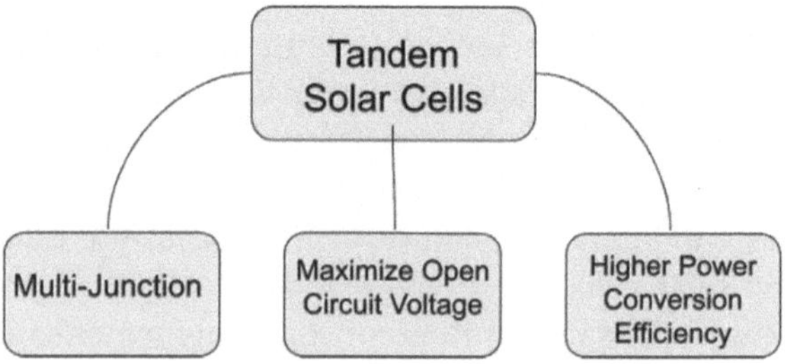

Figure 9.4 Tandem solar cells principle.

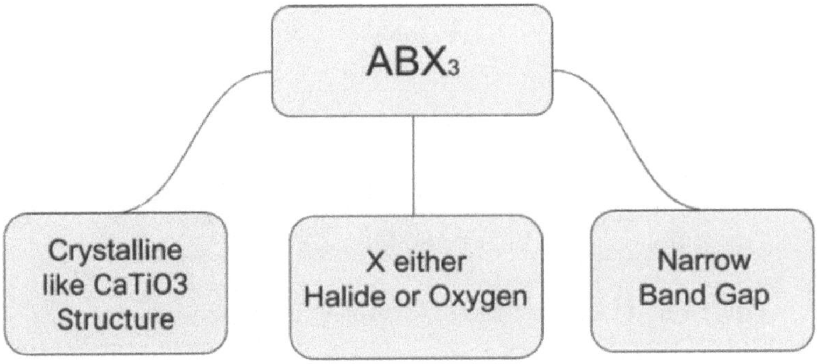

Figure 9.5 Main principles for perovskite solar cells.

structure of ABX3 with the same crystal structure as calcium titanite, where A and B elements act as cation X acts as an anion. X element must be either oxygen or halide. However, it is more commonly studied as oxygen than halide since oxides have higher electrical properties [23]. Figure 9.5 shows an exhibition for the fundamental points for the ABX3 molecular structure. It has excellent charge carrier transport based on its crystallinity and the length where the carrier can transport. The perovskite cell exhibits a narrower bandgap in 3D structures compared with the 2D structure [24]. The narrow bandgap is beneficial for PV devices [25]. Perovskite solar cells have power conversion efficiency more than conventional organic active materials.

9.3.5 Photonic Crystals

Photonic crystals (PCs) can be implanted in organic and inorganic solar cells. PCs are periodic structures established through the active material. For example, inorganic materials have low electron mobility compared with semiconductors. Also, their light absorption is much lower compared with silicon. Photonic crystals invoke photonic gaps in order to prevent light from spreading by them [26]. This leads to the containment of light in the active material for organic and inorganic solar cells. Control of trapped light in solar cells improves light absorption. The variation of the periodic sequence dimensions of photonic crystals determines the incident

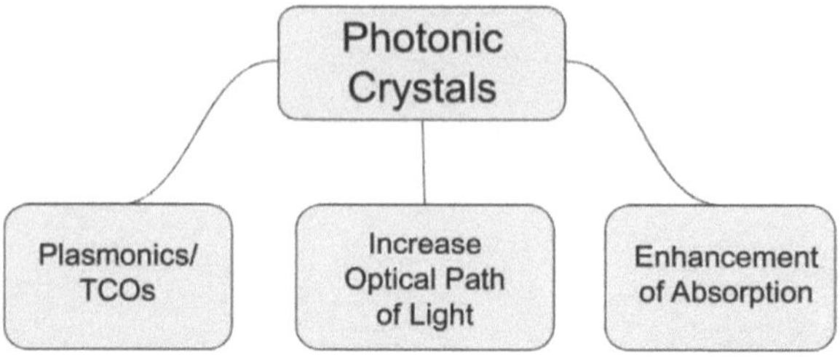

Figure 9.6 Main principles for photonic crystals-based solar cells.

trapped wavelength. Therefore, photonic crystals give a particular property for the required input wavelength [26]. Recent studies show the effect of the shape of the photonic crystals on the absorption magnitude.

The contrast between refractive indices of photonic crystals and active materials excites the manipulation of light. Thus, it increases the optical path of light and consequently causes an increase in absorption. Many plasmonic materials and transparent conducting oxides are used as photonic crystals. Zinc oxide, lithium fluoride, indium tin oxide, and other materials are commonly used to enhance the confinement of light in solar cells. These materials in each proposed structure cause the required refractive index contrast to confine light in the active material. A summary of the three crucial characteristics for the effect of the photonic crystal in the active material is illustrated in Fig. 9.6.

9.4 Polymer Nanocomposites

The production of organic solar cells with polymeric materials allows the production of flexible, ductile, thermally stable, lightweight, and quickly processed greener solar cells. The production costs of the organic solar cell decreased. However, the low-temperature resistance of polymers is a difficulty in the production of a highly efficient organic solar cell. There are certain defects in polymeric layers used in certain applications for example OPVs. These materials are unable to resist for a long time or keep their stability under high temperatures. Additionally, they are not mechanically firm. The involvement of nanoparticles through polymeric layers can solve these defects. The optoelectronic characteristics of the solar cells are significantly improved by the implantation of nanoparticles. The studies are concerned with the most adequate way to implant nanoparticles or nanofiller through polymeric layers. Recently, the most adequate form for such nanoparticles is PNCs [27]. On the other hand, recent research focuses on materials more applicable with low-cost fabrication techniques and can hold the required mechanical characteristics. Graphene is one of the most applicable materials in fabrication as it can be prepared using wrapping sheets of graphene. Additionally, graphene characteristics enhance the absorption due to its mechanical stiffness [28, 29].

The replacement of synthetic polymer with the natural polymer in organic solar cells will decrease the waste since natural polymers will degrade after using the organic solar cell. Natural polymers are not synthesized. However, it is created smoothly and naturally through proteins and polysaccharides as shown in Fig. 9.7. The availability of such natural sources is a crucial

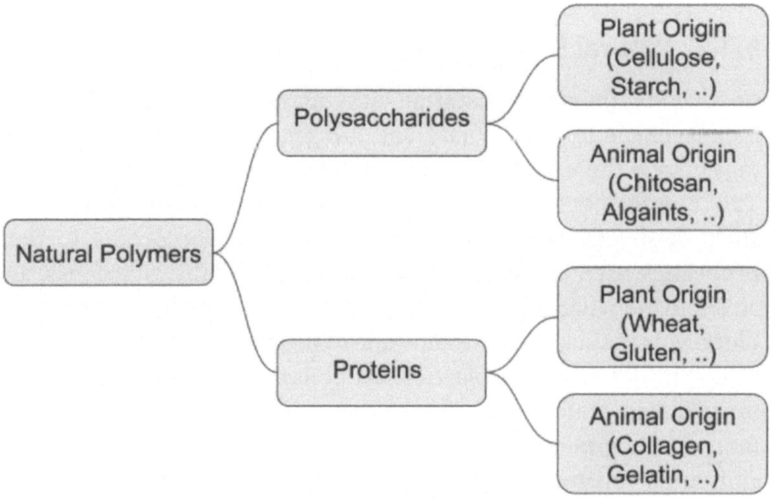

Figure 9.7 Natural polymers tree.

advantage for natural polymers. The comparison between natural and synthesized polymers represents the major difference between the effects of each of them on the surroundings. Natural polymers don't have severe effects compared with synthesized ones. That's because the preparation of the synthesized polymer requires chemical reactions with chemical exhausts. Such reactions for preparing specific carbonic polymers had a negative impact on the atmosphere.

As the natural polymers are formed naturally, they decay the same way in a natural behavior without any severe side effects. Based on these characteristics including safety information, decomposition, maintaining the characteristics for stiffness and enhancing absorbance in addition to its low cost, research focuses on improving the PNCs required for fabrication [30].

9.4.1 Natural Polymers

The primary sources of renewable biomass generated from an organic macromolecular matter of animals, plants, and microbes are the natural polymers.

9.4.2 Natural Biopolymer

Natural biopolymer material has many advantages, such as biodegradability, low toxicity, and low disposal costs. These advantages made the use of natural biopolymer materials very attractive. The biopolymers are classified according to their applications such as sheets, films, and fibers or classified according to their properties such as degree of crystallinity, solubility, hardness, and strength. Replacing synthetic plastic bags with bags based on natural biopolymers such as chitosan is one of the most promising solutions to dispose of synthetic plastic bags pollution. The biopolymers are formed from different resources as described in Fig. 9.8. However, many of the applications of chitosan are lost in the competitive world compared to synthetic polymers. Enhancement of the biopolymer properties is done by adding natural fillers such as cellulose fibers, and chitin fibers extracted from animal or plant waste are

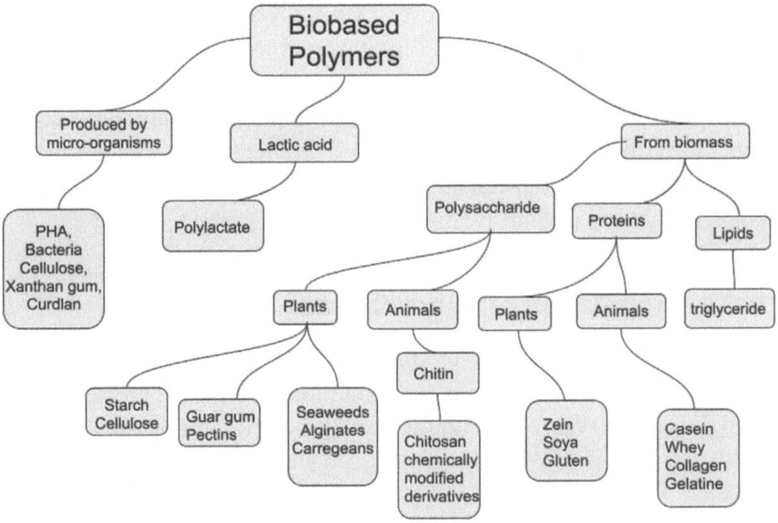

Figure 9.8 Bio-based polymers tree.

needed to overcome the limitations of chitosan. The formation of biocomposites from chitosan as matrices with cellulose fibers and chitin fibers as fillers do not pose environmental hazards [30].

9.4.3 Chitin and Chitosan

Chitin is the second abundant natural polymer in nature after cellulose. The primary sources of chitin are found in crustacean shells like shrimp and crab or insects but also some mushroom's envelopes, green algae cell walls, and yeasts [31]. Shrimp industries produce significant quantities of shrimp-coating waste during processing, around 45–55% of the weight of raw shrimps, but it can be used for the production of value-added items such as chitin [32]. Also, the yield of extracted chitin from shrimp shell waste is the highest in most of the polysaccharides, in which the yield of extracted chitin is around 30–40%.

The key derivative of chitin is chitosan, which consists of β-(1,4)-2-amino-2-deoxy-D-glucose, β-(1,4)-2-acetamido-2-deoxy-D-glucose, partly deacetylated chitin and thus formed by repeated units. The real difference is the acetyl content of the polymer between chitin and chitosan. Chitin and chitosan have been widely studied for various uses in pharmaceutical, agriculture, water treatment, and food packaging applications due to its biodegradability, biocompatibility, antimicrobial and antibacterial. Chitosan is the right candidate for manufacturing the polymeric layer in the organic solar cell due to its film-forming ability [33].

9.4.4 Preparation of Nanochitosan

Chitosan was dissolved in acetic acid by stirring the solution at 300 rpm, sodium tripolyphosphate crosslinker was added and dissolved into deionized water to produce nano green polymer [34].

9.5 Fabrication of Solar Cells

WBG gap structures like SiC based devices are constructed under a high-temperature environment. The wafer undergoes a thermal annealing process during bonding the layers of the structure. Additionally, deep etching is carried out on the SiC substrate to facilitate the extraction of gases during the annealing process. The device layers and SiC substrate are etched with the active acid, hydrofluoric acid (HF), to align the edges of the combined structure [17].

PNC and PCs have conventional fabrication techniques i.e., chemical vapor deposition (CVD), spin coating, and liquid phase deposition (LPD). Chemical vapor deposition can be carried out on a wide range of materials. Deposition can be maintained at high temperatures of up to 1000°C. The process is an excitement for a chemical reaction near the proposed film to be coated. The reaction occurs and is accompanied by the evolution of a vapor [35]. This resultant vapor is directed towards the layer and deposited uniformly on the thin film. Spin coating is a process of addition of droplets for a specific liquid on the substrate. Hence, a rapid rotational motion for the pre-installed substrate to distribute the liquid uniformly. The spin rate controls the thickness of the coating. The process ends with solvent evaporation and keeps the required

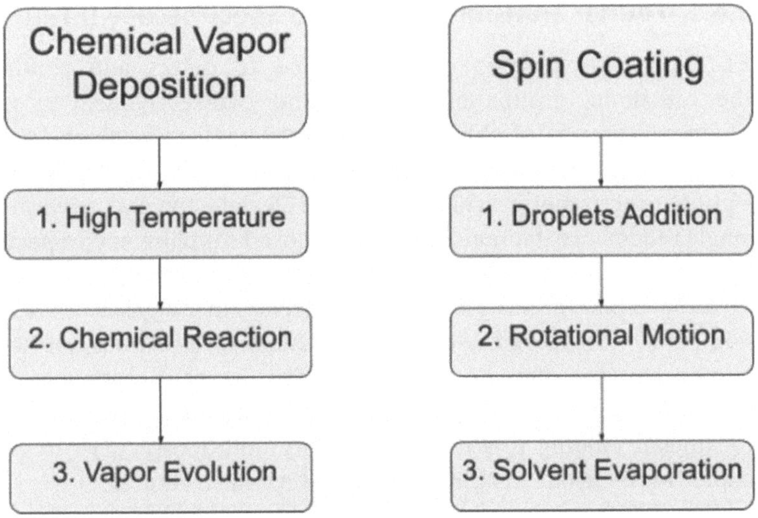

Figure 9.9 CVD and spin coating operation steps.

coated thin film on the substrate. Figure 9.9 highlights the main steps for the operation of chemical vapor deposition and spin coating.

9.6 Spectroscopic Characterization and Evaluation

The following techniques are suggested for the characterization of organic solar cells. Characterization techniques differ from each other according to their results, either qualitative or quantitative. The device typically results in an image that can show the sample elements or structure morphology. The source of the device can be either an electron beam or light source. The resolution of the sample image and the calibration procedures for the device are factors that must be taken into consideration before choosing the characterization technique. The following characterization techniques exhibit selectivity on the source, the resolution and calibration, and are applicable for the characterization of OPVs.

9.6.1 Fourier Transform–Infra-Red Spectroscopy (FT-IR)

FT-IR spectral analysis was performed to detect and confirm the functional groups of chitosan and detect changes in the functional groups of chitosan. Attenuated total reflectance (ATR) mid-infrared spectra were developed using a Tensor 27 (Bruker Optics) spectrometer armed with a DTGS detector and a warmed single reflectance diamond ATR with a closed sapphire anvil. Spectra were acquired with 128 scans at a 4 cm^{-1} resolution for samples at room temperature. The spectra wavelength ranges were 400–4000 cm^{-1}. Compared with other techniques, FT-IR does not have any destructive effect on the sample. The device components can be quickly stabilized mechanically and need no calibration by the technician. Adding this to its sensitivity and accuracy, FT-IR is a choice for different samples and sophisticated structures.

The FT-IR spectrum of chitin in Table 9.1 indicates a characteristic band at 3446 cm^{-1}, which is credited to the vibration of -NH and -OH groups. Moreover, there is a band 3102 cm^{-1}, which is allocated to the vibrational modes of NH in the amide. The characteristic carbonyl C=O of chitin at 1654 cm^{-1} is assigned to the vibrations of the amide I band. The band at 1560 cm^{-1} relates to the N-H distortion of amide II. The band at 1378 cm^{-1} is named according to the CH– deformation band. The vibration band at 1074 cm^{-1} indicated C–O–C vibration inside the chitin ring [36]. These peaks are crucial to distinguish the development between chitin and chitosan. Moreover, the vibration of $-CH_3$ group in chitin spectrum was absent in the FT-IR spectrum of chitosan after treating in

Table 9.1 FT-IR spectral value of chitin, chitosan

Assignment	Frequency Chitin	Frequency CS
Hydroxyl, O–H stretch	3446	—
Amine, N–H stretch	3102	3263
Aliphatic compound, $-CH_2$ stretch	—	2919
Secondary amide, C=O	1654	1643
Amide, N–H bend	1560	1541
Aliphatic compound $-CH_2$ bend	1378	1403
C–O–C glucosidic linkage	1074	1151

concentrated NaOH thus the $COCH_3$ in chitin backbone was removed during the deacetylation process and chitosan was manufactured [37].

9.6.2 Transmission Electron Microscopy (TEM)

TEM uses incident electron beams rather than light beams. The electron beam interacts with the sample during its path. The collision between electrons and the atom shells causes scattering of electrons [38]. This scattering, whether due to elastic collision or inelastic collision, reflects the existence of atoms. The output pattern of the transmitted electrons is collected, and consequently, it describes the structure of the sample [39]. TEM performs an accurate characterization; that's why it is used to characterize the WBG-based devices. The interface between the WBG wafer and the device layers is analyzed by TEM after the fabrication steps are finalized. Also, characterization for WBG based structures uses X-ray diffraction to test whether the layer has strains on the surface or not.

9.6.3 Scanning Electron Microscope (SEM)

SEM uses incident electron beams the same as TEM. SEM analysis gives information about the structure of the sample. Apart from TEM, SEM characterizes the structure of the sample by detecting the reflected, scattered electron beams from the surface [40]. The resultant image from the SEM detection is the morphology

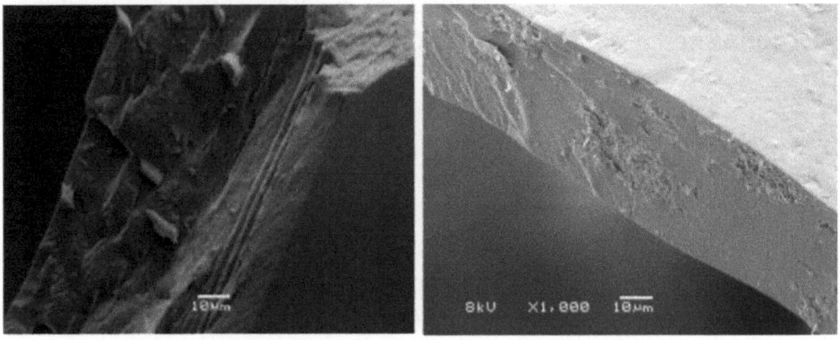

Figure 9.10 Chitosan and nanochitosan.

of the surface. However, it cannot represent the fundamental characteristics of the sample [41]. On the other hand, the visual image is very accurate, precise, and has a resolution that reaches about 1 nm [42]. SESEM, Leo Supra 55 – Zeiss Inc., Germany, high standard secondary electron (SE2) lens was used with a range of 6–8 kV to prevent the deteriorating of the polymer. The differences between the chitosan and nanochitosan SEM samples are shown in Fig. 9.10. The chitosan and nanochitosan showed homogeneity in the form of smooth surfaces without cracks or pores, indicating good miscibility.

9.7 Conclusion

The main goal of the proposed research is to tailor the natural nanopolymeric active layers within solar cells to improve the optoelectronic properties, and in turn, enhance the cell's efficiency. Adding of the conjugated polymer layer with nanoadditives will be modeled by computer software. After reaching the optimal results, the work will be designed and fabricated experimentally. The nanopolymer layer is manufactured so that the active materials can be patterned on lightweight, flexible substrates in wide areas employing low-cost printing and coating technologies. Upon the addition of the nanopolymeric layer, the cell's reflectivity will increase. Various parameters will be tested such as the tailoring and stacking of the solar cell with various polymeric materials, and the morphology of the stack (i.e. nanostructure) will be varied to highlight the optimum structure. A comparative study will be accomplished among the simulations and the experimental results to suggest future work.

References

1. G. Marland, T. A. Boden and R. J. Andres, Global, regional, and national fossil-fuel CO_2 emissions, Carbon Dioxide Information Analysis Center, Oak Ridge National Laboratory, U.S.
2. Migration of climate change, *The Intergovernmental Panel on Climate Change (IPCC)*, 2014.

3. World energy resources 2016, *World Energy Council*, 2016.
4. REN211, *International Renewable Energy Agency*, 2011.
5. N. S. Ouedraogo, Opportunities, barriers and issues with renewable energy development in Africa: a comprehensible review, *Current Sustainable/Renewable Energy Reports*, 2019, **6**(2), 52–60.
6. P. Bórawski, T. Yashchenko, A. Sviderskyi, J. Dunn, Development of renewable energy market in the EU with particular regard to solar energy, *Conference Proceedings of Determinants of Regional Development*, 2019.
7. W. Shockley and H. J. Queisser, Detailed balance limit of efficiency of p-n junction solar cells, *Journal of Applied Physics*, 1961, **32**, 510.
8. N. Brown, solar junction breaks concentrated solar world record with 43.5% efficiency, *Cleantechnica.com*, Apr. 19, 2011 [Online].
9. Z. Shahan, Sharp hits concentrator solar cell efficiency record, 43.5%, *Cleantechnica.com*, May 12, 2012 [Online].
10. Solaerotech, ZTJ space solar cell, datasheet, Apr. 2018.
11. W. Shockley and H. J. Queisser, Detailed balance limit of efficiency of p-n junction solar cells, *Journal of Applied Physics*, 1961, **32**, 510.
12. Y. Sun, K. H. Montgomery, X. Wang, S. Tomasulo, M. L. Lee and P. Bermel, Modeling wide bandgap GaInP photovoltaic cells for conversion efficiencies up to 16.5%, *IEEE 42nd Photovoltaic Specialist Conference*, PVSC 2015.
13. J. J. Wysocki and P. Rappaport, Effect of temperature on photovoltaic solar energy conversion, *Journal of Applied Physics*, 1960, **31**(3), 571.
14. K. H. Montgomery, C. Heredia and J. M. Woodall, Design and modeling of a high efficiency hybrid photovoltaic-photothermal concentrator (PVPTC) system, *IEEE 39th Photovoltaic Specialists Conference*, PVSC 2013.
15. G. Conibeer, Third-generation photovoltaics, *Materials Today*, 2007, **10**(11), 42–50.
16. M. A. Green, Photovoltaic principles, *Physica E: Low-dimensional Systems and Nanostructures*, 2002, **14**(1–2), 11–17.
17. M. Yamaguchi, T. Takamoto, K. Araki and N. Ekins-Daukes, Multi-junction III–V solar cells: current status and future potential, *Solar Energy*, 2005, **79**(1), 78–85.
18. F. Roccaforte, et al., Emerging trends in wide band gap semiconductors (SiC and GaN) technology for power devices, *Microelectronic Engineering*, 2018, **187–188**, 66–77.

19. B. Ozpineci and L. M. Tolbert, Comparison of wide-bandgap semiconductors for power electronics applications, Technical Report, Oak Ridge National Laboratory, Dec. 2003.

20. D. J. Burkea and D. J. Lipomi, Green chemistry for organic solar cells, *Energy & Environmental Science*, 2013, **6**, 2053–2066.

21. M. A. Green, K. Emery, Y. Hishikawa, W. Warta and E. D. Dunlop, NIMS sets a new world record for the highest conversion efficiency in dye-sensitized solar cells, *Progress in Photovoltaics: Research and Applications*, 2012, **20**(1), 12–20.

22. G. Dennler, M. C. Scharber, T. Ameri, P. Denk, K. Forberich, C. Waldauf and C. J. Brabec, Design rules for donors in bulk-heterojunction tandem solar cells-towards 15% energy-conversion efficiency, *Advanced Materials*, 2008, **20**(3), 579–583.

23. N. Park, Perovskite solar cells: an emerging photovoltaic technology, *Materials Today*, 2015, **18**(2), 65–72.

24. Y. Zou, Y. Cui, H.-Y. Wang, Q. Cai, C. Mu and J.-P. Zhang, Highly efficient and stable 2D-3D perovskite solar cells fabricated by interfacial modification, *Nanotechnology*, 2019, **30**(27), 275202.

25. A. Kojima, et al., Organometal halide perovskites as visible-light sensitizers for photovoltaic cells, *Journal of the American Chemical Society*, 2009, **131**(17), 6050–6051.

26. P. Bermel, C. Luo, L. Zeng, L. C. Kimerling and J. D. Joannopoulos, Improving thin-film crystalline silicon solar cell efficiencies with photonic crystals, *Optics Express*, 2007, **15**(25), 16986–17000.

27. M. D. Calisir, E. Stojanovska and A. Kilic, Polymer nanocomposites for dye-sensitized solar cells, in M. Jawaid and M. M. Khan, (eds.) *Polymer-Based Nanocomposites for Energy and Environmental Applications*. Woodhead Publishing, UK, 2018, pp. 361–396.

28. I. S. Fahim, H. G. Salem and W. Mamdouh, Chitosan nanocomposite mesoporous membranes: mechanical and barrier properties as a function of temperature, *Journal of Material Science Research*, 2015, **4**(4), 1.

29. I. S. Fahim, N. Marei, H. G. Salem and W. Mamdouh, Effect of graphene and fullerene nanofillers on controlling the pore size and physicochemical properties of chitosan nanocomposite mesoporous membranes, *Journal of Nanomaterials*, 2015, **2015**, 979561.

30. J. Huang, S. Fu and L. Gan, *Lignin Chemistry and Applications*. Elsevier, 2019.

31. S. Morooka and K. Kusakabe, Microporous inorganic membranes for gas separation, *MRS Bulletin*, 1999, **24**(3), 25–29.
32. T. P. White, et al., Tandem solar cells based on high-efficiency c-Si bottom cells: top cell requirements for >30% efficiency, *IEEE Journal of Photovoltaics*, 2014, **4**(1), 208–214.
33. A. Bose and T. W. Wong, Oral colon cancer targeting by chitosan nanocomposites, in *Applications of Nanocomposite Materials in Drug Delivery*. Woodhead Publishing, 2018, pp. 409–429.
34. A. Elhussieny, M. Faisal, G. D'Angel, N. Aboulkhair, N. M. Everitt and I. S. Fahim, Valorisation of shrimp and rice straw waste into food packaging applications, *Ain Shams Engineering Journal*, 2020, in press.
35. A. Mihi, C. Zhang and P. V. Braun, Transfer of preformed three-dimensional photonic crystals onto dye-sensitized solar cells, *Angewandte Chemie International Edition*, 2011, **50**(25), 5712–5715.
36. K. Prabu and E. Natarajan, Isolation and FTIR spectroscopy characterization of chitin from local sources, *Advances in Applied Science Research*, 2012, **3**(2), 1870–1875.
37. M. Norhidayah, F. Ahyat, A. Mohamad and A. Ahmad, Chitin and chitosan extraction from portunus pelagicus, *Malaysian Journal of Analytical Sciences*, 2017, **21**(4), 770–777.
38. P. J. Goodhew, F. J. Humphreys and R. Beanland, *Electron Microscopy and Analysis*. Taylor and Francis, 2001.
39. R. F. Egerton, P. Li and M. Malac, Radiation damage in the TEM and SEM, *Micron*, 2004, **35**, 399–409.
40. J. I. Goldstein, D. E. Newbury, P. Echlin and D. C. Joy, *Scanning Electron Microscopy and X-Ray Microanalysis* (2nd ed.). Plenum Press, New York, 1992.
41. J. I. Goldstein, et al., *Scanning Electron Microscopy and X-Ray Microanalysis*. Springer, 2018.
42. K. Akhtar, et al., Scanning electron microscopy: principle and applications in nanomaterials characterization, in *Handbook of Materials Characterization*. Springer International Publishing, 2019.

Chapter 10

Green Chemistry Assisted Synthesis of Metallic Nanomaterials Applications

Devi Radhika[a] **and Karthik Kannan**[b]

[a] *Department of Chemistry, School of Engineering and Technology, JAIN-Deemed to be University, Jakkasandra, Ramnagara 562112, Karnataka, India*

[b] *Center for Advanced Materials, Qatar University, P.O. 2713, Doha, Qatar*

karthikkannanphotochem@gmail.com

In green chemistry, nanoscience and technology unlock novel prospects in physical, chemical and biophysical revolutions. The metallic nanoparticles have built the arena of synthesis, catalysis, enzyme immobilizations, and molecular communications simpler, quick and well regulated. Metal oxide nanomaterials are deemed to be one of the majority attractive materials in an extensive range of applications due to their impressive optical, electronic, photocatalytic, and biological properties due to their non-toxic nature, large specific surface area, useful optical bandgap, and high biological activity. This chapter deals with the metallic nanomaterials using bioengineered routes from various sources of plants, microbes, and chemical routes via sonication and microwaving, etc. It gives advanced knowledge to the scientific community who intends to design efficient metal nanoparticles/composites using green chemistry principles.

Nanomaterials for Spectroscopic Applications
Edited by Kaushik Pal

ISBN 978-981-4877-69-5 (Hardcover), 978-1-003-16033-5 (eBook)
www.jennystanford.com

10.1 Introduction

Metal oxide based nanomaterials with extraordinary characteristics have been proposed, prepared, and used as main materials in the field of photocatalysis and biomedicine, due to their non-toxic nature, large specific surface area, useful optical band gap, and high biological activity. This chapter reveals the photocatalytic, antibacterial and anticancer performances of metallic nanoparticles for multifunctional applications prepared by green route synthesis. Consequently, we also discuss the encountered difficulties and prospects for the future application of metal oxide-based composites as promising candidates for the progress of highly efficient photocatalytic and biomedical systems.

Numerous metallic nanoparticles have a lot of applications such as sensors, catalysis, pigments, drug delivery, biomedicine, and biological field. Concerning to environmental remediation, metallic nanoparticles have extensive and become promising material owing to their narrow bandgap, low cost, non-toxicity, thermal, and chemical stability [1, 2]. It was found that mixed metal oxides could exhibit outstanding high electronic, ionic conductivity and catalytic activity, in turn, which may support to accomplish elevated power output [2–5].

While reducing the size to nanoscale, the surface-to-volume ratio and specific surface area rise extremely. These novel properties of semiconductor nanomaterials such as Ag, Au, Pd, Pt, CuO [6–10], NiO [11–16], TiO_2 [17–22], MgO [23], CdO [24] and SnO_2 [25], etc. have been described as extraordinary photocatalysts (organic pollutant degradation) and antimicrobial (foodborne pathogens) agents have applications in evolving technologies like energy conversion, optoelectronics, sensors, bioimaging devices, catalysis, solar cells, detectors, biomedicine [6, 16, 21–25]. Also, the metallic nanoparticles are mostly utilized for water splitting (H_2 production), odor control, and inactivation of bacterial and cancer cells. The pollutants released straight to the atmosphere have to be treated to photodegradation without the construction of added waste and byproducts into water and CO_2. It is renowned truth that water pollution is a large encounter for the nearby and upcoming years.

In this situation, dyes participate vital roles in the contamination of water bodies and are extremely toxic in nature, which is often used in several factories. As dyes showed a complex structure, it is difficult to degrade certainly. Consequently, various photocatalytic approaches have been approved to destroy the organic contamination integrated into the water sources [21–25].

10.2 Importance of Metallic Nanomaterials

A progressive idea in the present semiconductor technology is doping of impurities into the semiconductor lattices that is an effective way for monitoring the electrical conductivity and enhancement of the optical, magnetic and biological properties. Anticancer and antibacterial properties have been manipulated by the coupling of metal oxides such as Ag, Au, Pd, Pt, and CuO [6–10], NiO [11–16], TiO_2 [17–22], MgO [23], CdO [24] and SnO_2 [25] which are the best candidates for biomedical applications. Another ongoing trend in the development of semiconductor nanoparticles and nanocomposites is to achieve an expanded absorption in the visible region to be used in photocatalytic and solar cell applications. Nanocomposites have drawn much scientific attention in current research works because of the enhanced features as opposed to pure metal nanoparticles. A nanocomposite is a mixed or matrix material that involves various substances. This combination creates novel properties of the materials when the nanocomposite has size varying between 1–100 nm. Nanocomposite can be synthesized from different groups of materials which can be classified into three main building blocks: metals, ceramics, and polymers. A nanocomposite can have distinctly diverse electrical, thermal, electrochemical, mechanical, catalytic, and optical properties from the essential materials behind it. Taking into account the spatial structure, nanocomposites can be considered as zero-dimensional (0D) such as core-shell, one-dimensional (1D) like nanotubes and nanowires, two dimensional (2D) such as lamellar, and three-dimensional (3D) like metal matrix composites. Based on their properties related to structure, nanocomposites are categorized as nanofilamentary, nano-layered and nano-particulate composites.

Nanocomposite materials attract much more attention from researchers, scientists, and engineers. This had inevitably led to a rapid increase in the number of research articles examining these materials. Furthest, those nanocomposites have developed as materials of the 21st century that extend several technological and business developments in all segments of life.

10.3 Production of Various Metallic Nanomaterials Using Green Chemistry

There are two approaches available to prepare nanomaterials. Mechanical grinding is belonging to the top-down approach and the bottom-up mainly consists of generation of colloidal dispersion by wet chemical synthesis route. It is opted to achieve narrow size distribution of nanoparticles and this process mentions the build-up of material from the bottom-most: atom-by-atom. Wet chemical routes, such as co-precipitation process, ultrasound synthesis, hydrothermal method, sol-gel method, microwave-assisted, salt-assisted aerosol decomposition process, spray pyrolysis technique, and refluxing synthesis was prepared for nanomaterials. But the green route is very essential to produce eco-friendly nanoparticles and it's very important for the current world needs to utilize natural resources. Further, it has been described that green route synthesis shows excellent properties and this method has been used in recent years and was proved to be successful [10, 11, 14, 15, 22–25]. Up to now, there is a lot of work has been performed on the production of various metal oxide nanoparticles via the green route from various natural resources such as extracts from plant parts. Among the different techniques reported, the green route is a simple and promising process to yield homogeneous and small-sized powders [22–25]. Numerous metal oxide nanoparticles prepared from various plant part extracts have been discussed in Table 10.1. Metal nanoparticles produced from the plant by-products such as stem, leaf, seed, leaf, latex are broadly organisms applied for nanomaterials preparation. The biogenic production of nanoparticles gives good-yield, nano-sized materials with suitable

Table 10.1 Production of nanomaterials from various plant extracts

Plant species	Mode of synthesis	Nanomaterials	Size (nm)	References
Guilandina Moringa L.	Green route	CuO	12	[6]
Caloropsis procera	Green synthesis	CuO	40–46	[7]
Oak fruit hull extract	Green route	CuO	20–35	[8]
Punica granatum	Green synthesis	CuO	10–50	[9]
Andrographis paniculata	Green route	CuO	30	[10]
Guilandina moringa L.	Green route (hot plate combustion)	NiO	8–10	[11]
Agathosma betulina	Green route	NiO	15–23	[12]
Berchemia chanetii	Bioengineered route	NiO	18	[13]
Melaleuca viminalis	Green synthesis	NiO	20–40	[14]
Citrus fruit juice	Green synthesis	NiO	20	[15]
Aloe Barbadensis	Green synthesis	TiO_2	20	[16]
Vitex negundo Linn	Green synthesis	TiO_2	10	[17]
Curcuma longa	Green synthesis	TiO_2	43	[18]
Vigna unguiculata	Green synthesis	TiO_2	11	[19]
Moringa oleifera	Green synthesis	TiO_2	12	[20]
Ledebouria revoluta	Green synthesis	TiO_2	47	[21]
Andrographis paniculata	Green synthesis (microwave-assisted)	TiO_2	19	[22]
Andrographis paniculata	Green synthesis (microwave-assisted)	MgO	35	[23]
Andrographis paniculata	Green synthesis	CdO	22	[24]
Andrographis paniculata	Green synthesis (microwave-assisted)	SnO_2	27	[25]
Ocimum tenuif lorum	Green synthesis	Ag	25–40	[43]
Syzygium cumini	Green synthesis	Ag	29–30	[44]
Psidium guajava	Green synthesis	Au	25–30	[45]
Mangifera indica	Green synthesis	Au	17–20	[46]
Pinus resinosa	Green synthesis	Pd	16–20	[47]
Glycine max	Green synthesis	Pd	15	[48]

properties. The biomediated process of nanoparticle synthesis can be categorized into three steps such as activation level, growth level, and termination level. An extreme calcination temperature is needed to eliminate the impurities or forerunner to obtain crystalline phase pure materials.

10.4 Metallic Nanomaterials in Environmental and Biological Applications

Metal oxide nanoparticles attracted many young scientists due to their appreciable magnetic behavior, chemical stability, and biocompatibility [10, 15–17]. The metallic nanoparticles are also well-known in the usage of enzymatic or metabolic degradation, drug delivery, tissue repair, pathogen detection, hyperthermia, and antigen diagnosis (Fig. 10.1) [21–25].

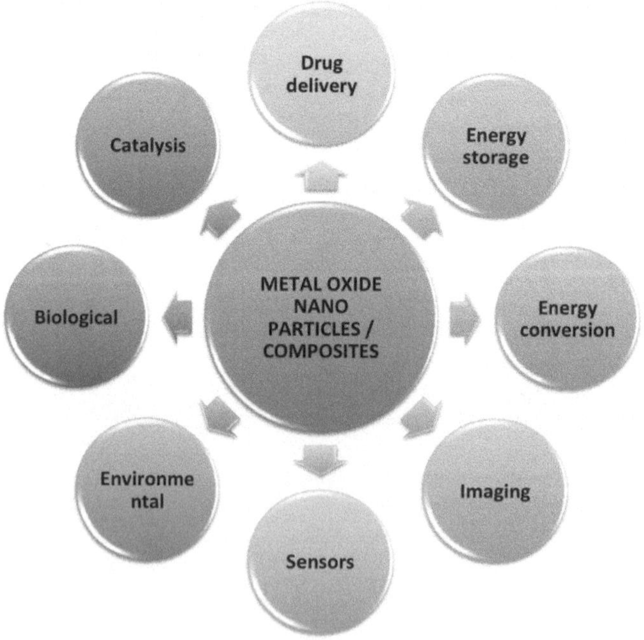

Figure 10.1 Applications of metal oxide nanomaterials.

10.4.1 Photocatalytic Performance

Organic dyes from industrial wastewater are huge pollutants which significantly affect water resources in the industrial surroundings. On a universal scale, over 0.7 million tons of unprocessed synthetic dyes are mass-produced every year mostly because of their usage in the fabric, industrial painting, fleece supplies, provisions, cosmetics, synthetics, and purchasers electronics sectors. Approximately 20% of the total global production of dyes is lost during the dyeing procedure and is emitted into the environment as textile effluent. The release of these dyes in the environment is a warning alarm due to the possible carcinogenic features of these chemicals. Moreover, weathering of these organic dyes through hydrolysis, oxidation, anaerobic decoloration or other chemical reactions happening in the wastewater, can provide probable carcinogens that are reported to cause adverse effects on human health as well as environmental quality [26, 27].

Nonetheless, with the help of several oxidants such as H_2O_2, O_3, Cl_2, and ClO_2 chemical oxidation does not experience difficulty even though they cannot achieve the whole mineralization of water contaminants. To overcome these issues, advanced oxidation processes (AOPs) have been presented to eradicate various toxic compounds that could not be efficiently eliminated by conservative treatment methods. Semiconductor photocatalysis has been filed in a lot of investigations related to environmental issues, such as water purification [27]. Semiconductor photocatalysis has gained much interest throughout the previous three decemvirates as an assuring elucidation for both energy production and ecological issues. Widespread works have been performed to yield hydrogen from water splitting by applying a variety of photocatalytic semiconductors. Currently, scientists and engineers gained much attention in mixed photocatalysis that has been also spotlighted on ecological applications such as water and air sanitization. Numerous researches focus on photocatalytic semiconductors in current literature. In photocatalytic semiconductors, electron-hole pairs are introduced after the bandgap excitation. When a photocatalyst is irradiated by light with energy identical to or superior to the bandgap energy, the valence band electrons can be stimulated to the

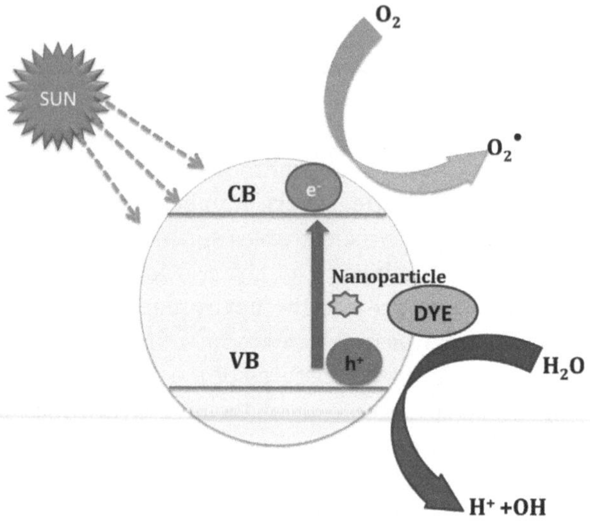

Figure 10.2 Photocatalytic activity of nanoparticles.

conduction band (CB), making a positive hole in the valence band (VB) as shown Fig. 10.2.

$$\text{Photocatalyst} + (h\nu) \rightarrow e_{\text{CB}}^{-} + h_{\text{VB}}^{+} \tag{10.1}$$

Metal oxides such as CuO, NiO, TiO_2 MgO, CdO, SnO_2, etc. are the most popular photocatalysts due to their relationally high activity, chemical stability, accessibility, low manufacture costs, and non-toxicity, has been extensively examined and developed to have the latent to entirely oxidize several unprocessed compounds in the target unprocessed contaminants [21–25]. There is a lot of research done in the arena of metal oxide nanoparticles/nanocomposites with effective photocatalytic performance and they are the photodegradation performance values reported in Table 10.2.

10.4.2 Antibacterial Activity of Metallic Nanomaterials

Generally, inorganic nanoparticles can interact with micro-organisms acting as a good antibacterial and antifungal agent. Though, before consuming micro-organisms as a means of transportation for bioactive molecules, like nanoparticles or drugs linked

Table 10.2 Photocatalytic degradation efficiency of various dyes against synthesized metallic nanomaterials via the green route

Nanomaterials	Method	Concentration (mg)	Degradation efficiency (%)	Reaction time (min)	Light source	Refs.
SnO_2	Microwave-irradiated green route	5.0	76	100	UV lamp	[28]
ZrO_2	Green route (microwave-assisted)	1.5	91	240	UV light	[29]
MgO	Microwave-assisted green synthesis	5.0	88	90	Solar light	[30]
SnO_2	Green synthesis	2	99	120	Sunlight	[25]
CuO	Green route (microwave-assisted)	5.0	98	120	Sunlight	[10]
$ZnFe_2O_4$	Green synthesis (sugarcane juice)	5.0	82	120	Sunlight	[31]
NiO	Green synthesis	5.0	88	120	Sunlight	[15]
NiO	Green synthesis	5.0	91	150	Sunlight	[16]
TiO_2	Green synthesis	5.0	99	80	UV light	[22]

to nanoparticles, it is necessary to examine how to incorporate these molecules within micro-organisms. It has already been demonstrated how gram-negative bacteria can be employed for constructing bacterial ghosts, signifying the latest advances in delivery and pointing vehicles appropriate for the distribution of hydrophobic or water-soluble drugs. Also, in pathogenic fungi like *Candida albicans*, numerous components related to the cell wall, have been recognized to play a key role in fungal–host interactions [32].

The gram-negative bacterial cell wall is structurally and chemically more complex. It comprises of a thin PG layer and over the surface membrane. This outer membrane frequently provides confrontation to hydrophobic species comprising detergents and covers as a rare component, lipopolysaccharide, which increases the negative charge of cell wall membranes and is vital for structural integrity and feasibility of the bacteria [33, 34].

Generally, the antibacterial performance of nanoparticles/composites is accomplished using various types of gram-negative and positive bacterial strains. The strength of metallic nanoparticles is varied among 50 and 100 μg/mL.

The antibacterial mechanism of the synthesized nanoparticles based on the subsequent features [35],

I. Creation of reactive oxygen species (ROS)
II. Heavy metal ions dispensation

The photogeneration of ROS on the metal oxide nanocomposite surface has been stated by Karthik et al. [10, 16]. Usually, the antibacterial performance mostly varies upon the ROS, specific surface area and size of the particle, etc. Few nanocomposites are providing ROS via the Fenton reaction mentoring to DNA damage, lipid peroxidation and protein oxidation can destroy bacteria without damaging nonbacterial cells. There are other possible mechanisms included in the antibacterial performance. Prepared nanocomposite obstructs the bacterial cell wall membrane and binds with the mesosome. These intracellular functional modifications are initiated by the oxidative stress influenced by ROS leading to cell death as shown in Fig. 10.3. Very few works of literature explain the electrostatic attraction mechanism.

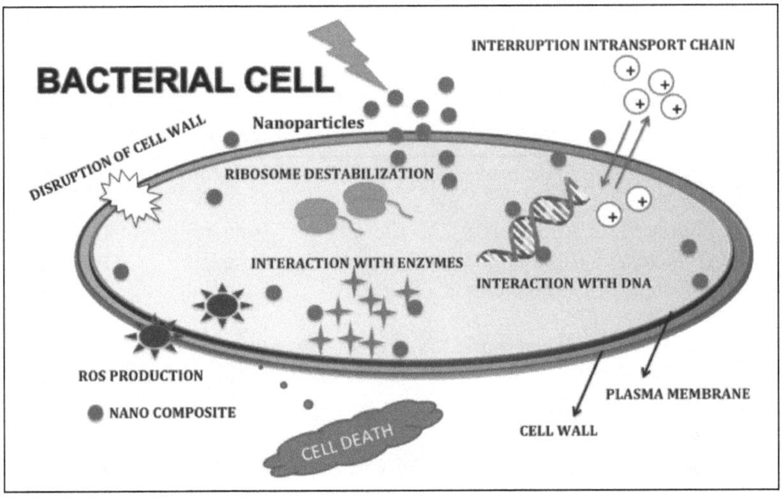

Figure 10.3 Antibacterial activity mechanisms of metallic nanoparticles.

M^{2+} is discharged from the surface of nanoparticle and communication with the membrane of the cell wall of the microbe. The cell wall (−ve) and M^{2+} (+ve) are equally fascinated and they reason denaturation of proteins which causes in thrashing of imitation competence of the DNA thus inducing the demise of the pathogen.

$$\mathrm{MO} + h\upsilon \rightarrow e^- + h^+$$
$$h^+ + \mathrm{H_2O} \rightarrow {}^\bullet\mathrm{OH} + \mathrm{H^+}$$
$$e^- + \mathrm{O_2} \rightarrow {}^\bullet\mathrm{O_2^-}$$
$${}^\bullet\mathrm{O_2^-} + \mathrm{H^+} \rightarrow \mathrm{HO_2^\bullet}$$
$$\mathrm{HO_2^\bullet} + \mathrm{H^+} + e^- \rightarrow \mathrm{H_2O_2}$$

The general mechanism of antibacterial activity by nanocomposite have described in Fig. 10.3. Karthik et al. [10, 16] have depicted the photocatalytic, antibacterial and anticancer performances of microwave-assisted green route CuO, NiO, MgO, and CdO nanoparticles [36]. The bioengineered nanocomposite of CdO-ZnO and their photocatalytic and biological performance have been described by Rajaboopathi et al. [37]. Numerous metal oxide nanoparticles/nanocomposite showed outstanding performance against both gram-positive and negative bacteria

Table 10.3 Antibacterial performance of various metal oxide nanoparticles synthesized via green route against different human pathogens

Nanomaterials	Method	Tested pathogens	Concentration	ZOI (mm)	Refs.
MgO (*P. Guvajava*)	Green route	*E. coli*	5 mg/mL	15	[38]
MgO (*Burm. f.*)	Green route	*E. coli*	5 mg/mL	12	[38]
MgO (*P. Guvaja*)	Green route	*S. aureus*	5 mg/mL	16	[38]
MgO (*Aloe Barbadensis*)	Green route	*S. aureus*	5 mg/mL	15	[38]
CdO (*A. paniculata*)	Green route	*E. coli*	100 μg/mL	16	[24]
CdO (*A. paniculata*)	Green route	*S. aureus*	100 μg/mL	13	[24]
CuO (*A. paniculata*)	Green route (microwave-assisted)	*E. coli*	100 μg/mL	20	[10]
CuO (*A. paniculata*)	Green route (microwave-assisted)	*S. aureus*	100 μg/mL	18	[10]
MgO (*A. paniculata*)	Green route	*E. coli*	100 μg/mL	20	[23]
MgO (*A. paniculata*)	Green route	*S. aureus*	100 μg/mL	18	[23]
NiO (*Limonia acidissima Christm*)	Green route	*S. aureus*	100 μg/mL	14.6	[16]
NiO (*Limonia acidissima Christm*)	Green route	*E. coli*	100 μg/mL	15.3	[16]
TiO_2 (*Ledebouria revoluta*)	Green route	*E. coli*	60 μg/mL	4	[21]
TiO_2 (*Ledebouria revoluta*)	Green route	*C. tetani*	60 μg/mL	5.4	[21]
Ag (Cordia dichotoma)	Green route	*E. coli*	1 mg/mL	6	[49]
Ag (*Cordia dichotoma*)	Green route	*E. coli*	2 mg/mL	8	[49]

accounted through several researchers as shown in Table 10.3 and the observed antibacterial performance in terms of zone of inhibition based on current literature displayed.

10.4.3 Anticancer Activity of Metallic Nanomaterials

The incorporation of green nanotechnology with the drug delivery arena has essentially begun an original empire of "green nanomedicine." Nanoparticles involved in the arena of biomedical applications increased day by day in numerous processes including drug delivery, bioimaging, and gene delivery. They can act as smart weapons against so many drug-resistant micro-organisms and as a good alternative for antibiotics. Nanoparticles using for drug delivery were derived from their two important fundamental characteristics. The first one is, due to their nano size can penetrate through very small capillaries and are absorbed by the cells, permitting an effective growth of drugs at the target sites. The second one is, the usage of biodegradable materials for the synthesis of nanoparticles permits the lengthy liberation of drugs within the site targeted over a particular period of time.

Cancer is an irregular, uncontrolled growth of cells triggered by genetic mutations. It leads to a dis-regulated balance of cell proliferation and is able to progress as a small population of cells invades tissues and metastasize to other parts of the body causing significant morbidity and death. The probable factors of initiating cancer in persons comprise genetic, lifestyle such as tobacco, diet, and lack of exercise, certain types of infections, and environmental exposures to distinctive types of chemicals and radiations.

Clinical cancer treatments include chemotherapy, surgery, radiation therapy, immunotherapy, gene therapy and cryotherapy. Conventional clinical treatments have their own advantages and several disadvantages. For example, chemotherapy has been expected to damage the cancerous cells foremost to tumor dissolution. Nevertheless, the clinical results show tumor resistance because of overcompensation from the parallel signaling pathways and side outcomes to the standard cells. Therefore, new therapeutic platforms are required to overcome the demerits of conventional clinical treatments [39].

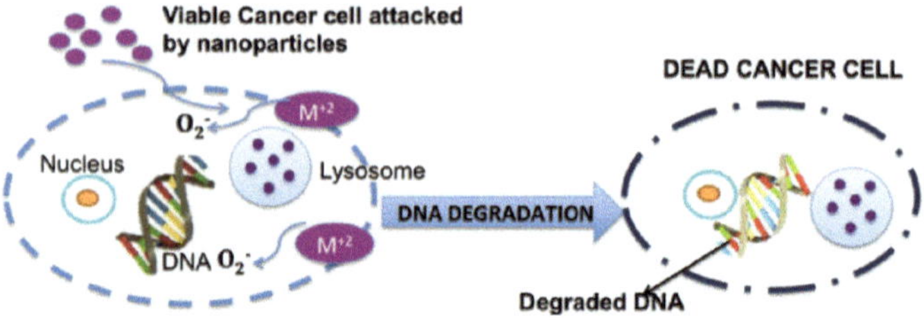

Figure 10.4 Anticancer activity mechanism of metallic nanoparticles.

Chemotherapeutic agents have poor water solubility, non-specificity, multidrug resistance, and less permeability across biological barriers. Nanomedicines have been extensively inspected to overcome the above drawbacks of anticancer drugs in the preceding few decades. The meaning of nanomedicine is the utilize of nanotechnology in the preclusion and treatment of disease in the human body. Nanomedicine may transform the medical procedures to simple and cellular level treatment that repairs the damaged cells or get inside cells and substitute or support damaged intracellular structures as shown in Fig. 10.4.

There are numerous improvements of nanomaterials employed for biomedical applications, for instance (1) nanostructures with a smaller size and various shapes having large surface areas will help to anchor and interact with cell membranes. This could guide to a rise in the mobility and communication with cell membranes that can effect in improved cellular uptake of nanoparticles, (2) the surface of the nanoparticles can be easily modified with hydrophilic polymers which are essential for escaping from macrophage capture, (3) nanocomposite surfaces can easily facilitate both in vitro and in vivo cell-specific objectifying, (4) nanoparticles be able to improve the intracellular concentration of the drugs in cancer cells whereas preventing needless toxicity in the direction of healthy cells and (5) chemotherapeutic drug molecules can be entrapped within the nanoparticles and/or conjugated on the surface which gives a high drug loading efficiency. Among several treatments for cancer therapy, very few nano medicine-based formulations have been supported for medical usage and many other formulations are under medical trials [40]. Metal oxide nanoparticles/composites

Table 10.4 Anticancer performance of metal oxide nanoparticles against different human breast and lung cancer cell lines

Sample	Method	Cancer cell lines	IC_{50} (μg/mL)	Refs.
CuO	Green route (*M. oleifera*)	*MCF-7*	26.7	[6]
CuO	Green route (*A. indica*)	*MCF-7*	25.5	[6]
MgO	Green route (*A. paniculata*)	*MCF-7*	15.7	[23]
NiO	Green route (*A. paniculata*)	*MCF-7*	44.91	[15]
ZnO	Green synthesis (*Dalbergia latifolia*)	*A549*	29.0	[41]
TiO_2	Green synthesis (*Ledebouria revoluta*)	*A549*	53.6	[21]
TiO_2	Microwave-assisted green route (*A. paniculata*)	*A549*	24.9	[22]
CuO	green route (*A. paniculata*) – microwave-assisted	*A549*	14.76	[10]
CuO	Green synthesis (*Ficus religiosa leaf extract*)	*A549*	200	[42]
Ag	Green synthesis (*Achillea biebersteinii*)	*MCF-7*	20	[50]
Ag	Green synthesis (*Alternathera tenella*)	*MCF-7*	42.5	[51]
Ag	Green synthesis (*Cymodocea serrulata*)	*A549*	100	[52]
Au	Green synthesis (*Antigonon letopus Hook. and Arn.*)	*MCF-7*	257.8	[53]

with anticancer property against different cancer cell lines have been reported in Table 10.4.

10.5 Conclusions, Outlook and Perspectives

The green synthesis of nanoparticles is a commercial and environmentally friendly alternative to physical and chemical processes. Biomediated synthesis of nanoparticles is a green chemistry approach that links nanotechnology with plants and delivers a sustainable, resource-effective and less cost preparation. It has great potential as in the production of nanoparticles as these techniques is free from hazardous and toxic chemicals used in usual chemical and physical processes and are economically cheap. The adaptability of metal oxide nanoparticles provides the possibility of these biodegradable, biocompatible and sustainable materials to be used in a diversity of applications. Bioengineered metal oxide nanoparticles and composites are becoming the heart of research and product growth as monitored from various kinds of researches carried out on metal oxide nanoparticles and composites in current years. As a result, one can predict that shortly, bioengineered metal oxide nanoparticles composites will be broadly and comprehensively utilized in the oil, gas, medical, biotechnology, electronics, and food industries. Green synthesized metal oxide nanocomposites have also been shown to be feasible choices of synthetic materials for impending replace of conventional materials and technologies in various functional applications in different industries. As metal oxide nanoparticles and composites constantly increase the huge attention from researchers all over the globe, the direction toward production and practical applications in advanced technology will quickly enlarge in a way encouraging the shift of novel green synthesized metal oxide nanocomposites from lab to industrial scale.

References

1. C. Artini, F. Locardi, M. Pani, I. Nelli, F. Caglieris, R. Masini, J. R. Plaisier and G. A. Costa, *Journal of Physics and Chemistry of Solids*, 2017, **103**, 59–66.
2. D. Radhika, K. Karthik, A. S. Nesaraj and R. Namitha, *Materials Research Innovation*, 2019, 1–7, doi:10.1080/14328917.2019.1686858.
3. Y. Liu, L. Fan and Y. Cai, *ACS Applied Materials & Interfaces*, 2017, **928**, 23614–23623.
4. S. Sa-nguanprang, A. Phuruangrat, K. Karthik, S. Thongtem and T. Thongtem, *Journal of the Australian Ceramic Society*, 2020, **56**, 1029–1041, doi:10.1007/s41779-019-00447-y.
5. R. Aswini, S. Murugesan and K. Kannan, *International Journal of Environmental Analytical Chemistry*, 2020, doi: 10.1080/03067319.2020.1718668.
6. D. Dilaveez Rehanaa, R. Mahendiran, S. Kumar and A. K. Rahimana, *Biomedicine & Pharmacotherapy*, 2017, **89**, 1067.

7. K. R. Reddy, *Journal of Molecular Structure*, 2017, **1150**, 553.
8. M. Sorbiun, E. S. Mehr, A. Ramazani and S. T. Fardood, *International Journal of Environmental Research*, 2018, **12**, 29–37, doi:10.1007/s41742-018-0064-4.
9. X. Fuku, N. Thovhogi and M. Maaza, *AIP Conference Proceedings*, 2018, **1962**, 040006.
10. K. Kannan, D. Radhika, S, Vijayalakshmi, K. K. Sadasivuni, A. A. Ojiaku and U. Verma, Facile fabrication of CuO nanoparticles via microwave-assisted method: photocatalytic and antimicrobial and anticancer enhancing performance, *International Journal of Environmental Analytical Chemistry*, 2020, doi:10.1080/03067319.2020.1733543.
11. A. Ezhilarasi, J. Judith Vijaya, K. Kaviyarasu, L. John Kennedy,R. Jothi Ramalingam and H. A. Al-Lohedan, *Journal of Photochemistry and Photobiology B: Biology*, 2018, **180**, 39–50, doi:10.1016/j.jphotobiol.2018.01.023.
12. F. T. Thema, E. Manikandan, A. Gurib Fakim and M. Maaza, *Journal of Alloys and Compounds*, 2016, **657**, 655–661.
13. A. T. Khalil, M. Ovais, I. Ullah, M. Ali, Z. K. Shinwari, D. Hassan and M. Maaza, *Artificial Cells, Nanomedicine, and Biotechnology*, 2018, **46**, 838–852.
14. B. Sone, X. Fuku and M. Mazza, *International Journal of Electrochemical Science*, 2016, **11**, 8204–8220.
15. K. Karthik, M. Shashank, V. Revathi and T. Tatarchuk, *Molecular Crystals and Liquid Crystal*, 2018, **673**(1), 70–80, doi:10.1080/15421406.2019.1578495.
16. K. Kannan, D. Radhika, M. P. Nikolova, K. K. Sadasivuni, H. Mahdizadeh and U. Verma, *Inorganic Chemistry Communications*, 2020, **113**, 10755, doi:10.1016/j.inoche.2019.107755.
17. K. G. Rao, C. H. Ashok, K. V. Rao, C. S. Chakra and P. Tambur, *Asian Pacific Journal of Tropical Medicine*, 2015, **2**, 28–34.
18. R. D. A. Jalill, R. S. Nuaman and A. N. Abd, *World Scientific News*, 2016, **49**, 204–222.
19. A. Chatterjee, M. Ajantha, A. Talekar, N. Revathy and J. Abraham, *Materials Letters*, 2017, **9**, 95–99.
20. V. Patidar and P. Jain, *International Research Journal of Engineering and Technology*, 2017, **4**, 470–473.
21. R. Aswini, S. Murugesan and K. Kannan, Bio-engineered TiO_2 nanoparticles using Ledebouria revoluta extract: larvicidal, histopathological,

antibacterial and anticancer activity, *International Journal of Environmental Analytical Chemistry*, 2020, doi:10.1080/03067319.2020.1718668.

22. K. Karthik, S. Vijayalakshmi, A. Phuruangrat, V. Revathi and U. Verma, *Journal of Cluster Science*, 2019, **30**, 965–972.
23. K. Karthik, S. Dhanuskodi, S. Prabukumar, C. Gobinath and S. Sivaramakrishnan, *Materials Letters*, 2017, **206**, 217–220.
24. K. Karthik, S. Dhanuskodi, C. Gobinath, S. Prabukumar and S. Sivaramakrishnan, *Journal of Materials Science: Materials in Electronics*, 2017, **28**, 7991–8001.
25. K. Karthik, V. Revathi and T. Tatarchuk, *Molecular Crystals and Liquid Crystals*, 2018, **671**(1), 17–23, doi:10.1080/15421406.2018.1542080.
26. World Health Statistics, 2011, Publications of the World Health Organization, 20 Avenue Appia, 1211 Geneva 27, Switzerland.
27. H. Zollinger, *Color Chemistry: Syntheses, Properties, and Applications of Organic Dyes and Pigments*, 2nd rev. ed., VCH, Weinheim, Germany, 1991.
28. S. Sathishkumar, M. Parthibavarman, V. Sharmila and M. Karthik, *Journal of Materials Science*, 2017, **28**, 8192–8196, doi:10.1007/s10854-017-6529-y.
29. H. M. Shinde, T. T. Bhosale, N. L. Gavade, S. B. Babar, R. J. Kamble, B. S. Shirke and K. M. Garadkar, *Journal of Materials Science*, 2018, **29**, 14055.
30. K. Karthik, S. Dhanuskodi, C. Gobinath, S. Prabukumar and S. Sivaramakrishnan, *Journal of Photochemistry and Photobiology B: Biology*, 2019, **190**, 8–20.
31. S. W. Lee, C. Mao, C. Flynn and A. M. Belcher, *Science*, 2002, **296**, 892–895.
32. A. Chwalibog, E. Sawosz, K. Mitura, M. Grodzik, P. Orlowski and A. Sokolowska, *International Journal of Nanomedicine*, 2010, **5**, 1085–1094.
33. F. Baquero, M. T. Coque and F. Cruz, *Antimicrobial Agents and Chemotherapy*, 2011, **55**, 3649–3660.
34. T. Jin and Y. He, *Journal of Nanoparticle Research*, 2011, **13**, 6877–6885.
35. M. A. Subhan, T. Ahamed, N. Uddin, A. K. Azad and K. Begum, *Spectrochimica Acta Part A: Molecular and Biomolecular Spectroscopy*, 2015, **136**, 824–831.
36. K. Karthik, S. Dhanuskodi, C. Gobinath, S. Prabukumar and S. Sivaramakrishnan, *Journal of Physics and Chemistry of Solids*, 2018, **112**, 106–118.

37. S. Rajaboopathi and S. Thambidurai, *Current Applied Physics*, 2017, **17**, 1622–1638.
38. L. Umaralikhan and M. Jamal Mohamad Jaffar, *Iranian Journal of Science and Technology, Transactions A: Science*, 2016, **1**, 1–9.
39. K. C. Zimmermann and D. R. Green, *Journal of Allergy and Clinical Immunology*, 2001, **108**(4), S99–103.
40. B. Summer and J. Gao, *Nanomedicine*, 2008, **3**(2), 17–140.
41. N. Elavarasana, S. Prakasha, K. Kokilaa, C. Thirunavukkarasu and V. Sujatha, *New Journal of Chemistry*, 2020, **44**, 2166.
42. R. Sankar, R. Maheswaran, S. Karthik, K. Subramanian and V. R. Shivashangari, *Materials Science and Engineering C*, 2014, **44**, 234.
43. R. S. Patil, M. R. Kokate and S. S. Kolekar, *Spectrochimica Acta Part A*, 2012, **91**, 234–238.
44. V. Kumar, S. C. Yadav and S. K. Yadav, *Journal of Chemical Technology and Biotechnology*, 2010, **85**(10), 1301–1309.
45. D. Raghunandan, S. Basavaraja, B. Mahesh, S. Balaji, S. Y. Manjunath and A. Venkataraman, *Nanobiotechnology*, 2009, **5**(1–4), 34–41.
46. D. Phillip, *Spectrochimica Acta Part A*, 2010, **77**(4), 807–810.
47. F. Coccia, L. Tonucci, D. Bosco, M. Bressan and N. d'Alessandro, *Green Chemistry*, 2012, **14**(4), 1073–1078.
48. R. K. Petla, S. Vivekanandhan, M. Misra, A. K. Mohanty and N. Satyanarayana, *Journal of Biomaterials and Nanobiotechnology*, 2012, **3**(1), 14–19.
49. B. Devaraj, V. Seerangaraj and V. Bhuvaneshwari, *Materials Research Express*, 2018, **5**, 055404.
50. J. Baharara, F. Namvar, T. Ramezani, M. Mousavi and R. Mohamad, *Molecules*, 2015, **20**(2), 2693–2706.
51. P. Sathishkumar, K. Vennila, R. Jayakumar, A. R. M. Yusoff, Hadibarata and T. Palvannan, *Bioprocess and Biosystems Engineering*, 2016, **39**(4), 651–659.
52. P. Palaniappan, G. Sathishkumar and R. Sankar, *Spectrochimica Acta Part A*, 2015, **138**(22), 885–890.
53. G. Balasubramani, R. Ramkumar, N. Krishnaveni, A. Pazhanimuthu, T. Natarajan, R. Sowmiya and P. Perumal, *Journal of Trace Elements in Medicine and Biology*, 2015, **30**(24), 83–89.

Chapter 11

Preparation of Metal Nanoparticles Extractions from Green Natural Products

Muthukumar Harshiny,[a] Sekar Aiswarya Devi,[b] Manickam Matheswaran,[b] and Muthukonda V. Shankar[c]

[a]*Department of Biotechnology, Indian Institute of Technology Madras, Chennai, Tamil Nadu 600036, India*
[b]*Department of Chemical Engineering, National Institute of Technology, Tiruchirappalli, Tamil Nadu 620015, India*
[c]*Nanocatalysis and Solar Fuels Research Laboratory, Department of Materials Science and Nanotechnology, Yogi Vemana University, Kadapa, Andhra Pradesh 516005, India*
matheswaran@nitt.edu

Metal nanoparticles (MNPs) are novel and inspiring material with unique traits and various applications. Multifunctional MNPs can be prepared using physicochemical and biological methods. Though, exploiting physicochemical routes requires high energy, and cost, it also leads to environmental damage via harmful chemical solvent utilization, consequently rising noxiousness and health concerns. In this regard, green extract mediated synthesis of MNPs has recently arisen as an eco-friendly method for synthesizing MNPs. Importantly covered, synthetic processes of MNPs using green natural products alike plant extracts and micro-organisms such as algae, bacteria, fungi, etc. Besides, the various parameters like

Nanomaterials for Spectroscopic Applications
Edited by Kaushik Pal

ISBN 978-981-4877-69-5 (Hardcover), 978-1-003-16033-5 (eBook)
www.jennystanford.com

concentration, pH, reaction time, temperature, and so on affect the green synthesis of MNPs are also explained. Finally, explored the biomolecules present in plant extracts interact with metal ions and transformed into nanostructures and challenges in MNPs green synthesis. Inclusively, this chapter highlights green chemistry, preparation of MNPs from green natural products, factors, preventive plant-based synthesis as an actual alternate to the conventional synthesis of MNPs. This chapter attempts to explain the challenge, pros, and cons, necessitate green synthesis and its multifunctional applications.

11.1 Introduction

Nanoscience and technology are the two curious fields implanting its footprint in technology research and opportunity's door to specific engineering on the atomic scale and influences from the medical to the environment [1]. Nanomaterials are cornerstones of nanoscience and nanotechnology; although it deemed to be one of the most important studies in recent sciences [2]. The novel materials in the nanoscale dimension by altering its physicochemical and biological traits with changing size, shape, composition and morphology make appropriate for various applications [3]. Amid, the metals for NPs such as aluminium, cobalt, copper, gold, iron, lead, silver and zinc (Al, Co, Cu, Au, Fe, Pb, Ag and Zn) have been widely used [4]. Most importantly, the MNPs have unique properties such as smaller sizes, surface traits such as pore size, crystalline structures, high surface area to volume ratio, charge, density and so on [5]. Ascribed to these extraordinary and attractive properties MNPs are gaining significant applications in various field alike catalysts, clinical, pharmaceutical, biomedical sciences, chemical engineering, space research, energy-environment and electrochemical [4, 6].

However, the synthesis of nanomaterials with requisite characteristics of major continuing challenges. The nanoscience and nanotechnology field focus on:

(i) The improvement of the fabrication process and analytical technique for synthesizing structures and materials.
(ii) Recognize the alteration in physicochemical traits due to miniaturization and formation process.
(iii) Utilization of the above properties in the improvement of innovative, efficient materials and devices [7, 8].

Mostly, nanoparticles are prepared in two ways the 'top-down' and 'bottom-up' method [3]. However, other than biological methods were throbbing from harmful and noxious elements, invest high cost, energy, ecological poisonousness, prolonged reaction times and toxic by-products [9]. The hazardous possessions of physicochemical and other approaches might deter large-scale consumption [10]. Moreover, other than biological methods conferred fewer yields. Therefore, alternative, cost-effective, safe and environmentally sound techniques for MNPs synthesis were highly needed [11].

It was noticed that amid, the bottom-up approach, the green synthetic route produces a non-toxic and consumes far less energy than other methods, thus reducing the environmental pollution [12]. Therefore, the eco-friendly green mediated synthesis of NPs has been considered a promising alternative. The bioreduction of MNPs using combinations of biotic things available naturally alike algae, fungi, plants, bacteria, yeast and other microbes are environmentally benign [13]. It has been reported that the choice of the synthesis method for metallic nanoparticles was very significant. As during NPs preparation steps, alike interaction of metal ions with reducing agent, the adsorption process of a stabilizing agent with MNPs and various environmental parameters impact on its morphology, stability and physicochemical traits [12, 14]. This perspective article, further presents synthesis methods and their potential adverse effects on the environment, the natural green extract used for synthesis, synthesis influencing parameters, challenges in synthesis methods. As significant and serious roles of green mediated for MNPs synthesis using plants have been discussed in six sections were shown in Fig. 11.1.

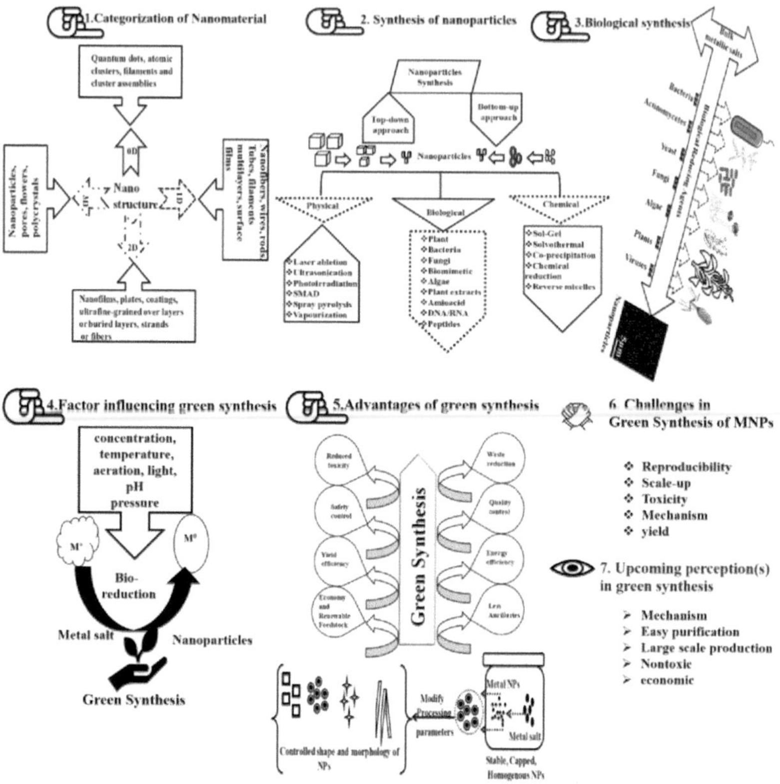

Figure 11.1 (i) Classification of nanomaterials and MNPs, (ii) synthesis of NPs, (iii) biological source mediated NPs synthesis, (iv) discuss advantages and green chemistry approaches in nanoparticles synthesis, (v) influencing parameters, and (vi) highlights challenges on the utilization of various green natural agents for the metal nanoparticles synthesis.

11.2 Classification of Nanomaterials

Nanotechnology is implanting its paw marks in the current epoch very hastily. Nanotechnology has the competence to determine, observe, manipulate and manufacture materials on an atomic or molecular scale, usually between one and 100 nanometers [5]. Nanomaterials have evoked ample attention in science mainly due to its novel properties, control design and fabricate materials at the nanoscale. At the nanoscale dimensions, the material traits transform extensively from there bulk counterparts [15].

Nanomaterials exhibit a new-fangled phenomenon coupled with the quantized properties and majority of surfaces. The quantization effect arises in the nanometer regime, as the total dimensions of substances are alike to the characteristic wavelength of excitations in materials [16]. Nanomaterials and nanostructures play an essential role in nanoscience and nanotechnology applications. It will inspire and promote the development of novel nanomaterials. As well, nanostructured materials are classified as zero, one, two and three dimensions. Amid the nanostructure materials, nanoparticles are playing an imperative role in the various fields [17].

11.2.1 Nanoparticles

The NPs are intermediates between the bulk materials, the atomic and molecular structures. NPs hold the novel size-dependent traits known as 'specific surface area,' which is related to the surface area to volume [18]. Electron transportation demonstrated in phenomena alike Coulomb blockade, the catalytic and thermodynamic traits of structures might be made-to-order after design materials rationally on this length scale. Thus, synthetic methods regulate the structure, nano size and NPs yield [19]. In addition, assembling nanoparticles generate new structures, which has unforeseen collective intrinsic physicochemical properties. The unique physicochemical and biological properties of NPs originate from several factors such as (i) large surface atom and energy, (ii) spatial confinement, and (iii) particle size, structure, formation and disparity [20].

11.2.2 Categorization of Nanoparticles

As the size of material approaches to the nanometer regime, enhanced physicochemical traits can be explored for various applications [21]. NPs exhibit in different types such as organic and inorganic NPs, from that it classified into pure metal NPs (Ag, Au, Cu, Fe, Ni, Co, etc.). Bimetallic colloids (Pt-Rt, Pt-Ni, Co-Mo, Pd-Fe/Ru), metal-organic (TiO_2, ZnO, Fe_2O_3, CrO_2, Meallates, etc.)

[22, 23]. Metal chalcogenides (PbS, CdS, ZnS, ZnSe, CdSe, CdTe, HgS, CuInSe, etc.) [24] and Specific traits NPs such as ferromagnetic shape memory alloys, magnetic NPs, conducting polymers NPs, quantum dots, liposomes and so on. These NPs holds interesting properties, which are unusual in the sense that they are not observed in their bulk counterpart [5].

Furthermore, the unavailability of qualified medicines composed with the advent of some antibiotics' adversative toxic properties, microbial resistance to antibiotics, environmental contamination, has been presently reaching a critical level. It has necessitated research on novel materials with enhanced properties like low cost, toxicity, easily available, fabrication and so on [25]. In exploring various options to address the above problem, nanomaterials, like metal nanoparticles, have emerged as promising material since; NPs possess greater permanence, lower toxicity, higher stability and selectivity when compared to organic NPs, drugs and chemicals agents [26].

11.2.3 Synthesis of Nanoparticles

The NPs physicochemical traits were found to be different based on the method of synthesis. Nanostructured materials are prepared either by 'bottom-up' or 'top-down' approach. In bottom-up methods initiate with atoms or small molecules as 'building blocks' as well accumulate nanoclusters and bulks from the materials as mentioned above. The bottom-up methods primarily engage a chemical and biological approach to fabricate NPs [27]. In, this method entails controlled condensation of solute molecules, where formed in a chemical reaction. The constraint of the condensation or growth direct particle formation in expected size and shape [28]. Though, distinct chemical synthesis of material with expected structure, the nanomaterials synthesis with even shape and size was difficult. Hence, large scale fabrication of nanostructured with precise structure, monodispersed size and uniform shapes were remnants difficult process [3].

The 'top-down' method for processing of nanostructured material comprises initiating with bulk material and gaining a nanostructure through structural decay. In the top-down approach,

NPs are synthesized by physically slicing or by abrasion of bulk material to attain the anticipated size [29]. In the course of evolution and talented to grasp the sub-micron level structure via various analytical techniques such as chemical etching by induced laser, ball milling and so on [30]. One of the primary limits in this process was material surface structure imperfections, as other corporeal traits of NPs are mainly dependent on the surface structure [31]. Schematic in the figure shows various strategies used for the synthesis of NPs.

The three main approaches for the synthesis of NPs are physical, chemical and biological were. However, the chemical mediated method has been extensively studied for the synthesis of nanostructures. But high-cost investment, the requirement of sophisticated equipment, stabilizing and capping agents [32]. In addition, environmental problems fade not only the application of NPs but also the need for a safe and alternate way; it includes a clean, non-toxic and eco-friendly method of NPs synthesis. Also, the recent developments in NPs synthesized using biological resources as a reducing agent are environmental being [33].

11.3 Biological Synthesis

Numerous literature have been reported biological syntheses of NPs were carried out by biological methods in those organic compounds and biomolecules reduce metal ions to NPs. Several biological resources have been assessed for MNPs preparation, including algae, bacteria, fungi, plants, viruses, and yeast [34]. As biosynthetic approaches for the synthesis of NPs have less cost and toxic effect than conventional physicochemical syntheses. Biological methods utilize natural resource alike micro-organisms and plant extracts as a reducing agent. As well, these bioreductants precluding the necessity of additional chemical reagents for capping and stabilizing agents. In the same way, biological methods can control the physicochemical properties of NPs [12]. As well, a green solvent medium, reducing agent and a non-hazardous substance for the stabilization makes the process eco-friendly and biocompatible procedures for the synthesis of NPs. Specifically, among the biosources for MNPs synthesis algae, bacteria, fungi, and plants were highly studied [10, 35].

11.3.1 Algae-Mediated Synthesis

Algae-mediated biosynthesis of NPs became an attractive branch also termed as phyconanotechnology. Algae are a photosynthetic organism existing in various atmospheres alike moisty surface rocks, fresh and marine water [36]. Heretofore, algae such as *Chlorophyceae, Phaeophyceae, Cyanophyceae, Rhodophyceae*, in addition some species alike *Chondrus crispus, Gracilaria edulis, Kappaphycus alvarezii, Sargassum wightii, Tetraselmis kochinensis, Spirogyra insignis, Rhizoclonium fontinales, Prasiola crispa, Chlorella vulgari, Cystoseira baccata, Klebsormidium flaccidum* [37]. Many other organisms like *Coelastrella, Anabaena, Sargassum, Turbinaria, Spirulina* species and others have been used for the synthesis of MNPs [38]. Algal extracts contain bioactive compounds such as polyphenols, tocopherol, pigments, phycobilin along with carbohydrates, fats, minerals, oil, proteins, and polyunsaturated fatty acids. These biocompounds in algae have the ability to accrue metals and reduce metal ions as well acted as a stabilizing agent [39].

The NPs synthesis was primarily stated to be intracellular, but then algae synthesis NPs by extracellular mode were exploited. For, 'intracellular' synthesis there is no necessity of microalgae pre-treatment as the process depends on metabolic pathways. The NADPH or NADPH dependent reductase may act as a reducing agent [34, 36]. In the 'extracellular' process cell metabolism encompassing of antioxidants, hormones, ions, metabolites, lipids pigments and various proteins may act as a reducing agent. As well, for extracellular process algal biomass were undergone pre-treatments alike washing and blending [39]. As of the existing reports, the synthesis of NMs by algae-mediated involves three steps. Firstly, algal extraction; secondly, preparation of metal precursor with a suitable solvent; finally, algal extract and the metal ion precursor were incubated. Moreover, algae are comparatively expedient and easy to handle for NPs synthesis at low temperature, less toxicity with better energy efficacy [32, 34].

11.3.2 Fungal-Mediated Synthesis

Presently, fungi are used for large scale synthesis of NPs. The formation and utilization of nanomaterials by fungal are myco-nanotechnology. Commonly, fungi-mediated MNPs synthesis can be via in vivo and in vitro procedures [40]. Many filamentous fungi have mycelia and the intracellular components synthesized NPs, but in this method, the extraction of the NPs from the mycelia is required. Consequently, biomass and extracellular materials from fungal are broadly applied to the NPs formation from metal ions [41]. In this process, disruption of the cells, autolysis incubation followed by the dissolution of membrane protein, improves the reduction of metal ions into MNPs. The phylum, ascomycetes and basidiomycetes fungal species such as *Fusarium oxysporum*, *Agaricus bisporus*, *Aspergillus niger*, *Fusarium semitectum*, *Aspergillus flavus*, *Aspergillus fumigatus*, *Pleurotus platypus*, *Fusarium solan*, *Volvariella volvacea*, *Penicillium* sp., *Verticillium* sp., *Trichoderma* sp., *Trichothecium* sp. was used to synthesized MNPs and bimetallic NPs [13, 42]. Filamentous fungi have a good binding ability and synthesize NPs extracellularly. The synthesized NPs were found to be highly stable and monodisperse. But, this process's problem is maintaining sterile conditions and fermentation media are found in the reaction. As well, the exact molecular mechanisms evolved in MNPs are unclear [43].

11.3.3 Bacteria Mediated Synthesis

Recently, various organisms have been stated to synthesize MNPs either intracellularly or extracellularly. Amongst micro-organisms, bacteria have alluring attributes. Bacteria such as *Aeromonas*, *Bacillus*, *Corynebacterium*, *Escherichia*, *Enterobacter*, *Klebsiella*, *Lactobacillus*, *Pseudomonas* and so on have been testifying for the synthesis of MNPs [13, 44]. Amid several methods, extracellular synthesis has been frequently studied. Moreover, the bacteria with enzymes, metal-resistant genes, peptides and reducing cofactors were important in acting as reducing agents [33]. Moreover, the above compounds were acting as a capping and anti-aggregating agent. The biosynthesis of MNPs occurred in two steps. Initially, at the cell wall metal ions were accrue by biosorption then reduced ions to MNPs [34].

11.3.4 Plant Mediated Synthesis

Recently, the plant-mediated MNPs have drawn more attention due to the physicochemical traits of MNPs. The MNPs were synthesized using different parts of hydrophytes, mesophytes and xerophytes plant alike callus, fruit, leaves, flower, root, peel, seed and stem [13]. As, plant extracts are rich in amino acids, as well as secondary metabolites, such as alkaloids, flavonoids, polysaccharides, polyphenols, steroids, terpenoids and vitamins. These metabolites have important parts reducing metal salt and capping, inhibit the aggregation and stabilization of synthesized NPs [10].

Mostly, organic compounds in the plant extract through chelation inactivate metal ions. Metal ions are grabbed and covered by organic compounds; as well as forms NPs from metal ions. The monodispersed morphology and yield of NPs can be altered by regulating the reaction environments [45]. The plant extract mediated MNPs synthesis is parted into three steps. Firstly, reduction and metal ion nucleation were happening. Secondly, NPs were combined together to form metal particles. Finally, NPs shape formed. The obtained NPs are centrifuged and obtained precipitates were washed with an appropriate solvent to remove impurities [34].

11.3.5 Other Bioresource Mediated Synthesis

Alike, viruses, tobacco mosaic virus (TMV), M13 bacteriophage were used to produce several nanoparticles, important for semiconductor NPs and quantum dot (QD) applications. Amid the eukaryotic microbe, yeast such as *Candida glubrata, Schizosaccharomyces pombe,* has been used to synthesize semiconductors in the peptidoglycan layer of the cell wall intracellularly. As well, actinomycetes, human cell lines and enzymes, secondary metabolites were also used to synthesize NPs [34, 35].

While amid the several biological sources were used for NPs synthesis, micro-organism intervened procedure was not of industrial achievable, owing to the necessities of the extremely sterile environment and preservation [46]. Thus the use of plant extracts as a reducing, stabilizing and capping agent is more beneficial over micro-organisms due to alleviate processing, less biohazard and maintaining cell cultures [46, 47]. It necessitates that syntheses of NPs would to be from noxious chemicals along with giving natural capping agents [48]. Currently, plant extract mediated NPs synthesis can be identified as an environmentally benign. The plant extracts are likely to influence the reduction process, stabilize NPs to prevent the agglomeration and demonstrate no toxicity of the NPs produced from biogenic [34, 47].

Moreover, plant extracts utilization also cuts the expenditure of micro-organisms isolation, growing media, purification and separation over NPs synthesis of other biological sources. Henceforth, this study amassed that bioinspired synthesis of NPs that conferred advancement over physicochemical methods that are simplistic, quick, toxic-free and more effectively applied in various applications. The easy and biocompatible process developments are needed for inexpensive synthesis, non-toxic NPs [49].

11.3.6 Why Plant Mediated Nanoparticles Synthesis?

Nanomaterials are cornerstones of Nanoscience and nanotechnology; although it is deemed to be one of the most important studies in recent sciences. Currently, many people were interested in plant extract mediated MNPs and nanomaterial. Due to its enhanced physicochemical traits and its novel application in numerous fields [50]. In this regard, applying green chemistry principles in NPs synthesis is a novel emergent issue relating to sustainability.

Currently, the 'green' synthesis has alluring attention for its ability to design alternate, energy-efficient, less toxic and safer methods towards NPs synthesis. Green chemistry was intended to eliminate hazardous constituents, trying to reduce the toxicity synthesized materials [51].

Green chemistry pacts with synthesis approaches based on the principles of mainly preventing waste and accidents, economy, avoiding harmful material and methods, design for energy efficiency, reduce derivatives, biodegradable and reduced toxicity. Therefore, the biocompatible, clean and biodegradable method revolving researchers to the biosynthesis of NPS by 'green' chemistry principles [52].

In recent eon, biosynthesis of nanometals using plant extracts has become faster and nontoxic technique. The conventional synthesis of NPs includes chemicals such as solvents, raw materials, reagents, and template materials. Such chemicals have engendered the noxious intermediates and products [13, 14]. Accordingly, many researchers have reported the biosynthesis of MNPs by plant leaf extracts to have ample benefits. In addition, to understand the properties of nanomaterials toxicity intensity and to plan nanoscale materials that can be integrated into better performance material that was less hazardous to the environment. To overcome the limitations associated with chemical and physical methods of MO NPs and to diminish the noxious products; it is necessary to move towards safer alternative methods of green synthesis [53].

Among the other biological synthesis methods, the use of plants for nanoparticle synthesis offers a wide range of benefits such as

(1) It proficiently exploits renewable resource materials, exterminate surplus waste and avoids toxic reagents and solvents.
(2) The active phytocomponents act as a reduction and stabilization agent, the consequent reduction in the synthesis method's overall cost.
(3) The high energy and pressure are not required, leads to an energy-saving process.
(4) In this process, the extravagant procedure such as cell cultures maintaining, product recovery, prolonged reaction time were eliminating.
(5) The plant extract concerned NPs can be aptly scaled up for large scale manufacture; also, the method can be economically viable [55].

Several reports show that several plants and plant-related materials have been used for easy, rapid synthesis of NPs, it confirmed to be compatible material for widespread application from biological to biomedical. Also, phytochemicals from green extract acted as capping, reducing, stabilizing agents and assisted in evading NPs aggregation [32].

In addition, following 'green' principles reduce the use and generation of hazardous substances. The aim of green usages is to make better, harmless materials, by preference the harmless source and to decrease waste. In addition, literature depicted that to achieve the sustainability of nanomaterial synthesis; more research is needed to explore [54, 55].

11.4 Factors Influencing Green Synthesis

In comparison to engineered NPs, few studies confirm that biosynthesized NPs are less toxic, stable, better physicochemical properties and enhanced application. However, a study of the relevant literature exposed challenging problems and limits the development of green mediated synthesis [56]. The main parameters such as plant species, concentration and stoichiometric ratio of plant extract and metal precursors, pH, reaction time, solvent, and temperature. Amongst, the amount of extract and reaction time highly influences the controlling the plant-mediated NPs size, size distribution, morphology, shape, stability and number of particles [57].

As well, process-engineering limitations, operational scalability, a lack of existence cycle assessment and identifying the bioactive components were also considered as major concerns. Thus, identification of the process as mentioned above and parameters optimization for the synthesis of NPs are exceedingly needed in green mediated synthesis. Besides the need for method optimization but, molecular mechanism as well as biochemical participation in the process remain unknown and would be discovered [8, 34]. Only green mediated NPs synthesis has been projected as an alternative to lesser use of dangerous materials and reaction conditions for the synthesis of MNPs the advantages over the traditional syntheses. Owing to differences in the preparation conditions of the NPs, alike precursors, reactants, processing temperature and pH. Moreover, even when using the same synthesis methods, nominal concentrations and conditions are sometimes not reproducible the same properties [58]. Some of the main factors that affect NPs biosynthesis are described below.

11.4.1 Impact of pH

The pH of the reaction medium has an important role in the synthesis of NPs. Several studies have reported that changing the reaction medium pH tends to form NPs in different shapes and sizes [8]. Many researchers have reported that in the acidic conditions the NPs peak become broader, the particle size increases also slow reaction rate was observed. It was ascribed to the electrostatic revulsion of anions that exist in the acidic reaction mixture. By increasing pH, NPs the absorption displays a narrow peak with monodispersed size. Because at higher pH functional groups ionization was occurring. So, it has been believed that the basic condition is favored for controlling the particle size. It was in concurrence with the reports in the literature [34].

11.4.2 Temperature

Temperature is another important factor in the synthesis of MNPs, as it controls the synthetic process the reaction kinetics. Several studies reported that temperature less than 30°C, the degree of growth beaten the nucleation; thus large particles were attained. As well synthesis of MNPs done at room temperature takes an extended time to form a MNPs [59]. Even so, the reaction mixture acceleration can be observed by increasing the temperature. High reaction temperature led to a fast reduction in the rate of metal ions and succeeding homogeneous nucleation and formation of small size MNPs [60]. In addition, literature studies showed that a rise in temperature leads to an increase in the plasmon band's intensity and a decrease in the mean diameter of MNPs. Also, a temperature above 100°C leads to a poor reduction process, uncontrolled and fast aggregation. Therefore, the synthesis of MNPs using green sources needs less than 100°C. The temperature of the reaction medium determines the properties of the MNPs formed thus; the temperature might be attuned between 30 and 100°C [15].

11.4.3 Time

The quality and nature of MNPs synthesized using natural extract are influenced by reaction time. Amid biosynthesis plant-mediated

MNPs are usually rapid compared to other bioextract [60]. It was noted that within 20 min color changes of the solution mixture were observed, but the SPR band of NPs was broadened because of the less conversion of metal ion to MNPs. Also, many researchers observed that the efficient MNPs formation rate at high reaction time also increases in contact time from 30 to 2 h enhances excellent plasmon band formation as a huge amount of metal ions has been converted to MNPs [34]. Though, further increase in the contact time decrease in the absorption intensity. It was due to the aggregation of NPs leading to changes in particle size. The green extract comprising the more phytochemicals will reduce the metal salt quickly. In some other plant have lesser reducing compounds that will take a prolonged time for reducing metal salt. The reaction time also reliant on reaction mixture pH, temperature, plant extract reducing power, secondary metabolites of plant extract used for NPs formation [60].

11.4.4 Concentration

The initial metal ion concentration of reaction solution and the concentration of biomolecules present in reactant extracts have a significant impact on the formation of MNPs.

Many researchers have found that an increase in reactant extracts dosage enhances the formation of MNPs as well as change in morphology of MNPs. Likewise, the initial metal ion concentration in the reaction was also pay a vital influence factor [14]. It was found that the optimum concentration of metal ions was 1 mM and it shows better MNPs. But, above 1 mM concentrations of metal ion solution caused aggregation of NPs. Thus, optimization of the metal ion solution to the extract concentration volume ratio is highly necessary for higher yield and better physicochemical properties of MNPs [58].

11.4.5 Other Influencing Parameter

Likewise, the synthesis of MNPs depends on some other operational parameters such as light, pressure, stabilizer concentration and so on influence the synthesis rate, the morphology of NPs [8]. Recently, the light-assisted synthesis of NPs has developed as a promising approach to synthesizing nanostructures with different shapes and sizes. It has been detected that synthesis in the presence of light, reduced the time for MNPs in a few minutes. Similarly, the pressure applied to the NPs synthesis medium influences the physio properties of the MNPs. Also, it has been found that the rate of reduction was faster at ambient pressure conditions [14].

The secondary metabolites and plant extraction procedures were also influencing the formation of MNPs. As the concentration of metabolites varies from plant to plant so that it affects the reducing rate and stabilization process in NPs synthesis. Moreover, the purification of synthesized NPs can influence quantity and quality. The stability and quality of NPs depend on various parameters and determine the physicochemical properties of NPs. Thus, before applied NPs from agriculture to medicine optimization, synthesis influence parameters are highly necessitated [34].

11.5 Challenges in Green Synthesis of MNPs

Globally, several types of research reported that MNPs prepared by green synthesis have admirable applications from nanosensing to Nanomedicine. As well, it was considered sustainable and safe technology for the future. But green nanotechnology has many challenges alike nonexistence of strong strategy procedures for NPs production, recycle/reuse/regeneration and toxicology [13]. Also, the still green synthesis faces several challenges in transforming from the idea of realism. Significant challenges are discussed below.

11.5.1 Reproducibility

In the green synthesis of MNPs attaining complete reproducibility is a challenging part of using plant sources. At this time, the plants are grown in various environmental conditions and plant age, genetically altered the plant metabolism [61]. So, primary and secondary metabolites are regulated according to the impetus. Also, it influences the concentration and existence of metabolites [45]. The metabolites present in a natural source are acting as reducing and stabilizing agents for the green synthesis of MNPs.

So, plant source collection and selection are vital necessities for the reproducibility of MNPs synthesis [62].

11.5.2 Scale-Up

In the preparation of MNPs, scale-up is one of the challenging sections. It was due to plant source, biomolecules composition, and plant extracts heterogeneity. So, the plant extraction should be done in the same conditions to obtain a homogeneous to produce MNPs with good physicochemical characteristics [63]. Due to temperature, humidity, living organisms and the environment homogeneity of plant extraction are impractical. In this regard, many researchers suggested plant tissue cultured for reproducibility and productivity of primary and secondary metabolites of plants for the synthesis of MNPs [45]. Inopportunely, plant metabolites were produced by a particular part of plants and lack of capacity to accumulate in dedifferentiated cells were still makes snags in homogenization. Thus, novel developments are needed for scalable and large-scale, reproducible synthesis of MNPs using plant extracts [55].

Likewise, toxicity, predictability of nanostructures formation and activity, utilization of MNPs attained by green synthesis by plants risk, cost-effective valuation and management challenges have to overcome in attaining greener nanotechnology [55]. Hence, design a procedure to implement novel synthesis of nanostructures and challenges to overcome before practical application of MNPs. Recent approaches alike emerging reliable and scalable synthesis methods, design of safer nanoparticles and maximizing the use of nanomaterials [61].

11.6 Upcoming Perception(s) in Green Synthesis

(1) The green mediated field has initial start-up issues alike morphology, crystal growth control and stability. Besides, the biologically synthesized NPs formation mechanism was not evidently stated. Hence, future studies have to focus on mechanisms.

(2) Next, the uncharted part is purification and the yield of synthesized NPs. Still, the plant-mediated synthesis of MNPs has been frequently synthesized at the laboratory level. Thus, more parameter optimization is essential for large-scale production.

(3) Likewise, non-toxic and cost-effectiveness is a factor that will be another revolutionary in the direction of sustainable development. Upcoming research must accentuate the rapid biosynthesis of NPs also is directed to be a massive dive into sustainability and 'green' material for upcoming generations.

11.7 Conclusion

The preparation of MNPs from green natural products has been a highly fascinating study. Several natural extracts such as algae, bacteria, fungi, yeast, and plant source have been used as effectual sources for NPs synthesis. Amongst, plant extract conferred high efficiency and NPs with good physicochemical traits and less toxic. Also, imperative parameters, concentration, time, temperature and pH have been highlighted. In conclusion, the upcoming green synthesis of NPs research should be reaching industrial scale through resolving contemporary problems. Then, green synthesized MNPs will be biocompatible and multifunctional material for advanced applications.

Acknowledgments

MH acknowledges SERB-DST for the National Post-Doctoral Fellowship (SO.No:PDF/2018/000795).

References

1. Dowling, A., Clift, R., Grobert, N., Hutton, D., Oliver, R., O'Neill, O., Pethica, J., Pidgeon, N., Porritt, J., Ryan, J., Seaton, A., Tendler, S., Welland, M. and Whatmore, R. (2004). *Nanoscience and Nanotechnologies: Opportunities and Uncertainties*. The Royal Society and The Royal Academy of Engineering, London, UK.

2. Hussain, C. M. and Palit, S. (2018). Nanomaterials, ecomaterials, and wide vision of material science. In Martínez, L., Kharissova, O. and Kharisov, B. (eds.) *Handbook of Ecomaterials*. Springer, Cham, pp. 1–29.
3. Ibrahim, K., Khalid, S. and Idrees, K. (2019). Nanoparticles: properties, applications and toxicities. *Arabian Journal of Chemistry*, **12**(7), 908–931.
4. Geeta, G. and Anirban, R. C. (2019). A review on the biosynthesis of metal and metal salt nanoparticles by microbes. *RSC Advances*, **9**, 12944–12967.
5. Jaison, J., Ahmed, B., Yen, S. C., Alain, D. and Michael, K. (2018). Review on nanoparticles and nanostructured materials: history, sources, toxicity and regulations. *Beilstein Journal of Nanotechnology*, **9**, 1050–1074.
6. Andreescu, S., Ornatska, M., Erlichman, J. S., Estevez, A. and Leiter, J. C. (2012). Biomedical applications of metal oxide nanoparticles. In *Fine Particles in Medicine and Pharmacy*. Springer, US, pp. 57–100.
7. Chetna, D., Neeraj, D., Xian, J. L., Alice, N. J. Y., Navin, K. V., Roger, W. B., Rajamani, L. and Seeram, R. (2015). Methods and strategies for the synthesis of diverse nanoparticles and their applications: a comprehensive overview. *RSC Advances*, **5**, 105003–105037.
8. Jayanta, K, P. and Kwang, H. B. (2014). Green nanobiotechnology: factors affecting synthesis and characterization techniques. *Journal of Nanomaterials*, **2014**, 12.
9. Aman, G. and Narendra, K. J. (2019). Advances in green synthesis of nanoparticles. *Artificial Cells, Nanomedicine, and Biotechnology*, **47**, 844–851.
10. Peralta-Videa, J. R., Huang, Y., Parsons, J. G., et al. (2016). Plant-based green synthesis of metallic nanoparticles: scientific curiosity or a realistic alternative to chemical synthesis? *Nanotechnology for Environmental Engineering*, **1**, 4.
11. Sharma, R. K., Yadav, S., Gupta, R. and Arora, G. (2019). Synthesis of magnetic nanoparticles using potato extract for dye degradation: a green chemistry experiment. *Journal of Chemical Education*, **96**(12), 3038–3044.
12. Kharissova, O. V., Kharisov, B. I., Oliva González, C. M., Méndez, Y. P. and López, I. (2019). Greener synthesis of chemical compounds and materials. *Royal Society Open Science*, **6**, 191378, http://dx.doi.org/10.1098/rsos.191378.

13. Singh, J., Dutta, T. and Kim, K. (2018). Synthesis of metals and their oxide nanoparticles: applications for environmental remediation. *Journal of Nanbiotechnology*, 16, https://doi.org/10.1186/s12951-018-0408-4.
14. Shah, M., Fawcett, D., Sharma, S., Tripathy, S. K. and Poinern, G. E. (2015). Green synthesis of metallic nanoparticles via biological entities. *Materials (Basel)*, **8**(11), 7278–7308, doi:10.3390/ma8115377.
15. Kolahalam, L. A., Kasi Viswanath, I. V., Diwakar, B. S., Govindh, B., Reddy, V. and Murthy, Y. L. N. (2019). Review on nanomaterials: synthesis and applications. *Materials Today: Proceedings*, **18**, 2182–2190, https://doi.org/10.1016/j.matpr.2019.07.371.
16. Adekoya, J. A., Ogunniran, K. O., Siyanbola, T. O., Dare, E. O. and Revaprasadu, N. (2018). Band structure, morphology, functionality, and size dependent properties of metal nanoparticles. In Seehra, M. S. and Bristow, A. D. (eds.) *Noble and Precious Metals: Properties, Nanoscale Effects and Applications*. IntechOpen, doi:10.5772/intechopen.72761.
17. Jitendra, N. T., Rajanish, N.T. and Kwang, S. (2012). Zero-dimensional, one-dimensional, two-dimensional and three-dimensional nanostructured materials for advanced electrochemical energy devices. *Progress in Materials Science*, **57**(4), 724–803
18. Neelu, C. (2018). Silver nanoparticles: synthesis, characterization and applications. In Khan, M. (ed.) *Silver Nanoparticles: Fabrication, Characterization and Applications*. IntechOpen, doi:10.5772/intechopen.75611.
19. Dmitri, V. T. and Elena, V. S. (2016). Introduction: nanoparticle chemistry. *Chemical Reviews*, **116**(18), 10343–10345.
20. Akhtar, M. S., Panwar, J. and Yun, Y.-S. (2013). Biogenic synthesis of metallic nanoparticles by plant extracts. *ACS Sustainable Chemistry & Engineering*, **1**(6), 591–602.
21. Rai, P. K., Kumar, V., Lee, S., Raza, N., Kim, K. H., Ok, Y. S. and Tsang, D. C. W. (2018). Nanoparticle-plant interaction: implications in energy, environment, and agriculture. *Environment International*, **119**, 1–19.
22. Sharma, G., Kumar, A., Sharma, S., Naushad, Mu., Dwivedi, R. P., ALOthman, Z. A. and Mola, G. T. Gaurav, S., Amit, Kumar., Shweta, Sharma., Mu.N., RamPrakash, D., Zeid A. A., Genene, T. M. (2019). Novel development of nanoparticles to bimetallic nanoparticles and their composites: a review. *Journal of King Saud University – Science*, **31**(2), 257–269.
23. Yong, X., Lei, C., Xuchun, W., Weitang, Y. and Qiao, Z. (2015). Recent advances in noble metal based composite nanocatalysts: colloidal

synthesis, properties, and catalytic applications. *Nanoscale*, **7**, 10559–10583.

24. Mal, J., Nancharaiah, Y. V., Van Hullebusch, E. D. and P. N. L., Lens (2016). Metal chalcogenide quantum dots: biotechnological synthesis and applications. *RSC Advances*, **6**, 41477–41495.
25. Harshiny, M., Matheswaran, M., Arthanareeswaran, G., Kumaran, S. and Rajasree, S. (2015). Enhancement of antibacterial properties of silver nanoparticles-ceftriaxone conjugate though Mukia maderaspatana leaf extract mediated synthesis. *Ecotoxicology and Environment Safety*, **121**, 135–141.
26. Patra, J. K., Das, G., Fraceto, L. F., et al. (2018). Nano based drug delivery systems: recent developments and future prospects. *Journal of Nanbiotechnology*, **16**, 71, https://doi.org/10.1186/s12951-018-0392-8.
27. Yuliang, W. and Younan, X. (2004). Bottom-up and top-down approaches to the synthesis of monodispersed spherical colloids of low melting-point metals. *Nano Letters*, **4**(10), 2047–2050.
28. Nguyen, T. K., Thanh, N. and Maclean, S. M. (2014). Mechanisms of nucleation and growth of nanoparticles in solution. *Chemical Reviews*, **114**(15), 7610–7630.
29. Zhang, X., Sun, C. and Fang, N. (2004). Manufacturing at nanoscale: top-down, bottom-up and system engineering. *Journal of Nanoparticle Research*, **6**, 125–130, https://doi.org/10.1023/B:NANO.0000023232.03654.40.
30. Murty, B. S., Shankar, P., Raj, B., Rath, B. B. and Murday, J. (2013). Synthesis routes. In *Textbook of Nanoscience and Nanotechnology*. Springer, Berlin, Heidelberg.
31. Kaushik, N. T., Snehit, S. M. and Rasesh, Y. P. (2010). Biological synthesis of metallic nanoparticles. *Nanomedicine: Nanotechnology, Biology and Medicine*, **6**(2), 257–262.
32. Deepali, S., Suvardhan, K. and Krishna, B. (2019). Biogenic synthesis of nanoparticles: a review. *Arabian Journal of Chemistry*, **12**(8), 3576–3600.
33. Thanusu, P., Manash, D. D. and Sujoy, K. D. (2019). Biofabrication of supported metal nanoparticles: exploring the bioinspiration strategy to mitigate the environmental challenges. *Green Chemistry*, **21**, 5469.
34. El-Seedi, H. R., El-Shabasy, R. M., Khalifa, S. A. M., Saeed, A., Shah, A., Shah, R., Iftikhar, F. J., Abdel-Daim, M. M., Omri, A., Hajrahand, N. H., Sabir, J. S. M., Zou, X., Halabi, M. F., Sarhann, W. and Guo, W. (2019). Metal

nanoparticles fabricated by green chemistry using natural extracts: biosynthesis, mechanisms, and applications. *RSC Advances*, **9**, 24539.

35. Das, R. K., Pachapur, V. L., Lonappan, L., et al. (2017). Biological synthesis of metallic nanoparticles: plants, animals and microbial aspects. *Nanotechnology for Environmental Engineering*, **2**, 18, https://doi.org/10.1007/s41204-017-0029-4.
36. Prerna, K., Amrit, K. and Dinesh, G. (2019). Algae-based metallic nanoparticles: Synthesis, characterization and applications. *Journal of Microbiological Methods*, **163**, 105656.
37. Shanmugam, R., Chelladurai, M., Kanniah, P., Mahendran, V., Gnanadas, G. and Gurusamy, A. (2014). Algae mediated green fabrication of silver nanoparticles and examination of its antifungal activity against clinical pathogens. *International Journal of Metals*, **2014**, 92643, https://doi.org/10.1155/2014/692643.
38. Felix, L. O., Sasikumar, V., Manivel, A., Nooruddin, T., Dharumadurai, D. and Chari, N. (2016). Algal nanoparticles: synthesis and biotechnological potentials. In *Algae - Organisms for Imminent Biotechnology*. IntechOpen, doi:10.5772/62909.
39. Si, A. D., Mourad, M., Kushlani, W., Carlos, D. M. F., Clémence, S., Dennis, A. B. and Clayton, J. (2017). Algae-mediated biosynthesis of inorganic nanomaterials as a promising route in nanobiotechnology: a review. *Green Chemistry*, **3**, 552–587.
40. Mahendra, R., Alka, Y., Bridge, P. and Aniket, G. (2009). Myconanotechnology: a new and emerging science. In *Applied Mycology*. Chapter 14, p. 258, doi: 10.1079/9781845935344.0258.
41. Molnár, Z., Bódai, V., Szakacs, G., et al. (2018). Green synthesis of gold nanoparticles by thermophilicfilamentous fungi. *Scientific Reports*, **8**, 3943, https://doi.org/10.1038/s41598-018-22112-3.
42. Khwaja, S. S. and Azamal, H. (2016). Fabrication of metal nanoparticles from fungi and metal salts: scope and application. *Nanoscale Research Letters*, **11**, 98.
43. Surajit, D. and Hirak, R. D. (2017). Fungi-mediated biosynthesis of nanoparticles and application in metal sequestration. In *Handbook of Metal-Microbe Interactions and Bioremediation*. CRC Press, Chapter 25, pp. 423–434.
44. Marooufpour, N., Alizadeh, M., Hatami, M., Asgari Lajayer, B. and Prasad, R. (2019). Biological synthesis of nanoparticles by different groups of bacteria. In *Microbial Nanobionics*, Nanotechnology in the Life Sciences. Springer, Cham.

45. Marslin, G., Siram, K., Maqbool, Q., Selvakesavan, R. K., Kruszka, D., Kachlicki, P. and Franklin, G. (2018). Secondary metabolites in the green synthesis of metallic nanoparticles. *Materials (Basel)*, **11**(6), E940, doi: 10.3390/ma11060940.
46. Gupta, D. and Chauhan, P. (2017). Green synthesis of silver nanoparticles involving extract of plants of different taxonomic groups. *Journal of Nanomedicine Research*, **5**(2).
47. Shakeel, A., Mudasir, A., Babu, L. S. and Saiqa, I. (2016) A review on plants extract mediated synthesis of silver nanoparticles for antimicrobial applications: a green expertise. *Journal of Advanced Research*, **7**, 17–28.
48. Hussain, I., Singh, N. B., Singh, A., et al. (2016). Green synthesis of nanoparticles and its potential application. *Biotechnology Letters*, **38**, 545–560, https://doi.org/10.1007/s10529-015-2026-7.
49. Poovathinthodiyil, R., Jie, F. S. and Wallen, L. (2003). Completely "green" synthesis and stabilization of metal nanoparticles. *Journal of the American Chemical Society*, **125**, 46.
50. Kalpana, V. N. and Devi, R. (2018). Review on green synthesis, biomedical applications, and toxicity studies of ZnO NPs. *Bioinorganic Chemistry and Applications*, **2018**, 3569758.
51. Haohong, D., Dingsheng, W. and Yadong, L. (2015). Green chemistry for nanoparticle synthesis. *Chemical Society Reviews*, **44**, 5778–5792.
52. Razavi, M., Salahinejad, E., Fahmy, M., Yazdimamaghani, M., Vashaee, D. and Tayebi, L. (2015). Green chemical and biological synthesis of nanoparticles and their biomedical applications. In Basiuk, V. and Basiuk, E. (eds.) *Green Processes for Nanotechnology*. Springer, Cham.
53. Jeevanandam, J., Chan, Y. S. and Danquah, M. K. (2016). Biosynthesis of metal and metal oxide nanoparticles. *ChemBioEngReviews*, **3**(2), 55–67.
54. Haohong, D., Dingsheng, W. and Yadong, L. (2015). Green chemistry for nanoparticle synthesis. *Chemical Society Reviews*, **44**, 5778–5792.
55. Siavash, I. (2011). Green synthesis of metal nanoparticles using plants. *Green Chemistry*, **13**, 2638–2650.
56. Umesh, K., Parida, S. D., Padan, K. J., Niranjan, R. and Birendra, K. B. (2016). Plant mediated green synthesis of metallic nanoparticles: challenges and opportunities. In *Fabrication and Self-Assembly of Nanobiomaterials: Applications of Nanobiomaterials*. William Andrew, Chapter 6, pp. 149–177.

57. Dauthal, P. and Mukhopadhyay, M. (2016). Noble metal nanoparticles: plant-mediated synthesis, mechanistic aspects of synthesis, and applications. *Industrial and Engineering Chemistry Research*, **55**, 9557–9577, doi: 10.1021/acs.iecr.6b00861.
58. Harshiny, M., Nivedhini Iswarya, C. and Matheswaran, M. (2015). Biogenic synthesis of iron nanoparticles using *Amaranthus dubius* leaf extract as a reducing agent. *Powder Technology*, **286**, 744–749.
59. Madhuree, K., Aradhana, M., Shipra, P., Satyendra, P. S., Vasvi, C., Mohana, R. M., Shatrunajay, S., Poonam, K. and Chandra, S. N. (2016). Physico-chemical condition optimization during biosynthesis lead to development of improved and catalytically efficient gold nano particles. *Scientific Reports*, **6**, 27575, doi:10.1038/srep27575.
60. Dipin, D. L. and Arti, S. S. (2020). *Phytonanofabrication: Methodology and Factors Affecting Biosynthesis of Nanoparticles*. IntechOpen, doi: 10.5772/intechopen.90918.
61. Silva, L. P., Reis, I. G. and Bonatto, C. C. (2015). Green synthesis of metal nanoparticles by plants: current trends and challenges. In *Green Processes for Nanotechnology*. Springer, Cham, pp. 259–275, doi:10.1007/978-3-319-15461-9_9.
62. Hussain, M. S., Fareed, S., Ansari, S., et al. (2012). Current approaches towards production of secondary plant metabolites. *Journal of Pharmacy and Bioallied Sciences*, **4**(1), 10–20, doi:10.4103/0975-7406.92725.
63. Srivastava, S. K., Ogino, C. and Kondo, A. (2015). Nanoparticle synthesis by biogenic approach. In Basiuk, V. and Basiuk, E. (eds.) *Green Processes for Nanotechnology*. Springer, Cham.

Chapter 12

Ostwald Ripening of $PDIF\text{-}CN_2$ Molecules on Bilayer and Multilayer Graphene

Jinta Mathew
Poothakuzhiyil, Rajamudy P.O. Idukki, Post-Code 685604, Kerala, India
jintamathew@gmail.com

The Ostwald ripening of $PDIF\text{-}CN_2$ molecules on exfoliated bilayer and multilayer graphene is depicted in this chapter. Spin coated $PDIF\text{-}CN_2$ solution on exfoliated graphene exhibits movement of $PDIF\text{-}CN_2$ islands across the graphene layer until the islands stabilize by lowering the surface energy. Decreased surface area and increased height of islands have been noted after the Ostwald ripening. The height of islands has risen from 1.4 nm to 2 nm by the change in time. The morphological change of $PDIF\text{-}CN_2$ islands on graphene was observed by atomic force microscopy (AFM).

12.1 Introduction

In 1897 Wilhelm Ostwald reported a study of the nucleation of crystals in droplets [1]. Nucleation of numerous small crystals takes place in the early stages then few of the crystals gradually enlarge by assembling the surrounding crystals. The surface area of a few

Nanomaterials for Spectroscopic Applications
Edited by Kaushik Pal

ISBN 978-981-4877-69-5 (Hardcover), 978-1-003-16033-5 (eBook)
www.jennystanford.com

large crystals is less than the area of numerous smaller crystals in the system. The surface area to volume ratio is less in larger crystals in comparison with smaller crystals. The driving force for the spontaneous process of coalescence and ripening is to lower the surface energy, which is proportional to the surface area so the Ostwald ripening is a process of growth of the crystals in a system to favor the energetic condition [1–4]. The two mechanisms of Ostwald ripening are diffusion-controlled and reaction controlled growth. Lifshitz and Slyozov have derived an expression for the diffusion-controlled growth, which is $R = ct^{1/3}$, where R, mean radius of the island, c is a constant which depends temperature, molar volume, surface energy and t is the time [5]. Lifshitz and Slyozov's theory states that the average diameter of the particle depends on the cube root of time. On the other hand, movements of molecules on the edges of a crystal cause a change in fractal dimension [6]. A fractal is a non-uniform geometric object with an infinite serpentine structure at all scales. Fractal dimension is defined as:

$$d_{\mathrm{f}} = \frac{\ln N(l)}{\ln(1/l)}$$

where d_{f} is the fractal dimension, $N(l)$ is the number of unit dimension and l is the scaling factor. It is an essential factor in the latest technology to control the Ostwald ripening for the solution synthesis of quantum dots [7]. Ostwald ripening is one of the reasons behind the formation of 1D nanostructure from nanoparticles of metals and semiconductors [8].

The layers of molecules on graphene may undergo different stages of evolution during the initial stages of growth. Here, we have studied the initial nucleation of PDIF-CN_2 and the formation of molecular clusters, which move on the surface of multilayer graphene (MLG).

12.2 Experiments

Graphene has exfoliated from highly oriented pyrolitic graphite (HOPG) by scotch-tape method as shown in Fig. 12.1. The exfoliated graphene on scotch tape was transferred onto 300 nm thick SiO_2, which was thermally grown on Si (001) wafer. SiO_2 substrate were

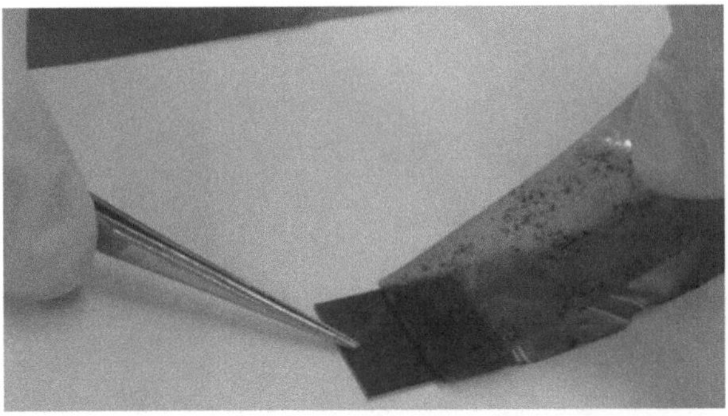

Figure 12.1 Exfoliation of graphene by using adhesive tape.

rinsed in acetone and treated by oxygen plasma in prior to the transfer of graphene onto it [9, 10]. The substrates were then heated to 100°C for 5 min to transfer graphene flakes with large area on SiO_2 [11]. Eventually, the samples were rinsed with acetone to remove the glue that have deposited on the substrates from the scotch tape. After exfoliation, graphene transferred onto 300 nm SiO_2 and checked the height using AFM. PDIF-CN_2 solution was made using toluene and was spin coated on exfoliated graphene/SiO_2 at 1000 rpm for 1 min [12]. N,N′-1H,1H-perfluorobutyl dicyanoperylenecarboxydiimide (PDIF-CN_2) is an n-type semiconductor exhibiting high electron mobility and an excellent air stability. This sample after the deposition of the molecules on graphene was kept for few days in ambient conditions and has been analyzed periodically by AFM. Transmission electron microscopy (TEM) measurements have been done on PDIF-CN_2 on suspended few layer graphene on a fresh sample and after 15 days of preparation, the sample are kept in ambient condition as same as the samples of PDIF-CN_2 on exfoliated graphene.

12.3 Results and Discussion

By observing the morphology of PDIF-CN_2 on graphene intermittently using AFM showed that the size of the islands changed periodically. Ostwald ripening has been seen in PDIF-CN_2 after few

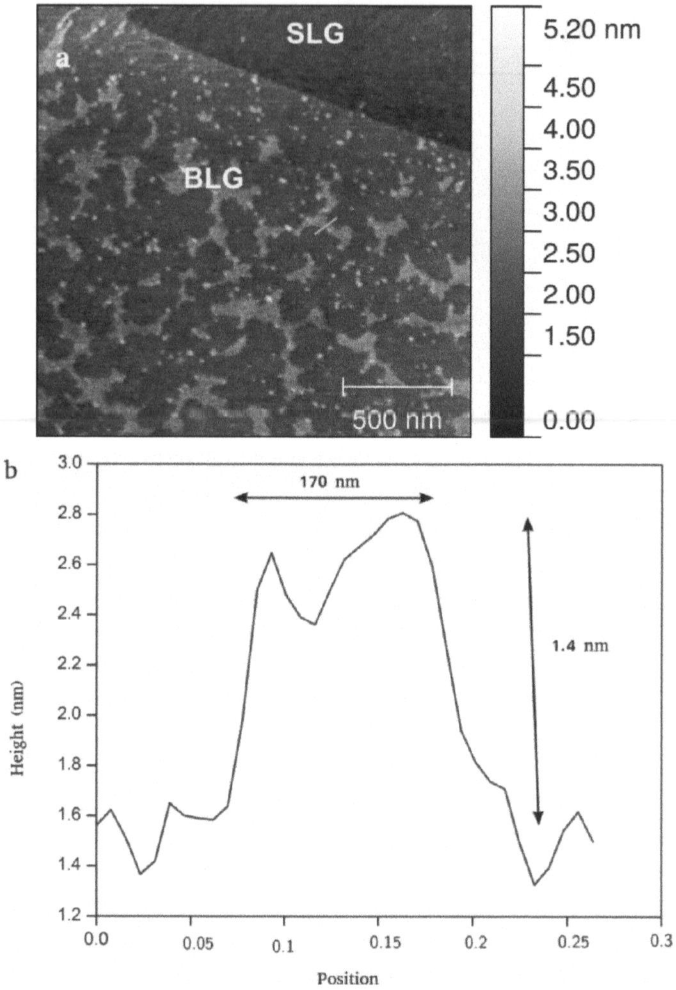

Figure 12.2 (a) AFM image of PDIF-CN_2 islands on bilayer graphene (BLG) after sample preparation. Scan size is 5 μm × 5 μm. (b) Profile along the PDIF-CN_2 islands indicated by a green line in Fig. 12.2a.

days of depositing the molecules on bilayer and multilayer graphene without any special treatment. The small islands move together to form big clusters and leave other regions empty of PDIF-CN_2.

Figure 12.2a shows the AFM image of PDIF-CN_2 islands on bilayer graphene immediately after the sample preparation. PDIF-

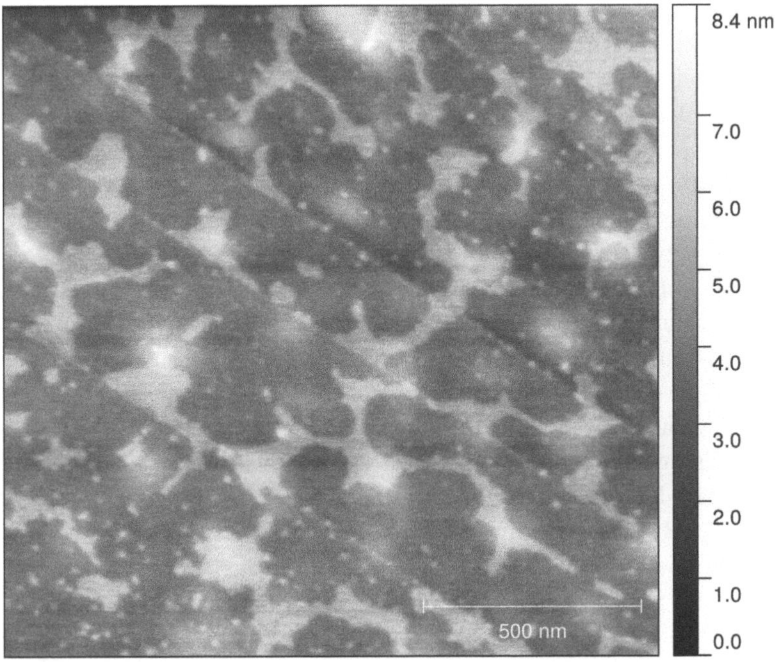

Figure 12.3 The image shows the coarsening of PDIF-CN_2 islands on multilayer graphene, which has 1 nm high graphene steps and the islands are preferentially positioned at the steps of graphene flakes and extended over the terrace.

CN_2 islands are 1.4 nm high and their width is 150–200 nm as illustrated by the line scan performed along the line shown in Fig. 12.2b. The surface area of PDIF-CN_2 islands on bilayer graphene is 0.66 μm^2. TEM studies of PDIF-CN_2 islands on graphene demonstrate that the molecules are inclined to 43° angle relative to the substrate surface [12].

Figure 12.3 is an AFM image of PDIF-CN_2 islands on multilayer graphene after 3 days of sample preparation. The islands on the steps of graphene start coarsening. The scan size is 1.6 μm × 1.6 μm. The image is taken in non-contact AFM mode. These molecules move around in the graphene flake and it moves till all the small islands move to form big clusters. Ripening of OS islands has been reported before and was depending on the type of the substrate and temperature. For example, Ribič et al. has studied the ripening of rubrene on sapphire steps, where they observed the changes of the island size at different times and after 48 h the arrangement of rubrene islands becomes stable [13]. The ripening of PDIF-CN_2 molecules takes place more slowly than rubrene on sapphire steps. As for the temperature dependence of ripening M. Dohr et al. have studied morphology and structure of dioctyl[1]benzothieno[3,2-b][1]benzothiophene (C_8-BTBT) on silica surfaces at different temperature. Wetting process and layer formation are interpreted and explain the thermal stability at a given temperature and dewetting occurs upon cooling [14].

Figure 12.4 illustrates two stages of PDIF-CN_2 layer morphology evolution with time. Figure 12.4a we obtained by AFM 4 days after deposition. We see that some of the islands have aggregated and formed larger islands with area 1.7 μm^2 and fractal dimension of d_f 2.3. These clusters do not overlay across different layers of graphene. Figure 12.4b is the AFM image of graphene flake after 11 days under ambient conditions. The bigger islands have 1.5 μm width, 4 μm length and 2 nm height. The surface area of one of the islands is 4.7 μm^2 and its fractal dimension d_f is 2.2. The height of the islands changed from 1.4 nm to 2 nm after Ostwald ripening. From this height variation, we can conclude that molecules become more upright oriented than before the completion of the ripening. In addition, from the lower d_f after 11 days, we conclude that molecules exhibit also edge mobility resulting in a smoother perimeter of the islands. There is no difference in island size on bilayer and multilayer graphene. The total area of the PDIF-CN_2 islands measured on fresh samples, after 4 days and 11 days are 23%, 21.77% and 20%, respectively. The decrease in area is due to the change in orientation of PDIF-CN_2 molecules. The apparent surface area is high when the molecules are slanted on graphene and it decreases when the molecules are upright. The ratio of coverage is $\frac{0.20}{0.23} = 0.87$. This is in reasonable agreement with a simple model of change in the projected area seen by AFM of PDIF-CN_2 unit cell on graphene at an angle of 43° and the upright molecule after ripening. The calculation yields a change in the surface area of 0.78. A schematic illustration

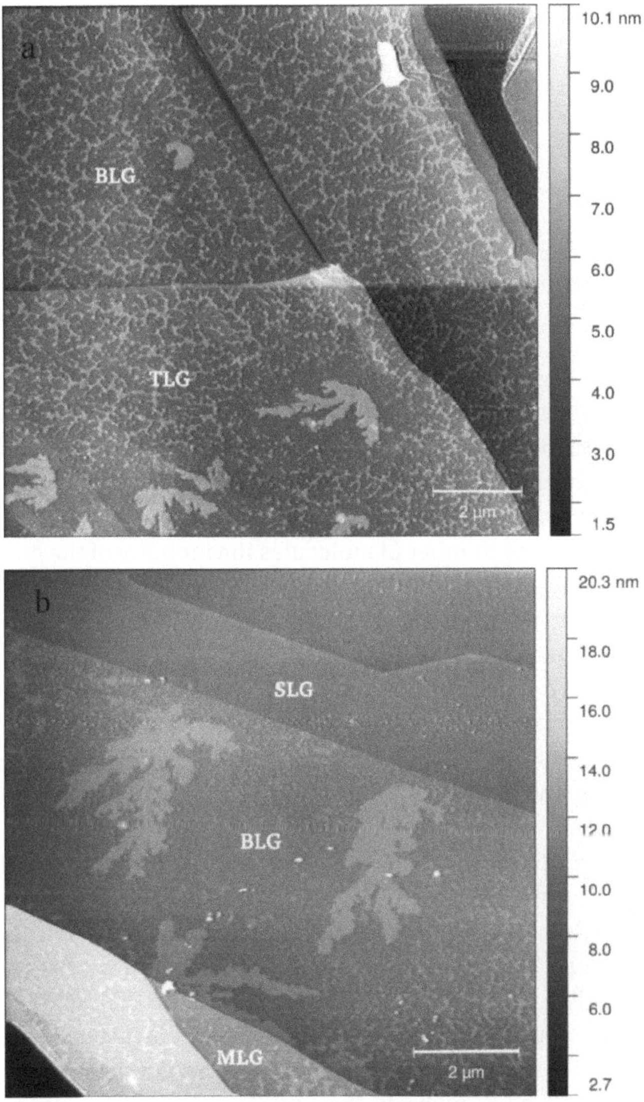

Figure 12.4 (a) AFM image of PDIF-CN_2 ripening on bilayer graphene after 4 days of sample preparation. (b) AFM image shows the Ostwald ripening of PDIF-CN_2 on bilayer and multilayer graphene after 11 days. Single-layer, bilayer and trilayer are denoted as SLG, BLG, and MLG. Scale bar shows 2 μm. These two images are taken in non-contact AFM mode.

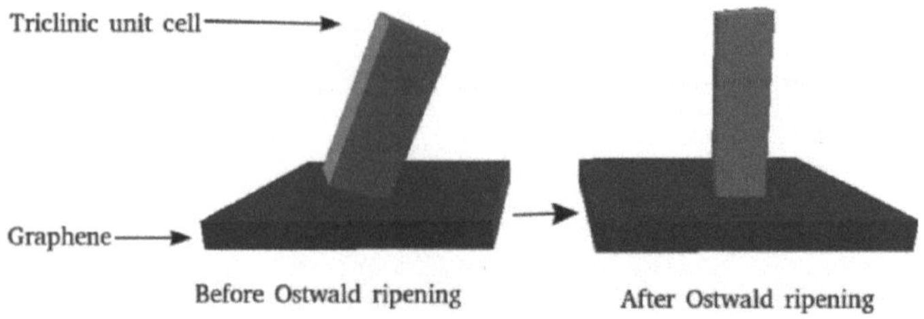

Figure 12.5 Schematic illustration of a PDIF-CN_2 unit-cell before and after Ostwald ripening.

is given in Fig. 12.5. PDIF-CN_2 molecules move on graphene layers and create bigger clusters to a decrease the surface area that leads to decrease in surface energy so that they can have equilibrium.

The movement of incoming molecules over a surface is controlled by the substrate–molecule and molecule–molecule interactions. In the case of a large number of molecules the mobility of the molecules deposited on the surface is impeded by the intermolecular interactions that result in the formation of clusters. The higher the density of incoming molecules, the higher is the probability of a molecule depositing on the substrate surface to encounter a new molecule, which effectively reduces the surface mobility of the molecules [13, 15]. The diffusion of PDIF-CN_2 molecules also depends on the density of the molecules and the molecule–molecule interaction.

Figure 12.6a shows the AFM image on a fresh sample of PDIF-CN_2 islands on graphene flake. There are no PDIF-CN_2 islands on single-layer graphene (SLG) but PDIF-CN_2 islands are seen on BLG and MLG. In Fig. 12.6b we can see the Ostwald ripening of PDIF-CN_2 on BLG and MLG, which is observed after 11 days on the same graphene flake shown in Fig. 12.6a. We have done TEM measurements to check the Ostwald ripening of PDIFCN_2 islands on suspended CVD graphene. We have analyzed the sample one day after preparation and 15 days after preparation but there was no ripening as we have seen on exfoliated graphene. We could see micron size clusters of PDIF-CN_2 molecules on exfoliated graphene but there were no such bigger changes of PDIF-CN_2 molecules on suspended CVD graphene.

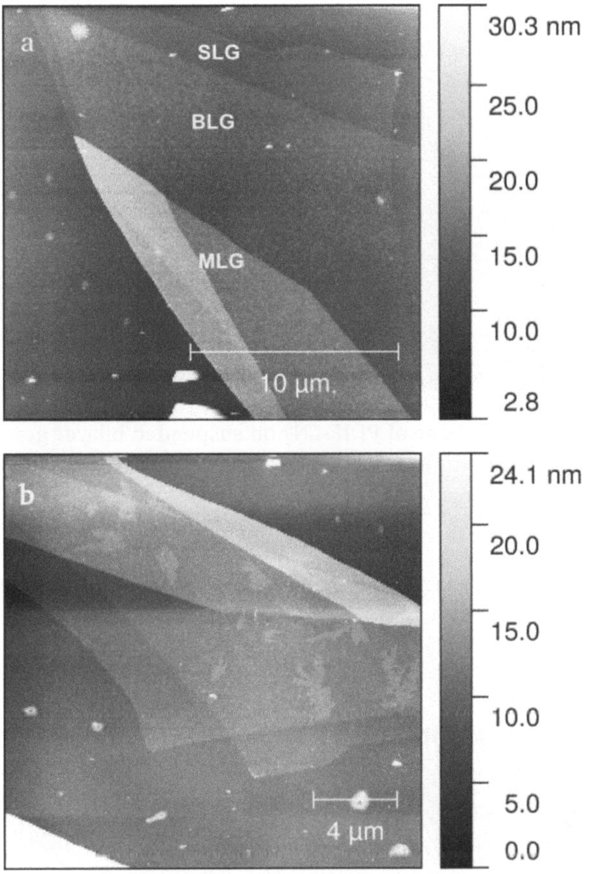

Figure 12.6 (a) The image shows the AFM topography of PDIF-CN_2 on graphene flake on a fresh sample. (b) The image shows the AFM topography of PDIF-CN_2 after 11 days of preparation on the same graphene flake shown in Fig. 12.6a. Bigger clusters of PDIF-CN_2 are visible on bilayer and multilayer graphene. Scale bar shows 2 μm.

Figure 12.7 is a TEM image of PDIF-CN_2 on suspended graphene after 15 days of sample preparation and the diffraction pattern is shown in the inset. We kept the sample in atmospheric conditions in the same way as on exfoliated graphene. But the surface energy of exfoliated and suspended graphene is different and due to this, Ostwald ripening is not observed in PDIF-CN_2 on suspended

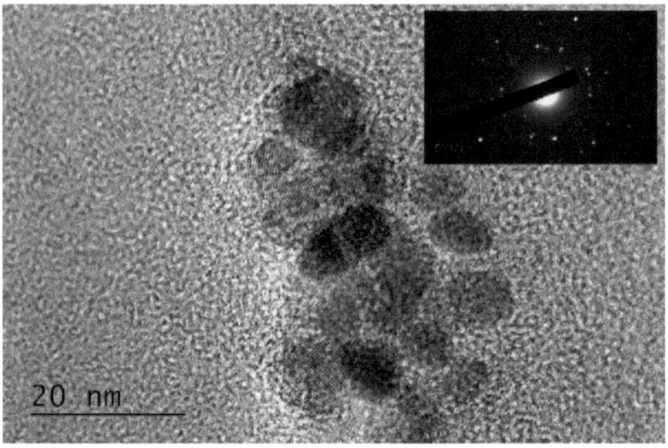

Figure 12.7 TEM image of PDIF-CN_2 on suspended bilayer graphene after 15 days of preparation. Scale bar is 20 nm. Inset shows the diffraction pattern.

graphene. The average size of the crystalline structures of PDIF-CN_2 is 10–20 nm for fresh samples and after 15 days.

12.4 Conclusion

We have found the Ostwald ripening of PDIF-CN_2 on exfoliated bilayer and multilayer graphene supported on SiO_2. PDIF-CN_2 molecules start to move around the graphene layer and form big clusters within 2 weeks. We couldn't see the ripening across the graphene layers. The molecules prefer to move inside the layer until it forms big clusters by using all the molecules in that layer. We have not observed Ostwald ripening of PDIF-CN_2 on suspended graphene.

References

1. W. Ostwald, Studien über die Bildung und Umwandlung fester Körper. *Zeitschrift für Physikalische Chemie*, 1897, **22U**(1) [Online]. Available on https://www.degruyter.com/view/journals/zpch/22U/1/article-p289.xml [accessed on 16 Apr, 2020].

2. T. Kraska, Direct observation of single ostwald ripening processes by molecular dynamics simulation, *Journal of Physical Chemistry B*, 2008, **112**(39), 12408–12413.
3. S. R. Forrest, Ultrathin organic films grown by organic molecular beam deposition and related techniques, *Chemical Reviews*, 1997, **97**(6), 1793–1896.
4. A. Baldan, Progress in Ostwald ripening theories and their applications to nickel-base superalloys. Part I: Ostwald ripening theories, *Journal of Materials Science*, 2002, **37**(11), 2171–2202.
5. Z. Hu, D. J. Escamilla Ramírez, B. E. Heredia Cervera, G. Oskam and P. C. Searson, Synthesis of ZnO nanoparticles in 2-propanol by reaction with water, *Journal of Physical Chemistry B*, 2005, **109**(22), 11209–11214.
6. P. Streitenberger and D. Förster, The effect of surface fractal dimension of particles on Ostwald ripening, *Physica Status Solidi B*, 1991, **164**(2), K65–K68.
7. R. D. Vengrenovich, Y. V. Gudyma and S. V. Yarema, Ostwald ripening of quantum-dot nanostructures, *Semiconductors*, 2001, **35**(12), 1378–1382.
8. G. M. Whitesides and B. Grzybowski, Self-assembly at all scales, *Science*, 2002, **295**(5564), 2418–2421.
9. K. S. Novoselov, et al., Electric field in atomically thin carbon films, *Science (80-.)*, 2004, **306**(5696), 666–669.
10. P. Blake, et al., Making graphene visible, *Applied Physics Letters*, 2007, **91**(6), 063124.
11. Y. Huang, et al., Reliable exfoliation of large-area high-quality flakes of graphene and other two-dimensional materials, *ACS Nano*, 2015, **9**(11), 10612–10620.
12. J. Mathew, S. Emin, E. Pavlica, M. Valant and G. Bratina, Interface-controlled growth of organic semiconductors on graphene, *Surface Science*, 2017, **664**, 16–20.
13. P. R. Ribič and G. Bratina, Initial stages of growth of organic semiconductors on vicinal (0 0 0 1) sapphire surfaces, *Surface Science*, 2008, **602**(7), 1368–1375.
14. M. Dohr, et al., Reversibility of temperature driven discrete layer-by-layer formation of dioctyl-benzothieno-benzothiophene films, *Soft Matter*, 2017, **13**(12), 2322–2329.
15. A.-L. Barabási and H. E. Stanley, *Fractal Concepts in Surface Growth*. Cambridge University Press, 1995.

Index